UCF
99.95

Coastal and Estuarine Risk Assessment

Environmental and Ecological Risk Assessment Series

Series Editor
Michael C. Newman
College of William and Mary
Virginia Institute of Marine Science
Gloucester Point, Virginia

Forthcoming Titles

Species Sensitivity Distributions in Ecotoxicology
Edited by
Leo Posthuma, Glenn W. Suter II, and Theo Traas

Risk Assessment with Time to Event Models
Edited by
Mark Crane, Michael C. Newman, Peter F. Chapman, and John Fenton

Coastal and Estuarine Risk Assessment

Edited by
Michael C. Newman
Morris H. Roberts, Jr.
Robert C. Hale

College of William and Mary
Virginia Institute of Marine Science
Gloucester Point, Virginia

LEWIS PUBLISHERS

A CRC Press Company
Boca Raton   London   New York   Washington, D.C.

Library of Congress Cataloging-in-Publication Data

Coastal and estuarine risk assessment / editors, Michael C. Newman, Morris H. Roberts, Jr., Robert C. Hale.
    p. cm. — (Environmental and ecological risk assessment)
    Includes bibliographical references (p.).
    ISBN 1-56670-556-8 (alk. paper)
    1. Marine pollution—Environmental aspects. 2. Estuarine pollution—Environmental aspects. 3. Ecological risk assessment. 4. Coastal animals—Effect of water pollution on. 5. Estuarine animals—Effect of water pollution on. I. Newman, Michael C. II. Roberts, Morris H. III. Hale, Robert C. IV. Series.

QH545.W3 C59 2001
577.7′27—dc21
                                                                                                            2001037702

This book contains information obtained from authentic and highly regarded sources. Reprinted material is quoted with permission, and sources are indicated. A wide variety of references are listed. Reasonable efforts have been made to publish reliable data and information, but the authors and the publisher cannot assume responsibility for the validity of all materials or for the consequences of their use.

Neither this book nor any part may be reproduced or transmitted in any form or by any means, electronic or mechanical, including photocopying, microfilming, and recording, or by any information storage or retrieval system, without prior permission in writing from the publisher.

All rights reserved. Authorization to photocopy items for internal or personal use, or the personal or internal use of specific clients, may be granted by CRC Press LLC, provided that $1.50 per page photocopied is paid directly to Copyright Clearance Center, 222 Rosewood Drive, Danvers, MA 01923 USA. The fee code for users of the Transactional Reporting Service is ISBN 1-56670-556-8/02/$0.00+$1.50. The fee is subject to change without notice. For organizations that have been granted a photocopy license by the CCC, a separate system of payment has been arranged.

The consent of CRC Press LLC does not extend to copying for general distribution, for promotion, for creating new works, or for resale. Specific permission must be obtained in writing from CRC Press LLC for such copying.

Direct all inquiries to CRC Press LLC, 2000 N.W. Corporate Blvd., Boca Raton, Florida 33431.

Trademark Notice: Product or corporate names may be trademarks or registered trademarks, and are used only for identification and explanation, without intent to infringe.

Visit the CRC Press Web site at www.crcpress.com

© 2002 by CRC Press LLC
Lewis Publishers is an imprint of CRC Press LLC

No claim to original U.S. Government works
International Standard Book Number 1-56670-556-8
Library of Congress Card Number 2001037702
Printed in the United States of America 1 2 3 4 5 6 7 8 9 0
Printed on acid-free paper

Preface

Experts working in diverse areas of coastal and estuarine risk assessment contributed to this first volume of the CRC Press Environmental Risk Series. Contributors were asked to address a comprehensive series of important topics including the regulatory context for coastal and estuarine risk assessment, emerging contaminants of concern, effects to marine mammals, bioavailability and exposure of marine organisms to inorganic and organic contaminants, and effects of contaminants on ecological entities ranging from biomolecules to landscapes.

Coastal and Estuarine Risk Assessment is the first book to address the application of the current risk assessment paradigm to coastal marine environments. Chapter authors bring together experiences from academia, private consultancies, and government agencies, resulting in a blending of diverse experiences, insights, and vantages.

The Editors

Michael C. Newman, Ph.D., is a Professor of Marine Science and the Dean of Graduate Studies at the College of William and Mary's Virginia Institute of Marine Science/School of Marine Science (VIMS/SMS). After receiving B.A. and M.S. (Zoology with marine emphasis) degrees from the University of Connecticut, he earned M.S. and Ph.D. (1981) degrees in Environmental Sciences from Rutgers University. He joined the faculty at the University of Georgia's Savannah River Ecology Laboratory (SREL) in 1983, becoming a SREL group head in 1996. He left SREL to join the VIMS/SMS faculty in 1998, and became the Dean of Graduate Studies for the School of Marine Science in 1999.

Dr. Newman's research interests include quantitative methods for ecological risk assessment and ecoepidemiology, population responses to toxicant exposure including genetic responses, QSAR-like models for metals, bioaccumulation and toxicokinetic models for metals and radionuclides, toxicity models including time-to-death models, and environmental statistics. He has published more than 85 articles on these topics. He has co-edited three books: *Metal Ecotoxicology: Concepts and Applications* (1991); *Ecotoxicology: A Hierarchical Treatment* (1996); and *Risk Assessment: Logic and Measurement* (1998). He has written three additional books: *Quantitative Methods in Aquatic Ecotoxicology* (1995); *Fundamentals of Ecotoxicology* (1998); and *Population Ecotoxicology* (2001).

Morris H. Roberts, Jr., Ph.D., is a Professor of Marine Science and Chair of the Department of Environmental Sciences at the College of William and Mary's VIMS/SMS. Upon completing undergraduate work at Kenyon College, he earned M.A. and Ph.D. (1969) degrees at the College of William and Mary in the School of Marine Science. He joined the faculty of Providence College to teach invertebrate zoology to upper-level students. In 1971, he became first Director of Invertebrate Studies and later Director of Research at the former Aquatic Sciences, Inc. He returned to VIMS/SMS as a faculty researcher in 1973, assuming his present role as department chair in 1994.

Dr. Roberts' current research focuses on the determination of ambient toxicity in estuarine and tidal freshwater systems. This work includes laboratory tests of ambient waters and sediments, and *in situ* tests of ambient waters. Earlier research involved evaluating the acute toxicity of chlorine and bromine chloride, Kepone, and TBT to estuarine species, bioaccumulation of PAH, Kepone and TBT in selected species, and the effects of PAH-contaminated sediment on fish and invertebrate species. He and his students have published over 60 articles and made over 35 presentations on these topics. He co-edited the book, *Water Chlorination Volume 6* (1985). Consistent with his interest in the methods of toxicology, he chaired the Aquatic Toxicology Subcommittee of ASTM Committee E47 and was responsible for preparing a compilation of all methods developed by the Committee (first edition, 1993; second edition, 1999).

Robert C. Hale, Ph.D., is an Associate Professor of Marine Science at the College of William and Mary's VIMS/SMS. His interest in pollutants and aquatic environments led him to obtain a B.S. in Chemistry and B.A. in Biology from Wayne State University, and then a Ph.D. in Marine Science from the College of William and Mary in 1983. He then joined Mobil Corporation's Environmental and Health Sciences Laboratory in Princeton, NJ. His activities as an Environmental Research Chemist at Mobil included pesticide fate and metabolism, complex mixture behavior, and ecotoxicology. He joined the VIMS/SMS faculty in 1987, receiving tenure in 1993 in the Department of Environmental Science.

Dr. Hale's current research interests focus on the bioavailability of organic contaminants in aquatic environments, and their fate and effects. Dr. Hale has also been active in the area of progressive analytical techniques (e.g., supercritical fluid extraction) for the determination of trace contaminants, as well as emerging contaminants, such as brominated flame retardants and non-ionic detergents. His group has published more than 30 articles and made more than 70 scientific presentations on these topics while at VIMS/SMS.

Acknowledgments

This books results from the Coastal and Estuarine Risk Assessment Forum (CERAF) which was held at the College of William and Mary (Williamsburg, Virginia, U.S.A.; July 20–21, 2000). Presentations and discussions from that meeting were incorporated into the chapters of this volume. We are grateful to the presenters and other participants of that forum for their engaging dialogue. The following individuals also provided invaluable assistance in the form of anonymous chapter reviews: A. Aguirre, Tufts University; S. Bartell, Cadmus Group; A. Bernhoft, National Veterinary Institute (Oslo); G.P. Cobb, Texas Tech University; M. Crane, University of London, Royal Holloway; N. Denslow, University of Florida; R.T. Di Giulio, Duke University; W.S. Douglas, Symbiosis Environmental; A.A. Elskus, University of Kentucky; N.S. Fisher, SUNY, Stony Brook; S. Ferson, Applied Biomathematics; M.E. Hahn, Woods Hole Oceanographic Institution; R.F. Lee, Skidaway Institute of Oceanography; M. Lydy, Wichita State University; J.S. Meyer, University of Wyoming; G.L. Mills, University of Georgia, Savannah River Ecology Laboratory; J.E. Perry, College of William and Mary–Virginia Institute of Marine Science; J.T. Phinney, American Society of Limnology and Oceanography; P.S. Rainbow, The Natural History Museum (London); J. Reinfelder, Rutgers University; C. Richards, University of Minnesota – Sea Grant College Program; E.J. Scollon, Texas Tech University; C. Strojan, University of Georgia, Savannah River Ecology Laboratory; M.A. Unger, College of William and Mary–Virginia Institute of Marine Science; and P.A. Van Veld, College of William and Mary–Virginia Institute of Marine Science.

Contributors

Gary Bigham
Exponent, Inc.
15375 SE 30th Place
Bellevue, WA 98007
U.S.A.

Alan L. Blankenship
Entrix Inc.
and
Michigan State University
East Lansing, MI 48824
U.S.A.

Jenee A. Colton
Exponent, Inc.
15375 SE 30th Place
Bellevue, WA 98007
U.S.A.

Mark Crane
University of London
Royal Holloway
School of Biological Sciences
Egham, Surrey
TW20 0EX
U.K.

David A. Evans
Virginia Institute of Marine Science
College of William and Mary
Route 1208 Greate Road
Gloucester Point, VA 23062
U.S.A.

John P. Giesy
Michigan State University
Department of Zoology
Institute for Environmental Toxicology
National Food Safety
  and Toxicology Center
East Lansing, MI 48824
U.S.A.

Timothy R. Gleason
U.S. Environmental Protection Agency
National Health Environmental Effects
  Research Laboratory
Atlantic Ecology Division
27 Tarzwell Drive
Narragansett, RI 02882
U.S.A.

Albania Grosso
Environmental Resources Management
Wallbrook Court
North Hinksey Lane
Oxford
OX2 0QS
U.K.

Ruth Gutjahr-Gobell
U.S. Environmental Protection Agency
National Health Environmental Effects
  Research Laboratory
Atlantic Ecology Division
27 Tarzwell Drive
Narragansett, RI 02882
U.S.A.

Robert C. Hale
Virginia Institute of Marine Science
College of William and Mary
Route 1208 Greate Road
Gloucester Point, VA 23062
U.S.A.

Marina Huber
U.S. Environmental Protection Agency
National Health Environmental Effects
  Research Laboratory
Atlantic Ecology Division
27 Tarzwell Drive
Narragansett, RI 02882
U.S.A.

Timothy J. Iannuzzi
BBL Sciences
326 First Street, Suite 200
Annapolis, MD 21403
U.S.A.

Paul D. Jones
Michigan State University
Department of Zoology
Institute for Environmental Toxicology
National Food Safety
  and Toxicology Center
East Lansing, MI 48824
U.S.A.

Kurunthachalam Kannan
Michigan State University
Department of Zoology
Institute for Environmental Toxicology
National Food Safety
  and Toxicology Center
East Lansing, MI 48824
U.S.A.

Mark J. La Guardia
Virginia Institute of Marine Science
College of William and Mary
Route 1208 Greate Road
Gloucester Point, VA 23062
U.S.A.

Byeong-Gweon Lee
Chonnam National University
Department of Oceanography
Kwang Ju
South Korea

Richard F. Lee
Skidaway Institute of Oceanography
10 Ocean Science Circle
Savannah, GA 31411
U.S.A.

Kenneth M. Y. Leung
Royal Holloway
University of London
School of Biological Sciences
Egham, Surrey
TW20 0EX
U.K.

Dave Ludwig
BBL Sciences
326 First Street, Suite 200
Annapolis, MD 21403
U.S.A.

Samuel N. Luoma
U.S. Geological Survey
Water Resources Division,
  Mail Stop 465
345 Middlefield Road
Menlo Park, CA 94025
U.S.A.

Christopher E. Mackay
Exponent, Inc.
15375 SE 30th Place
Bellevue, WA 98007
U.S.A.

Robert P. Mason
University of Maryland
Center for Environmental Science
Chesapeake Biological Laboratory
P.O. Box 38
Solomons, MD 20688-0038
U.S.A.

David Morritt
Royal Holloway
University of London
School of Biological Sciences
Egham, Surrey
TW20 0EX
U.K.

Wayne R. Munns, Jr.
U.S. Environmental Protection Agency
National Health Environmental Effects
  Research Laboratory
Atlantic Ecology Division
27 Tarzwell Drive
Narragansett, RI 02882
U.S.A.

Diane E. Nacci
U.S. Environmental Protection Agency
National Health Environmental Effects
  Research Laboratory
Atlantic Ecology Division
27 Tarzwell Drive
Narragansett, RI 02882
U.S.A.

Michael C. Newman
Virginia Institute of Marine Science
College of William and Mary
Route 1208 Greate Road
Gloucester Point, VA 23062
U.S.A.

Morris H. Roberts, Jr.
Virginia Institute of Marine Science
College of William and Mary
Route 1208 Greate Road
Gloucester Point, VA 23062
U.S.A.

Christian E. Schlekat
Environmental and Health Science
U.S. Borax
26877 Tourney Road
Valencia, CA 91355
U.S.A.

Neal Sorokin
Royal Holloway
University of London
School of Biological Sciences
Egham, Surrey
TW20 0EX
U.K.

James R. Wheeler
University of London
Royal Holloway
School of Biological Sciences
Egham, Surrey
TW20 0EX
U.K.

Paul Whitehouse
WRc-NSF
Henley Road
Medmenham, Marlow,
  Buckinghamshire
SL7 2HD
U.K.

Table of Contents

Chapter 1
Overview of Ecological Risk Assessment in Coastal
and Estuarine Environments 1
Morris H. Roberts, Jr., Michael C. Newman, and Robert C. Hale

Chapter 2
European Approaches to Coastal and Estuarine Risk Assessment 15
*Mark Crane, Neal Sorokin, James R. Wheeler, Albania Grosso,
Paul Whitehouse, and David Morritt*

Chapter 3
Emerging Contaminants of Concern in Coastal
and Estuarine Environments 41
Robert C. Hale and Mark J. La Guardia

Chapter 4
Enhancing Belief during Causality Assessments: Cognitive Idols
or Bayes's Theorem? 73
Michael C. Newman and David A. Evans

Chapter 5
Bioavailability, Biotransformation, and Fate of Organic Contaminants
in Estuarine Animals 97
Richard F. Lee

Chapter 6
The Bioaccumulation of Mercury, Methylmercury, and Other Toxic
Elements into Pelagic and Benthic Organisms 127
Robert P. Mason

Chapter 7
Dietary Metals Exposure and Toxicity to Aquatic Organisms:
Implications for Ecological Risk Assessment 151
Christian E. Schlekat, Byeong-Gweon Lee, and Samuel N. Luoma

Chapter 8
Endocrine Disruption in Fishes and Invertebrates: Issues for Saltwater
Ecological Risk Assessment 189
Kenneth M.Y. Leung, James R. Wheeler, David Morritt, and Mark Crane

Chapter 9
The Use of Toxicity Reference Values (TRVs) to Assess the Risks That
Persistent Organochlorines Pose to Marine Mammals 217
*Paul D. Jones, Kurunthachalam Kannan, Alan L. Blankenship,
and John P. Giesy*

Chapter 10
Effects of Chronic Stress on Wildlife Populations: A Population Modeling
Approach and Case Study 247
*Diane E. Nacci, Timothy R. Gleason, Ruth Gutjahr-Gobell, Marina Huber,
and Wayne R. Munns, Jr.*

Chapter 11
Structuring Population-Based Ecological Risk Assessments
in a Dynamic Landscape 273
Christopher E. Mackay, Jenee A. Colton, and Gary Bigham

Chapter 12
Incremental Chemical Risks and Damages in Urban Estuaries:
Spatial and Historical Ecosystem Analysis 297
Dave Ludwig and Timothy J. Iannuzzi

Chapter 13
Ecological Risk Assessment in Coastal and Estuarine Environments 327
Michael C. Newman, Robert C. Hale, and Morris H. Roberts, Jr.

Index 337

1 Overview of Ecological Risk Assessment in Coastal and Estuarine Environments

Morris H. Roberts, Jr., Michael C. Newman, and Robert C. Hale

CONTENTS

1.1 Introduction 1
1.2 Application of Risk Assessment in Estuaries...5
    1.2.1 Water Quality Criteria...5
    1.2.2 Sediment Quality Guidelines..6
    1.2.3 Toxics Characterization...7
    1.2.4 Relative Risk Assessment 8
    1.2.5 A Case Study of Risk Assessment in an Estuary 9
1.3 Forum Organization 10
References..11

1.1 INTRODUCTION

Ecological risk assessment is a logical process for objectively defining the probability of an adverse effect to an organism or collection of organisms when challenged with an environmental modification such as climatic change, xenobiotic exposure, infection with a disease organism, or some other potential stressor. The link between the parameter producing the effect (i.e., the stressor) and the organism(s) responding (i.e., the receptor) is accessible mainly by inference in the absence of full and detailed knowledge. Long used in economics and health sciences, risk assessment has a shorter history of application to ecological systems and even a shorter history for estuarine and coastal systems.

The risk assessment paradigm in general, and in ecological risk assessment in particular, involves (1) problem formulation, (2) parallel analyses of exposure and effects, and (3) risk characterization[1-5] (Figure 1.1). Each aspect of the paradigm involves an integrative process connecting various inputs or data to one or more outputs or conclusions. Initial assessments are often imprecise, pointing to data needs

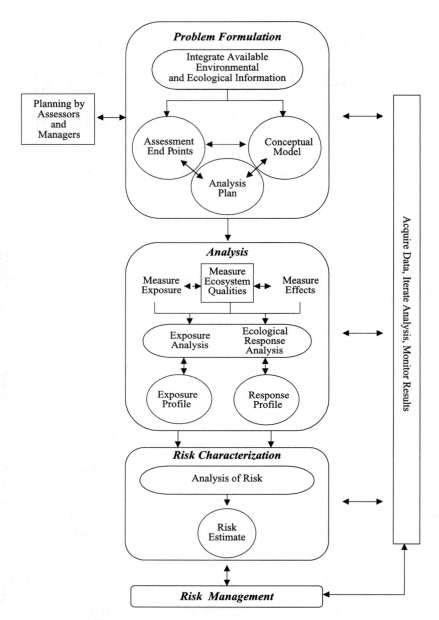

FIGURE 1.1 The environmental risk assessment paradigm as presented by the U.S. EPA. Information inputs are shown in boxes, processes in ovals, and outputs in circles. (Modified from Reference 5.)

that, if fulfilled, allow more rigorous assessments of risk with improved certainty in the evaluation. Problem definition includes a determination of a valued resource such as a population of commercially important species or keystone species, or community structure. In this step, one determines what assessment end point(s) will be used, develops a conceptual model that links the end point(s) to exposure, and an analysis plan that incorporates both the end point(s) and the model.

The analysis phase involves collection of data on exposure, effects, and characteristics of the ecosystem of interest. In coastal and estuarine situations, these parameters are particularly dynamic and the resultant gradients in salinity and other variables present special challenges to risk assessment. The analyses of exposure and ecological responses occur in parallel, leading to exposure and response profiles. Attention must be paid to environmental conditions in both analyses if the ultimate risk estimation is to be credible.

The final stage in the process is risk characterization, resulting ultimately in a risk estimate. To the extent possible, this characterization should be a quantitative probability statement, but this is not always achievable in complex and dynamic systems. As depicted in Figure 1.1, the risk characterization may be deemed inadequate, resulting in the need for additional data, reanalysis, and monitoring of the outcome for plausibility.

Risk assessment is a practical management tool rather than a purely scientific endeavor. It is useful for predicting the outcomes of activities such as chemical use or disposal, or species introductions, or for attributing observed effects to potential causes in a retrospective analysis. Risk assessment can generate enough understanding to allow informed decision making relative to choosing among several remedial actions. It is an especially important tool if the resources needed to reduce risk are limited, if competing options exist, or if the relative value of each action is not obvious.

This edited volume was developed to focus primarily on chemical exposure, effect measurement, and risk characterization in estuarine and coastal environments. The chemicals of concern are those that are or may become toxic in these environments. This focus is important not only because of intense historical uses, but also because human use of the estuarine and coastal system continues to expand rapidly. Historical human use has resulted in relic accumulations of chemical contaminants in localized areas, and present human activities in coastal areas produces continued introductions of chemical contaminants either directly (e.g., from discharge pipes) or indirectly (e.g., polycyclic aromatic hydrocarbon, or PAH, from atmospheric deposition as a result of vehicular emissions). Enhancing the importance of risk assessment in estuaries is the current rapid migration of people into the coastal zone.[6] For example, the Chesapeake Bay watershed had a 19.3% increase in population between 1970 and 1990. Smaller coastal watersheds such as St. Johns River (Florida), Tampa Bay (Florida), and San Diego Bay (California) had even larger population increases of 43 to 47%.[7] With this population growth, the potential for substantial increases in adverse impacts on estuarine and coastal environments is dramatically enhanced. Conversely, as these populations grow, the resources in estuaries and coastal systems that serve as an inducement to human immigration are among those most likely to deteriorate as a result of human activity.

Despite the focus on chemicals as the stressors of interest in most of this volume, one must recognize that ecological risk assessment is a tool with broader application, providing an objective approach to evaluate simultaneously the effects not only of chemical stressors, but also other stressors such as habitat loss (e.g., submerged aquatic vegetation, salt marshes, or shoreline green ways), fishing pressure, or reduced freshwater inputs (resulting from reservoir construction). This is reflected in two recurrent themes in the book, i.e., the need to apply a landscape context for assessment and the need to relate effects to stressors across all levels of ecological organization from bacteria and primary producers to top carnivores, and from individual organisms to communities.

Estuarine and coastal systems develop at the interface between land, fresh water, and the oceans. As such, they are characterized by gradients, most obviously a persistent salinity gradient. These systems are also physically dynamic over a variety of timescales. Tidal cycles modify mixing regimes over different, but predictable, periods in many coastal areas; wind varies aperiodically and interacts with tide to further modify the flow of water and associated materials. Seasonal changes in air temperature and precipitation are reflected in annual changes in temperature and salinity of the water. As a result, these systems must be looked at in an ecotonal context, as a habitat in which there are unique groups of organisms, as well as those from freshwater or euhaline systems that have sufficient physiological tolerances to handle conditions within these mixing zones and that utilize estuaries as sources of food or as refuges from predators. Biological communities are dynamic and often ephemeral during the year although recurrent over cycles of years.

These variable conditions affect both stressor and receptor. For there to be an impact, there must be co-occurrence of the stressor and receptor. The degree of exposure must be sufficient in intensity and duration to produce an effect. In less dynamic systems, the presumption of stressor and receptor co-occurrence is often reasonable over relatively long timescales. And equally important as a presumption, the receptors must remain within the volume influenced by the stressor. Some species inhabiting estuaries are highly mobile, and can move outside the region where a stressor occurs. Other species may have evolved behavioral mechanisms to cope with the natural variability of the system by isolating themselves from a stressor until the stressor intensity has decreased; e.g., bivalves can close their shells, and some organisms can encase themselves in mucus or another protective agent. Many organisms have enzymatic mechanisms that allow them to detoxify chemicals, mechanisms that are induced by the presence of the chemical.

The tools of risk assessment have been developed for optimal use in less dynamic systems. Therefore, the various steps in ecological risk assessment may need to be modified to address dynamic estuarine and coastal conditions effectively. Is proper consideration given to the effects of estuarine conditions on exposure concentrations during assessments of exposure? Specifically, is sufficient attention given to the effects of salinity, binding to sediment-associated or dissolved organic material, and isolation in water masses by density gradients? Is the temporal and spatial sampling utilized appropriate to the temporal and scalar variation in these dynamic systems? Is the ecosystem context inherent in framing most risk assessments adequate for coastal and estuarine systems that are physical boundaries and ecotones receiving large amounts of materials, energy, and species from outside of the system for which risk is to be estimated? Does the evaluation consider adequately all ecological components including primary producers, secondary consumers, and saprophytes?

In assessing hazard, is sufficient consideration given to variability in degree of exposure or in rates of stressor input? Does the approach consider the inherent seasonal variations in the presence of a species, reproductive activity, and growth? Some of these issues are not unique to estuarine and coastal systems, and obviously have persistent and collective implications about the validity of any risk assessment. Nevertheless, the implications of the estuarine environment on risk assessment need careful consideration.

1.2 APPLICATION OF RISK ASSESSMENT IN ESTUARIES

Although the use of ecological risk assessment in aquatic, and especially estuarine, systems is a relatively new development, exposure and hazard evaluations have been used in these systems for some time. These evaluations have been applied in setting water quality criteria and standards, definition of sediment quality guidelines, characterization of various segments of the estuaries with regard to toxicological effects, and the so-called relative risk assessment. In each of these cases, frequent mention is made of managing "risk" or the process being "risk-based," yet many times no actual determination of risk probabilities is conducted.

1.2.1 WATER QUALITY CRITERIA

Water quality criteria are primarily based on laboratory exposure and effects data. In collecting exposure and effects data, standard test designs require high consistency of exposure over time and effects are monitored over timescales of days to months. The intent is to achieve reproducible and precise data, even though these controlled conditions may not accurately reflect field scenarios. From the laboratory data developed for an array of species, a statistical estimate is made of the concentration of a material that will be protective of the "most sensitive species." Account is taken of the interaction between a limited number of environmental variables (hardness, acidity, or salinity) and the chemical of concern by modeling the relationships, again assuming time constancy in all these parameters. A safety factor is then applied to account for environmental variability and uncertainty.[8]

Data for freshwater and saltwater species can be treated separately in this type of analysis, leading to distinct water quality criteria for fresh water and salt water. The derivation of distinct freshwater and saltwater criteria is clearly important for many inorganic and organic materials that exhibit different bioavailabilities in these two media. Examples include chlorine and cadmium,[9,10] but there are many others.

The analysis of exposure and effects data yields a concentration that notionally should not be exceeded in a specific aqueous medium if the specified acute or chronic effect is to be avoided. This process is a valuable first estimate of the hazard concentration, but is not a risk assessment as defined herein, even though both exposure and effect information are central to any environmental risk assessment. The risk of adverse impact cannot be determined from the criteria.

The same effects data can be used to conduct a risk assessment as demonstrated by Hall and his colleagues.[11] In the problem formulation, they selected two specific chemicals, copper and cadmium, based on use patterns in Maryland waters. They then applied the broadest array of laboratory toxicity data available to determine the concentration of each of these metals that affected the most sensitive 10th percentile of species, thus defining a concentration protective of 90% of the species. If one desired a more protective concentration, one could derive that from the same data. This value is conceptually the same as the criterion value derived by the U.S. Environmental Protection Agency (U.S. EPA), but obtained in a way that involves all data. Measured concentrations in samples from specific locations can then be compared with the concentration that produces a specified degree of protection. The concentrations may also be compared with the effect curve to estimate the probability of an adverse effect at specific locations as a result of exposure to the specific compound for which the analysis was performed.

The issue of multiple stressors including effects of unidentified stressors was not addressed in this example although the authors noted the concurrence of both stressors in some locations. Hall et al.[11] discussed the strong preponderance of fish and benthos information in the available data set and the paucity of plant and zooplankton data. The importance of this observation cannot be assessed objectively at this point, but if these groups are more sensitive than fish or benthos, the analysis is clearly biased by the omission of such information. This observation points to a definite need to expand the array of organisms routinely used for effects evaluations to be more inclusive of important ecological components.

1.2.2 SEDIMENT QUALITY GUIDELINES

A variety of sediment quality guidelines (SQG) have been proposed in the past two decades.[12-17] Different guidelines yield different answers about the concentration that may be protective. All guidelines are based on exposure and effects data, and some incorporate both laboratory and field data. Clearly, many geochemical parameters can affect bioavailability and will vary from place to place. The statistically derived values are considered guidelines rather than criteria because spatial variability in underlying conditions at any specific location may be different from those embodied in the data from which the guidelines were derived. Despite the imprecision of these guidelines, if carefully applied, these values can be useful to managers faced with decisions involving sediment quality issues. Exceeding guidelines is *prima facie* evidence of a risk of sediment quality reduction and an attendant substantial, if not quantified, risk of adverse effects.

With further manipulation, these guidelines can also provide a basis for establishing remediation criteria. Based on a sediment quotient value (SQV) (i.e., the sum of the quotient between a measured value and the appropriate SQG for each contaminant of concern), one can identify one or more levels of exposure that will, with some degree of confidence, reduce adverse effects below a selected level(s).[18,19] The ratios identified at one site may then be cautiously applied as cleanup goals in other locations. The approach has been used recently by the Army Corps of Engineers in a feasibility study of sediment remediation in the Elizabeth River, Virginia.[20]

A risk analysis approach in this application analogous to that used for water would be a valuable endeavor. The use of guidelines does not allow an estimation of risk probability. At present, the risk of an adverse effect resulting from exceeding a guideline such as an effects range median (ERM) is undefined. If the concentration was two or more times above the ERM, the increase in risk is at present indeterminate. An analysis that includes all available data to estimate risk would provide valuable extensions to the present approaches.

Just as in the previous example related to water column impacts, risk analysis for sediments as portrayed here would be accomplished one material at a time. In reality, simultaneous exposure to multiple materials and other potential stressors is typical. Better tools are needed to deal with this complexity.

1.2.3 TOXICS CHARACTERIZATION

The sediment quality triad method is one approach used to characterize the impact of chemical contaminants in estuaries. It was developed originally for Puget Sound[12,21] and subsequently applied in San Francisco Bay[13] and elsewhere. In this method, there is synoptic determination of chemical contaminants in the sediment, measurement of toxic end points in laboratory tests of the sediment, and an assessment of the benthic community. To understand bioavailability of chemicals in the sediment, total organic carbon, acid-volatile sulfides, and other geochemical parameters are measured on sediment samples. From these three types of measures, one can assess the health of a system on empirical grounds. The power of this approach lies in the contemporaneous nature of the measurements.

The U.S. EPA Chesapeake Bay Program has, for a decade or more, been working on a toxics characterization of various strata within the bay and its tributaries. The program has been concerned with both water column and sedimentary effects. The method for characterizing these strata as a region of concern, region of emphasis, region with low probability for adverse effect, or region of insufficient or inconclusive data has been described in detail.[22] The characterizations are based on available ambient chemical concentration data for water, sediment, and finfish/shellfish tissue and on effects information such as laboratory toxicity data for ambient water and sediment samples and indices of biological integrity for benthic invertebrates. Data collected under the auspices of the U.S. EPA Chesapeake Bay Program Office[23–30] with contemporaneous data collection were used along with other available, although noncontemporaneous, data of certifiable high quality.

These determinations are risk based in that strata are ranked based on an estimate of adverse effect by a panel of experts using professional judgment to assess the likelihood of adverse impacts. This approach can resolve "regions of concern" from "regions of no probable effect," but does not allow objective evaluations of greater resolution.

These characterizations have nevertheless proved useful in prioritizing specific regions for cleanup and in identifying the need for additional data in regions that cannot currently be characterized. However, these characterizations alone do not provide a clear strategy for the remediation of specific regions. Once a strategy has been developed, quantitative cleanup goals must be established. Again, the characterizations do not provide objective criteria for cleanup. Finally, if resolution among regions is imprecise using this approach, one must conclude that there is insufficient resolution to recognize the benefits derived from various cleanup options. Thus, the use of these characterizations to guide management decisions is limited.

An alternative approach to characterize strata, developed and used by Ian Hartwell in Maryland, involves a risk-ranking approach based on ambient toxicity results using a standard battery of water column and benthic tests, and some measure of community health.[31–33] A score is calculated for each sample site based on severity of effect, degree of response, variability of each test, consistency among various effects measures, and number of effects measured. This approach allows one to compare objectively data from multiple sites even if there were differences in parameters measured, number of effects studied, and species tested.

In the sediment triad method, laboratory and field effects data are not always consistent with one another. This inconsistency may be due to test conditions modifying bioavailability, or a parameter other than toxicity producing an adverse effect on communities. Further, the sediment triad method deals only with the benthic component of the ecosystem. The risk-ranking method of Hartwell is an imperfect, but nevertheless useful, way to address some of these concerns. The method can utilize water column or benthic information, or both. Site scores can be compared with measures of community health such as fish or benthic indices of biotic integrity. Sites that show poor community health but minimal toxicity can be examined for other risk factors such as physical conditions, food availability, or infectious disease impacts. A site that is scored "toxic" can be subjected to further study to provide a chemical characterization, with the goal of identifying specific chemicals of concern. Thus, the outcome of analysis may lead to another round of data collection, analysis, and risk characterization. Such iteration is an inherent strength of the risk assessment process.

1.2.4 RELATIVE RISK ASSESSMENT

Another tool in environmental management applied recently has been relative risk assessment.[34] To implement this method, an assessment group is formed of environmental experts (e.g., research scientists, regulatory agents, environmental managers from business) and nonscientists (e.g., homeowners, educators, businesspeople). The group is organized into committees representing differing perspectives and expertise with the goal of differentiating the importance of various issues of concern among stakeholders. All committees are charged with identifying the problems and issues that need to be addressed from their perspective, and then a consensus is sought across the stakeholders.

To help build consensus, experts in each group are asked to compile available data on each issue and to inform the panels. For example, what chemicals are present in water or sediment? Is toxicity associated with water or sediment? Is rainfall directed from roads and parking areas treated to remove contaminants before release to waterways? Is the air polluted and, if so, with what pollutants? What are the human health implications? What effects are likely known to occur in fish and other wildlife? The specific list of questions for which data are accumulated will depend on the issues collectively identified and the technical concerns attendant to each. Armed with this technical information, digested into a form that is understandable to everyone, the group seeks to reach consensus about relative risks associated with each issue, and to rank these in terms of perceived risk.

Although based ultimately on opinion, the consensus about relative risk for each issue is founded on the technical information accessible to the panelists. All stakeholders have participated in the evaluation process and have a sense, if the process worked well, that their concerns were appropriately considered.

The next step in the process is to identify and evaluate possible steps that can be taken to reduce the perceived risks. Remediation methods used in other locations may provide applicable solutions, but many times these cannot be used without further consideration of local conditions. For example, capping of contaminated estuarine sediments, as was done at Eagle Harbor in Puget Sound, is not possible in a shallow estuary such as the Elizabeth River in Virginia with shoals less than 3 m deep, and the need to maintain deepwater shipping channels. Simple removal of surface sediments in shoal areas may be an unacceptable resolution if the contamination penetrates deeply into the sediment. However, removal of contaminated sediment to some depth and then restoring the bottom contours by capping with clean sediment combines two technologies to isolate any residual contamination under a cap without drastically changing the bottom contour of the system. Modifications of waste discharge processes for both point and non-point sources may provide a long-term benefit, but must focus on the most significant sources of contaminants and fit the landscape and socioeconomic structure. The assessment group, representing a broad array of stakeholders, can identify by consensus a cleanup strategy that is acceptable to the community. Environmental managers can then further evaluate and implement the strategy, with some confidence that there is acceptance of the proposed solutions, provided that the assessment group remains active and informed.

This relative risk approach, although subjective and imprecise, can be an effective tool to guide management decisions. It has the inherent advantage that data on exposure and effects are evaluated by all constituencies, which then participate in prioritizing factors of greatest importance and selecting among courses of action based on perceived risks. This model was used by the Elizabeth River Project, Inc., in developing its Strategic Plan for the restoration of the Elizabeth River, an urban river and "region of concern" in the Chesapeake Bay in Virginia.[35,36]

1.2.5 A Case Study of Risk Assessment in an Estuary

In 1993, the U.S. EPA Office of Water and National Center for Environmental Assessment cosponsored five prototype case studies to demonstrate the utility of the risk assessment paradigm as a tool for environmental management of entire watersheds. One of these pilot studies addressed an estuarine system, the Waquoit Bay watershed in Massachusetts. This small estuary is a shallow coastal feature with a diverse surrounding landscape. It was small enough to allow a manageable exploratory study, yet of sufficient scale that the application of formal risk assessment procedures in this system is likely to have relevance to larger estuarine systems.[37] Waquoit estuary is a National Estuarine Research Reserve in which extensive partnering is possible to enhance research. Stressors of concern in this estuary include nutrient enrichment, suspended solids, changes in water circulation patterns, toxic chemical inputs, and habitat alterations. Of particular interest are effects on submerged aquatic vegetation in the system. Models are used to assess the risks associated with each of the stressors both singly and in combination. Although reports on risk assessments in this estuary have not been identified, this is a work in progress that bears watching.

1.3 FORUM ORGANIZATION

The forum from which this book was developed was focused on scientific aspects of contaminant exposure and effects, with emphasis on an estuarine and coastal context. Presenters dealing with chemical exposures were asked to address not only issues such as analyte type and efficacy of analytical approaches, but also bioavailability in estuarine and coastal settings. One concern that is often overlooked is that of new use or otherwise unregulated chemicals. History teaches that such chemicals may become significant contaminants before they are detected in water or sediment if there are no overt impacts on biota, e.g., Kepone. History also teaches that there may be chemicals periodically entering the estuary at high concentrations that may be difficult to detect and evaluate for effects. Therefore, a paper was solicited specifically on these issues.

Presentations were also solicited concerning effects at various levels of biological organization from biochemical to cellular to whole organism to population to landscape levels. At each level of organization, there are concerns over the sensitivity and precision of measurement tools and the predictive value of measurements at any given level for effects at higher levels of organization. Can laboratory methods adequately portray effects in the environment? What role do various biomarkers play in risk assessment? Rarely, does one objectively study the consequences of exposure to multiple stressors. Is this important to a meaningful risk assessment? And ultimately, although individual species are important because they have commercial or recreational value, aesthetic value, or key ecological roles, the real issue is the preservation of functional assemblages at a landscape level. Does one incorporate all important ecosystem components into data collection and analysis? Where possible, presenters were asked to show how their data could be applied in a risk assessment. Through such case studies, the forum hoped to show how one can include issues overlooked in less comprehensive analyses.

This forum examined the science and tools of risk assessment, used both in the laboratory and in watersheds, but it also included the logical tools by which risk is evaluated in an objective fashion. The intent was to juxtapose different approaches and views in a way leading to a better understanding of the critical features of estuaries that need to be considered in a successful environmental risk assessment. The summary chapter examines how well this intent was met.

This forum was a meeting of scientists speaking to peers using their own jargon and shorthand. Readers of this volume will most likely be scientists and managers familiar with this language. Yet ultimately, this tool will be used in a very public risk management and regulatory context. As scientists, we must always recognize that the people we serve may not see the world as we do. People often place more value on their own perception of risk than on technically sound quantitative estimates of risk obtained by methods they do not understand. Our job is simultaneously to collect the best scientific information possible, improve the science undergirding risk assessment, enhance the formal methods of synthesis used, and, of great importance, make our estimates of risk meaningful both to managers and to our larger constituency, the general public. This means that we must make our science and thinking accessible to nonexpert decision makers and the communities they serve, so that the most informed and balanced decisions possible can be made and implemented. We must learn to communicate in nontechnical, but meaningful ways, while avoiding misleading oversimplification.

REFERENCES

1. National Research Council (NRC), Risk Assessment in the Federal Government: Managing the Process, National Academy Press, Washington, D.C., 1983, 191.
2. National Research Council (NRC), Building Consensus through Risk Assessment and Management of the Department of Energy's Environmental Remediation Program, National Academy Press, Washington, D.C., 1994, 108.
3. National Research Council (NRC), Science and Judgment in Risk Assessment, National Academy Press, Washington, D.C., 1994, 651.
4. Stern, P.C. and H.V. Friebirg, Understanding Risk. Informed Decisions in a Democratic Society, National Academy Press, Washington, D.C., 1996, 249.
5. U.S. EPA, Framework for Ecological Risk Assessment, U.S. EPA/630/R-92/001, Risk Assessment Forum, Washington, D.C., 1992, 41.
6. National Oceanic and Atmospheric Administration (NOAA). 1998 (online). "Population: Distribution, Density and Growth" by Thomas J. Culliton. NOAA's State of the Coast Report. NOAA, Silver Spring, MD, available at http://state-of-coast.noaa.gov/bulletins/html/pop_01/pop.html
7. National Oceanic and Atmospheric Administration (NOAA) 2000 (online). "Population and Development in Coastal Areas," available at http://www-orca.nos.noaa.gov/projects/population/population.html
8. Stephan, C.E. et al., Guidelines for Deriving Numerical National Water Quality Criteria for the Protection of Aquatic Organisms and Their Uses, U.S. EPA, OWRS, CSD, PB85-227049, National Technical Information Service, Springfield, VA, 1985, 98.
9. U.S. EPA, Ambient Aquatic Life Criteria for Chlorine-1984, U.S. EPA 440/5/84-030, Washington, D.C., 1985, 57.
10. U.S. EPA, Ambient Aquatic Life Criteria for Cadmium-1984, U.S. EPA 440/5-84-032, Washington, D.C., 1985, 127.
11. Hall, L.W., Jr., Scott, M.C., and Killen, W.D., Ecological risk assessment of copper and cadmium in surface waters of Chesapeake Bay watershed. *Environ. Toxicol. Chem.*, 17, 1172, 1998.
12. Long, E.R. and Chapman, P.M., A sediment quality triad: measures of sediment contamination, toxicity and infaunal community composition in Puget Sound. *Mar. Pollut. Bull.*, 16, 405, 1985.
13. Chapman, P.M., Dexter, R.N., and Long, E.R., Synoptic measures of sediment contamination, toxicity and infaunal community composition (the sediment triad) in San Francisco Bay. *Mar. Ecol. Prog. Ser.*, 37, 75, 1987.

14. Long, E.R. and Morgan, L.G., The potential for biological effects of sediment-sorbed contaminants tested in the National Status and Trends Program, NOAA Technical Memorandum NOS OMA 52, National Oceanic and Atmospheric Administration, Seattle, WA, 1990, various pagings.
15. Long, E.R. et al., Incidence of adverse biological effects within ranges of chemical concentrations in marine and estuarine sediments. *Environ. Manage.,* 19, 81, 1995.
16. Long, E.R., Field, L.J., and MacDonald, D.D., Predicting toxicity in marine sediments with numerical sediment quality guidelines. *Environ. Toxicol. Chem.,* 17, 714, 1998.
17. MacDonald, D.D., Ingersoll, C.G., and Berger, T.A., Development and evaluation of consensus-based sediment quality guidelines for freshwater ecosystems. *Arch. Environ. Contam. Toxicol.,* 39, 20, 2000.
18. Holland, J.L., van Dolan, R.F., and Snoots, T.R., Predicting stress in benthic communities of southeastern U.S. estuaries in relation to chemical contamination of sediments. *Environ. Toxicol. Chem.,* 18, 2557, 1999.
19. McGee, B.L. et al., Assessment of sediment contamination, acute toxicity, and population viability of the estuarine amphipod *Leptocheirus plumulosus* in Baltimore Harbor, Maryland, USA. *Environ. Toxicol. Chem.,* 18, 2151, 1999.
20. U.S. Army Corps of Engineers, Formulation Analysis Notebook. Elizabeth River Basin, Virginia. Elizabeth River Environmental Restoration — Feasibility Study, U.S. ACE, 2000, 173 pp. + appendices.
21. Chapman, P.M., Dexter, R.N., and Goldstein, L., Development of monitoring programmes to assess the long-term health of aquatic ecosystems. A model from Puget Sound, USA. *Mar. Pollut. Bull.,* 18, 521, 1987.
22. U.S. EPA, 1999. Targeting Toxics: A Characterization Report. A Tool for Directing Management and Monitoring Actions in the Chesapeake Bay's Tidal Rivers, Chesapeake Bay Program, U.S. EPA 903-R-99-010, 1999, 49 pp. + appendices.
23. Hall, L.W. et al., A Pilot Study for Ambient Toxicity Testing in Chesapeake Bay. Vol. 1, Year 1 Report, CBP/TRS 64/91, Environmental Protection Agency, Chesapeake Bay Program Office, Annapolis, MD, 1991, 141 pp. + appendices.
24. Hall, L.W. et al., A Pilot Study for Ambient Toxicity Testing in Chesapeake Bay. Year 2 Report, CBP/TRS 82/92, Environmental Protection Agency, Chesapeake Bay Program Office, Annapolis, MD, 1992, 145 pp. + appendices.
25. Hall, L.W. et al., A Pilot Study for Ambient Toxicity Testing in Chesapeake Bay. Year 3 Report, CBP/TRS 116/94, Environmental Protection Agency, Chesapeake Bay Program Office, Annapolis, MD, 1994, 99 pp. + appendices.
26. Hall, L.W. et al., A Pilot Study for Ambient Toxicity Testing in Chesapeake Bay. Year 4 Report, CBP\TRS 172/97 (U.S. EPA 903-R-97-011), Environmental Protection Agency, Chesapeake Bay Program Office, Annapolis, MD, 1997, 82 pp. + appendices.
27. Hall, L.W., Anderson, R.D., and Alden, R.W., III, Ambient Toxicity Testing Chesapeake Bay — Year 5 Report, U.S. EPA 903-R-98-008, Environmental Protection Agency, Chesapeake Bay Program Office, Annapolis, MD, 1998, 68 pp. + appendices.
28. Hall, L.W. et al., Ambient Toxicity Testing Chesapeake Bay — Year 6 Report, U.S. EPA 903-R-98-017, Environmental Protection Agency, Chesapeake Bay Program Office, Annapolis, MD, 1998, 79 pp. + appendices.
29. Hall, L.W. et al., Ambient Toxicity Testing Chesapeake Bay — Year 7 Report, U.S. EPA 903-R-00-006, Environmental Protection Agency, Chesapeake Bay Program Office, Annapolis, MD, 2000, 88 pp. + appendices.
30. Hall, L.W. et al., Ambient Toxicity Testing Chesapeake Bay — Year 8 Report, U.S. EPA 903-R-00-012, Environmental Protection Agency, Chesapeake Bay Program Office, Annapolis, MD, 2000, 103 pp. + appendices.

31. Hartwell, S.I., Demonstration of a toxicological risk ranking method to correlate measures of ambient toxicity and fish community diversity. *Environ. Toxicol. Chem.,* 16, 361, 1997.
32. Hartwell, S.I. et al., Correlation of measures of ambient toxicity and fish community diversity in Chesapeake Bay, USA, tributaries — urbanizing watersheds. *Environ. Toxicol. Chem.,* 16, 2556, 1997.
33. Hartwell, S.I., Empirical assessment of an ambient toxicity risk ranking model's ability to differentiate clean and contaminated sites. *Environ. Toxicol. Chem.,* 18, 1298, 1999.
34. Anonymous, Michigan's Environment and Relative Risk, Michigan Department of Natural Resources, Lansing, MI, 1992, 49.
35. Regional Action Team, Elizabeth River Restoration. A Watershed Action Plan to Restore the Elizabeth River, Elizabeth River Project, Inc., Norfolk, VA, 1996, 103.
36. Fisher, C.W., The Elizabeth River Project. A grassroots, watershed approach to restoring the river, in *Workshop Proceedings—AMSA/U.S. EPA 1997 Pretreatment Coordinators Workshop,* 1997, 6 pp. + 3 figures.
37. U.S. EPA, Watershed Ecological Risk Assessment, U.S. EPA/822-F-97-004, Washington, D.C., 1997, 4.

2 European Approaches to Coastal and Estuarine Risk Assessment

Mark Crane, Neal Sorokin, James Wheeler, Albania Grosso, Paul Whitehouse, and David Morritt

CONTENTS

2.1 Introduction 16
2.2 Legislative Procedure in The European Union 16
2.3 Principles of Chemical Risk Assessment in the EU 18
2.4. Prospective Risk Assessment in the EU.. 19
    2.4.1 New Chemical Substances... 19
    2.4.2 Existing Chemical Substances 20
    2.4.3 The Technical Guidance Document... 20
    2.4.4 The Precautionary Principle... 21
        2.4.4.1 What Is the Precautionary Principle?................................. 21
        2.4.4.2 When and How Should the Precautionary Principle Be Applied?.. 21
        2.4.4.3 Remaining Problems with the Precautionary Principle 22
    2.4.5 Prospective Risk Assessment for Saltwater Environments in the EU 23
        2.4.5.1 Perceived Problems with Marine Risk Assessment in the EU 23
        2.4.5.2 Estimating a Saltwater PNEC 23
        2.4.5.3 Factors Potentially Affecting Correlations between Freshwater and Saltwater Toxicity Data............... 24
        2.4.5.4 Saltwater Species Sensitivity Distributions........................ 25
2.5 Retrospective Risk Assessments 26
    2.5.1 The Dangerous Substances Directive and Other Marine Regulations 26
    2.5.2 The Water Framework Directive.. 30
        2.5.2.1 Principles of the Water Framework Directive 30

    2.5.2.2   What Is "Good Status" for Marine Waters? 31
    2.5.2.3   Direct Biological Assessment of the Tees Estuary
              — A Case Study 34
2.6 Conclusions 35
Acknowledgments 37
References 37

2.1 INTRODUCTION

The European Union (EU) is neither a national legislature, such as the U.S. federal government, nor an international organization, such as the United Nations (UN). European Union members are completely sovereign states that have surrendered some law-making and enforcing powers, so that the powers of the EU go considerably beyond those of international organizations such as the UN, but not as far as those of the U.S. government.[1] This chapter briefly describes the political structure of the EU and the environmental regulations that have emerged from this structure. Much of the environmental legislation from the EU has been fragmentary, addressing single issues, or has focused on freshwater environments. There is now an increasing move toward an integrated approach to the environment, epitomized by the Water Framework Directive (WFD). Marine and estuarine waters, although not entirely ignored by earlier legislation, are now explicitly considered within the WFD. However, there is a recurrent practical problem. What are the fate, behavior, and toxicity of the many thousands of chemicals used in the EU for which we have little or no data for saltwater species? Must we test every chemical and every taxonomic group, or can we extrapolate between chemicals, biological species, and ecosystems? This is currently an important debate within the EU, with some environmental regulators proposing that toxicity to marine species must always be tested, while others believe that toxicity to freshwater species can be used to predict toxicity to marine species. There is also a recurring conceptual problem in most EU legislation. How can we define "high ecological status" for estuaries and coastal waters when we have difficulties in defining this for freshwater systems that have been more intensively studied? This chapter discusses these problems and provides some examples of ways in which they are being addressed by European researchers.

2.2 LEGISLATIVE PROCEDURE IN THE EUROPEAN UNION

After the devastation of two World Wars in the space of just over 30 years, the leaders of continental Europe finally recognized that new political structures were needed to avoid further bloodshed. The 1957 Treaty of Rome was a tool to establish a common market, expand economic activity, promote living standards, and, perhaps most importantly, encourage political stability in Western Europe. The European Community formed by the Treaty of Rome unified the European Economic Community, the European Atomic Energy Community, and the European Coal and Steel

Community.[2] This common market has expanded in both membership and aims over the decades since its formation. The postwar organization, set up primarily for economic and security reasons, has now evolved into the EU, a body with legislative powers that penetrate deeply into the daily life of every member state. Currently, the EU has 15 member states: Belgium, Luxembourg, the Netherlands, France, Germany, Italy, the United Kingdom, Ireland, Denmark, Greece, Spain, Portugal, Finland, Sweden, and Austria. More countries, principally from the former Soviet bloc, are likely to join the EU in the near future, with "Accession States" such as Poland due to join within the next 5 years.

Four major EU institutions are responsible for legislation under the Treaty of Rome: the European Commission, European Parliament, Council of Ministers, and European Court. The Commission is the supreme EU executive, comprising 20 independent members appointed by individual member states. Members of the Commission are charged with operating in the interests of the community as a whole, not as national representatives. Each commissioner has responsibility for an area of community policy, which includes a commissioner for the environment, and their main function is to propose EU legislation. The Commission civil servants are divided among 25 different directorates that report to the Council of Ministers. The most important directorates involved in chemicals and environmental legislation are Directorate General (DG) III (Enterprise), DGVI (Agriculture), and DG XI (Environment).[2] Legislation generally emerges from the commission in the form of proposals for Directives. Once accepted by the Council and Parliament these are usually enforceable across all member states and are the main basis for statutory controls in EU environmental legislation. Directives empower the commission to define objectives, standards, and procedures, but allow member states flexibility in implementation, so they can use their own national legislative processes. A Directive is therefore binding about the ends to be achieved, but leaves the means to member states. This has led to a variety of national methods for achieving environmental objectives as defined by Directives.

The European Parliament was originally a consultative and advisory body, but is gaining increasing legislative powers, and is the only part of the EU legislature that is truly open to public scrutiny. Its function is to assess proposals for legislation by commenting on the Commission's proposals. It also has some control over EU budgets, and the Council of Ministers must consult with the European Parliament over all new legislation. Increasingly, the Parliament has to agree to legislation as part of a "co-decision" (Council and Parliament). The Council of Ministers comprises government ministers from each member state and is the primary decision-making body in the EU. Its main function is to consider proposals from the Commission. Voting in the Council must be unanimous to be accepted on some issues (e.g., tax and defense), although a form of majority voting is now frequently used in other areas. The main function of the European Court is to interpret and apply all community law. All judgments are binding and member states must ensure that national legislation is compatible with EU law.

This then is the political framework that currently generates Europe-wide environmental policy and regulation. These policies generally comply with several principles that have been agreed upon by member states.

2.3 PRINCIPLES OF CHEMICAL RISK ASSESSMENT IN THE EU

Environmental laws in the individual countries of the European Union date to the 1800s, with the Rivers Pollution Prevention Act 1876 in the United Kingdom.[1] Over the following 90 years, several pieces of environmental legislation were enacted in various European countries with, arguably, only limited success. In the 1960s, widespread public concern about environmental issues was aroused in Europe as well as in North America by books such as Rachel Carson's *Silent Spring,* and by several dramatic environmental accidents, particularly oil spills at sea. As a result of this, the environment became accepted as a serious political issue in the 1970s.

In 1972, the year of the first UN Conference on the Environment, the European Community established fundamental principles that were to guide future policies for a wide range of environmental problems. These principles have been carried forward and enhanced in subsequent programs.[1,3] The more familiar principles are as follows:

- Polluters should pay for damage that they cause (the "Polluter Pays Principle");
- Prevention of environmental damage is more cost-effective and environmentally beneficial than dependence on subsequent remediation ("prevention is better than cure");
- Environmental action should be taken at the most appropriate level (regional, national, or international: the "Subsidiarity Principle"); and
- Action to prevent environmental damage can be taken in the absence of complete scientific knowledge (the much-debated "Precautionary Principle" discussed later in Section 2.4.4).

Aquatic environmental legislation adopted by the EU over the past two decades and relevant to this chapter can be divided into four broad categories:

1. Directives that impose rules and obligations on the supply of chemicals;
2. Directives that try to limit or prohibit discharges of dangerous substances into waters by industrial plants;
3. Directives and regulations that set water quality objectives for various uses; and
4. Geographically specific regulations on marine pollution to help protect the North, Baltic, and Mediterranean Seas.

Before turning to ways in which the EU performs chemical risk assessments, it is important to distinguish between the two main types of risk assessment that may be performed. Chemical risk assessments for any environmental medium may be prospective or retrospective.[4] Prospective, or predictive, risk assessments are usually performed to assess the future, usually generic, risks from releases of chemicals into the environment. In contrast, retrospective risk assessments are performed when sites have been contaminated historically, and such assessments are therefore necessarily site specific. Prospective risk assessments tend to have received more attention at the Commission because of the need to remove trade barriers through harmonization of all aspects of product testing. However, individual member states in the EU spend considerable resources on emission control and environmental monitoring within retrospective risk assessment frameworks.

Section 2.4 discusses ways the EU addresses prospective risk assessment, and Section 2.5 looks at ways retrospective risk is assessed. Throughout both of these sections the reader should bear in mind that the term *risk assessment* is not used in the strict sense of "the probability of an adverse event occurring." Instead, and in common with much of the rest of the world, the term is used rather loosely in the EU to cover regulatory processes that address chemical hazards and concentrations of chemicals in the environment, without necessarily combining them probabilistically.

2.4. PROSPECTIVE RISK ASSESSMENT IN THE EU

2.4.1 NEW CHEMICAL SUBSTANCES

The 1967 Council Directive 67/548 on the Classification, Packaging and Labelling of Dangerous Substances was enacted to classify chemicals for dangerous properties, thereby ensuring adequate labeling at the point of supply. The approach taken was to use hazard labeling for substances over a certain threshold volume when these were supplied in member states. This was so EU citizens would be aware of any dangers if the chemical were released deliberately or accidentally, but also so that a market without trade barriers could be established in the EU. However, it was only with the sixth amendment to Council Directive 67/548 in 1979 (79/631/EEC) that a series of biological, physical, and chemical tests became a requirement before new chemicals could be marketed. The seventh amendment (92/32/EEC) in 1992 introduced hazard classification for the environment, and risk assessment was introduced through a daughter Directive (93/67/EEC) in 1993.

The amount and type of information required for each chemical depends on its production volume. Testing within this scheme comprises three levels. Level 0 (or base level) is for production volumes of up to 10 tonnes/year and requires short-term toxicity data for the invertebrate *Daphnia magna* and for fish (e.g., rainbow trout, *Oncorhynchus mykiss*). Algal growth inhibition tests were mandated by the seventh amendment in 1992. For higher production volumes of up to 1000 tonnes/year (Level 1), and greater than 1000 tonnes/year (Level 2), more thorough toxicological testing is required, such as long-term toxicity studies with *Daphnia* and freshwater fish.

Marketing and environmental safety of plant protection products (pesticides) was treated separately in the 1991 Plant Protection Products Directive (PPPD, EU Directive 91/414/EEC). Many of the data requirements for the PPPD are similar to those required for new substances, e.g., toxicity profiles for fish, invertebrates, and algae. A deterministic risk assessment of product impacts on humans and wildlife must be carried out for both new and existing pesticides. This combines hazard assessments with rather simple environmental fate models to estimate toxicity exposure ratios. There is also a directive covering biocides, which contains many similar elements to the PPPD, including deterministic risk assessments.

2.4.2 Existing Chemical Substances

Of course, there are many thousands of existing substances in use within the EU that were never subject to testing under the Directives described above. Because of this, Council Regulation 793/93 on Existing Chemicals was developed to harmonize the different national systems for risk assessment within the member states. The regulation came into effect in 1993 and covers about 100,000 chemical substances that are thought to be used in the EU. Under this regulation, information must be provided to the Commission by industry for substances produced or imported into the EU in quantities over 1000 tonnes/year. The information collected by the Commission is then used for setting priorities on the basis of a preliminary risk assessment.

The priority substances are assessed by distributing them among member states for evaluation by national experts. The experts in these member states can ask for additional information from manufacturers and importers if the substance is suspected to be dangerous. On the basis of these data and more refined risk assessments, a substance may be deemed as dangerous by the Commission and be banned or restricted. The risk assessment methodology that should be used is described in the Technical Guidance Document,[5] described in more detail below. The EU existing substances regulations are coordinated with the Organisation for Economic Cooperation and Development (OECD) Chemicals Program. The OECD is the main coordinating body for the development of new ecotoxicity testing strategies and agrees on tests that then become mandatory data requirements in EU Directives.

2.4.3 The Technical Guidance Document

EU directives and regulations generally state some of the basic principles of risk assessment for new and existing chemicals, but lack detail. Because of this, the European Commission, the member states, and the European chemical industries produced a Technical Guidance Document (TGD) on risk assessment. This rather lengthy set of guidelines condenses to a series of deterministic equations for estimating chemical hazard and exposure in various environmental compartments, and the comparison of these using a quotient approach. A predicted environmental concentration (PEC) is estimated from data or models on chemical emissions and distribution in the environment.

A predicted no effect concentration (PNEC) is estimated by adding a safety factor, usually 10, 100, or 1000 depending on the level of uncertainty and availability of data, to ecotoxicity data, although for marine and estuarine systems, a further assessment factor of 10,000 has been suggested.[11] In the risk characterization phase, PEC and PNEC values are compared to decide whether there is a risk from a substance or whether further information and testing are needed to refine the risk quotients. The TGD requires the calculation of PEC/PNEC ratios for aquatic ecosystems, terrestrial ecosystems, sediment ecosystems, top predators, and microbes in sewage treatment systems. The TGD has also been implemented in a computerized system: the European Union System for the Evaluation of Substances (EUSES), which includes algorithms for the deterministic equations plus conservative default values that can be used in the absence of data. This means that EUSES calculations

European Approaches to Coastal and Estuarine Risk Assessment   21

can be made with limited data, such as the base set for new chemicals,[6] four physicochemical properties of the substance under consideration, and the tonnage that is likely to be present in the EU.[7]

PEC values are derived for local as well as regional situations, each based on a number of time- and scale-specific emission characteristics. As a consequence, several different exposure scenarios are estimated, leading to different PEC/PNEC ratios, some of which may exceed a threshold of 1 and some of which may not. If the PEC/PNEC ratio is greater than 1, the substance is considered to be of concern and further action must be taken. This may be through consulting with industry to see whether additional data on exposure or toxicity can be obtained to refine the risk assessment. If the PEC/PNEC ratio remains above 1 after the generation of further information, risk reduction measures will be imposed.

This risk assessment procedure should be performed within the spirit of the Precautionary Principle.

2.4.4 THE PRECAUTIONARY PRINCIPLE

2.4.4.1 What Is the Precautionary Principle?

The conventional form of the Precautionary Principle states: "preventative action must be taken when there is reason to believe that harm is likely to be caused, even when there is no conclusive evidence to link cause with effect: if the likely consequences of inaction are high, one should initiate action even if there is scientific uncertainty."[8] This principle has been increasingly adopted in Europe over recent years, as in the 1987 Ministerial Declaration on the North Sea, and the 1992 Convention for the Protection of the Marine Environment of the North East Atlantic. Although the European Commission has embraced the Precautionary Principle in its approach to environmental regulation, it recently felt that there was a need to explain exactly what it meant by *precaution*. This is because of criticisms from scientists that the Precautionary Principle, as defined by some environmentalists, seems to be an illogical tool with no place in science-based decision making. On the other hand, the commission has suffered criticism from environmentalists that it was acting in an insufficiently precautionary manner by following a slow risk assessment process when there was *a priori* evidence of damage caused by the substance being assessed.

According to the Commission, the Precautionary Principle should be considered as part of a structured approach to the analysis of risk. It assumes that the potentially dangerous effects of a chemical substance have been identified through scientific procedures, but scientific evaluation does not allow the risk to be quantified with sufficient certainty.

2.4.4.2 When and How Should the Precautionary Principle Be Applied?

The Commission wishes to apply the Precautionary Principle when there is evidence for a potential risk, even if this risk cannot be fully demonstrated or quantified because of insufficient scientific data. It should be triggered when scientific evaluation has shown a potential danger, and efforts have been made to reduce scientific uncertainty and fill gaps in knowledge that could allow a scientifically based decision to be made.

When the Precautionary Principle is invoked, the view of the Commission is that measures should be:

- Proportional (measures should be chosen to provide a specific level of protection to humans or wildlife);
- Nondiscriminatory (comparable situations should not be treated differently, and different situations should not be treated in the same way, unless there are objective grounds for doing so);
- Consistent (measures taken should be of a comparable scope and nature to those already taken in equivalent areas);
- Cognizant of costs and benefits (the overall cost of action and lack of action should be compared, in both the long and short term);
- Subject to review (measures based on the Precautionary Principle should be maintained so long as scientific information is incomplete or inconclusive, and the risk is still considered too high to be imposed on society); and
- Capable of assigning responsibility for producing the scientific evidence necessary for altering conclusions based upon the Precautionary Principle (for chemicals this will usually be the company that wishes to manufacture or market the chemicals).

2.4.4.3 Remaining Problems with the Precautionary Principle

Despite the Commission's valiant attempts to define the Precautionary Principle accurately and sensibly, and thereby defuse arguments about its applicability, these arguments still remain and are likely to gather force whenever a decision about a potentially dangerous chemical needs to be made. Santillo et al.[10] summarize the views of many European environmentalists who believe that the spirit of the Precautionary Principle is not being implemented fully in the EU. The main disagreement appears to be over *when* the Precautionary Principle should be invoked, rather than over *whether* it should be invoked. Santillo et al.[10] argue that too often a decision to regulate a substance is deferred until the results of further research become available. In their view this approach is based upon two flawed assumptions:

1. That greater understanding of the system under study will always result from further scientific study, allowing risks to be more accurately defined and quantified; and
2. That risks (usually commercial) arising from precautionary action taken now are greater than the currently undefined risks (usually human health or environmental) of inaction until the results of further investigations become available.

Of relevance to the subject of this book is the Santillo et al. view that the Precautionary Principle is in opposition to approaches based upon risk assessment.

This is because of the inherent uncertainty of risk assessment approaches based upon limited data, usually from laboratory tests on only a few species, and the use of arbitrary safety factors. The views of environmentalists such as Santillo and coworkers arguably carry more weight in Europe than in North America, because Green political parties have over the past decade been quite successful in European elections. However, it is still not entirely clear what factors should, in their view, trigger the invocation of the Precautionary Principle and, perhaps more importantly, what types of scientific evidence would allow such a decision to be reversed.

2.4.5 Prospective Risk Assessment for Saltwater Environments in the EU

2.4.5.1 Perceived Problems with Marine Risk Assessment in the EU

Despite the theme of this book, readers of this chapter will so far have encountered rather few references to coastal or estuarine systems. This is in part because there are problems in using the approaches described in the TGD for saltwater prospective risk assessments, as that document deals mostly with freshwater and terrestrial habitats. For example, the TGD provides guidance for the calculation of local PECs for several different environmental compartments. However, releases into coastal or estuarine waters are not specifically considered. There is a view, in Europe at least, that experience of risk assessment in the marine environment is insufficient to give sound practical guidance in the TGD. There are fears that large dilution factors, low biodegradation rates, and possible long-term exposure with consequent prolonged effects on saltwater organisms may produce quite different scenarios in marine systems when compared with freshwater systems.[11] Furthermore, information on releases to saltwater systems is scarce for some substances, making it difficult to estimate a PEC. Despite this, modifications of EUSES to permit its use for risk assessment for the marine environment have been proposed.[12] These are being developed by the OSPAR (Oslo Paris Commission) DYNAMEC group, which has also developed a modification of the COMMPS (Combined Monitoring-based and Modelling-based Priority Setting) approach to prioritize those substances for which marine risk assessment is urgently required.

The remaining part of this section outlines some of the problems that researchers and environmental regulators in the EU currently have in attempting to estimate the toxicity of chemicals to saltwater biota.

2.4.5.2 Estimating a Saltwater PNEC

A key step in chemical risk assessment is the estimation of a PNEC. In practice, risk assessors must extrapolate from a relatively small data set, usually containing data from fewer than ten species, to estimate the PNEC.[13] This is normally achieved by applying safety factors to the lowest effects concentrations from reliable studies, although species sensitivity distribution models are considered by some regulatory authorities.[14] There are generally fewer data available for saltwater species than for freshwater species, especially for organic compounds,[15] largely because there are fewer standard test methods in the EU for saltwater species and because aquatic risk assessments have traditionally tended to focus on freshwater systems. Because of this paucity of data, many saltwater PNECs rely on extrapolations from freshwater data. This surrogate approach assumes that freshwater species respond like marine species, and that the distributions of freshwater and saltwater species sensitivities are similar — assumptions that are only now being addressed in current research programs.

2.4.5.3 Factors Potentially Affecting Correlations between Freshwater and Saltwater Toxicity Data

The degree of correlation between freshwater and saltwater toxicity data could potentially be influenced by two main factors.

1. *Biological differences between saltwater and freshwater animals.* For example, saltwater and freshwater invertebrates differ in their physiology, phylogeny, and life histories, which has implications for their sensitivity to toxicants. Most important is the greater phylogenetic diversity in marine environments compared with freshwater environments, with some components of marine assemblages absent from fresh waters (e.g., Echinodermata, Cephalopoda, and Ctenophora). Conversely, freshwater data sets may of course include insects, higher plants, and amphibians that are generally absent from the marine environment. Differences in physiology may also be responsible for differences in the uptake and toxicity of certain chemicals to freshwater and marine crustaceans and fish.[16-19] Many more saltwater species have pelagic planktonic stages that can exhibit markedly different sensitivities to chemicals.[20,21] Finally, reproductive strategies of marine invertebrates are less responsive to changing environmental conditions, which might lead to differences in sensitivity to toxicants.[22] Some studies show good correlations between the sensitivities of particular freshwater and saltwater invertebrates, fish, and algae.[26,27] After reviewing the European Chemical Industry ECETOC Aquatic Toxicity Database, Hutchinson et al.[22] concluded that freshwater to saltwater toxicity could be predicted with greater confidence for fish than for invertebrates.
2. *Differences in chemical behavior, especially speciation and bioavailability.* Differences in bioavailability in fresh and salt waters can be expected for a number of inorganic substances and can have a major impact on toxicity.[23] When reviewing toxicity data, it is important to recognize the possible differences between total concentrations of a substance and concentrations that are bioavailable or are biologically active.[24] Differences in solubility of organic chemicals between fresh water and salt water can influence partitioning between water and tissues, with the effect that differences in uptake or the time required to attain a critical body burden may occur. Such differences are often acknowledged in water quality standards (which may be regarded as PNECs), with different standards for salt and fresh waters.

2.4.5.4 Saltwater Species Sensitivity Distributions

Useful progress in discovering whether there are any systematic differences in the responses of freshwater and saltwater organisms can be made by comparing the sensitivities of freshwater and saltwater species to the same chemicals. The construction of species sensitivity distributions (SSDs), using toxicity data for different species, provides information about the range of species sensitivities to individual chemicals, and allows estimation of the concentration predicted to affect only a small proportion (typically 5%) of species. Recent evidence shows that these distributions may be strongly influenced by the mode of toxic action of a chemical, which can influence both the range and the complexity of the distribution.[25] Thus an understanding of the mode of toxicity of a chemical is important when comparing the distributions of species sensitivities. In addition, species sensitivity distribution models assume a random selection of test species, which is clearly not always the case.[13] A possible cause of any differences between sensitivity distributions of freshwater and saltwater organisms may therefore be due to differences in the taxonomic compositions of the data sets.

The U.S. EPA AQUIRE database was used to construct SSDs for both freshwater and saltwater species (e.g., Figure 2.1). To maximize the data set, short-term (acute) EC_{50} values were chosen, with a minimum of six species as recommended by the Ecological Committee on FIFRA Risk Assessment Methods,[28] although we recognize that in deriving PNECs environmental regulators would demand use of only long-term (chronic) test results. The database yielded 22 substances that had sufficient data (acute EC_{50} values for at least six freshwater and six saltwater species) to construct distributions based on log–logistic responses. Data were summarized using the regression coefficients and a point estimate referred to in Europe as the HC_5 (hazardous concentration that will exceed no more than 5% of species toxicity threshold values), otherwise known as the 95% protection level.[29] For the purposes of comparison, ratios of the HC_5 values for freshwater and saltwater species

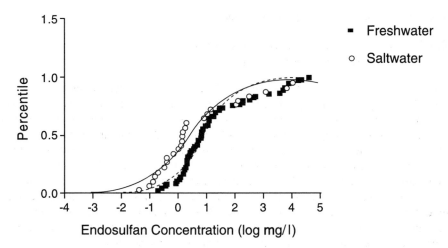

FIGURE 2.1 Freshwater and saltwater species sensitivity distributions for endosulfan.

responses are compared in Table 2.1. Despite the taxonomic differences alluded to earlier, differences in freshwater and saltwater HC_5 estimates were within a factor of 10 for 18 of the 22 chemicals examined in this way. Risk assessors in the EU would consider that this was sufficiently similar to suggest no difference in sensitivity between freshwater and saltwater organisms (S. Robertson, Environment Agency of England and Wales, personal communication). Three of the heavy metals (chromium, lead, and zinc) were more toxic to freshwater organisms, with HC_5 estimates for freshwater organisms more than ten times lower than for saltwater organisms. These results were likely due to the greater bioavailability of metals in fresh water. Ammonia was also more toxic to freshwater species, although this almost certainly resulted from a preponderance of data for fish (the taxon most sensitive to ammonia) in the freshwater data set. In other cases, a tendency to greater sensitivity by saltwater species was noted in the case of pesticides, although, again, differences in species composition of the available data sets may have had an influence.

The SSD approach described above can be used to identify species or groups to be recommended for further practical toxicity testing, and it uses the available toxicity data more effectively than simply relying upon the most sensitive species to generate a PNEC. From the limited analysis presented here, there is at least some evidence that freshwater toxicity data could in most cases be used to predict saltwater toxicity. If information is available on the distribution of chemical concentrations in the environment, then SSDs can be combined probabilistically with this exposure information to provide a richer perspective on the probability and severity of environmental harm.

However, SSDs are currently suitable only for regulation of individual substances. The next section shows that, for retrospective and site-specific risk assessments, the EU has also historically adopted a substance-by-substance approach. This is now changing, with more explicit recognition that complex mixtures of chemicals in complex environments may best be assessed by appropriate combinations of chemical analysis, biological sampling, and direct bioassay of environmental samples.

2.5 RETROSPECTIVE RISK ASSESSMENTS

2.5.1 THE DANGEROUS SUBSTANCES DIRECTIVE AND OTHER MARINE REGULATIONS

The Dangerous Substances Directive of 1976 (76/464) provided the first framework for eliminating or reducing historical water pollution by particularly dangerous substances, in both freshwater and saltwater systems. Member states were required to take appropriate steps to eliminate pollution by 129 toxic and bioaccumulable "List I" substances (otherwise known as "black list" substances) and to reduce pollution by "List II" substances (otherwise known as "gray list" substances). List I contains organohalogen and organophosphorus compounds, organotin compounds, carcinogenic substances, mercury and cadmium compounds, whereas List II includes biocides not included in List I; metalloids/metals and their compounds; toxic or organic compounds of silicon; inorganic compounds of phosphorus, ammonia, and nitrites; cyanides and fluorides; nonresistant mineral oils, and hydrocarbons of petroleum origin. List II compounds are regarded as less dangerous to the environment

TABLE 2.1
Summary Statistics from Log-Logistic Species Sensitivity Distributions Generated with Data on the Toxicity of 22 Chemicals to Freshwater and Saltwater Organisms

Chemical	n	STD (log)	α	β	r^2	HC_5 (log µg/l)	HC_5 (µg/l)	Lower 95% CI (µg/l)	Ratio FW:SW HC_5
Ammonia									
FW	27	0.895	3.609	0.492	0.958	2.493	311.17	301.12	
SW	14	0.729	4.460	0.401	0.984	3.207	1610.66	1592.58	0.19
Benzene									
FW	28	0.520	4.893	0.286	0.987	3.815	6531.30	6513.54	
SW	6	0.863	4.572	0.475	0.926	3.227	1686.55	1682.97	3.87
Cadmium									
FW	42	1.284	2.960	0.706	0.989	0.352	2.25	nc	
SW	31	1.339	3.083	0.737	0.972	1.075	11.89	0.005	0.19
Chlordane									
FW	25	0.778	1.760	0.428	0.962	0.470	2.95	2.68	
SW	8	0.666	0.837	0.367	0.969	−0.358	0.44	0.42	6.7
Chlorpyrifos									
FW	90	1.226	0.928	0.674	0.987	−1.422	0.038	0.001	
SW	19	1.566	0.872	0.861	0.938	−1.816	0.015	0.014	2.53
Chromium									
FW	15	1.139	3.002	0.626	0.976	0.66	4.57	2.25	
SW	7	0.887	4.204	0.488	0.930	1.913	81.85	78.34	0.056
Copper									
FW	42	1.014	2.368	0.558	0.974	0.521	3.32	2.09	
SW	24	0.602	2.333	0.331	0.992	1.091	12.33	11.53	0.27

(*continued*)

TABLE 2.1 (CONTINUED)
Summary Statistics from Log-Logistic Species Sensitivity Distributions Generated with Data on the Toxicity of 22 Chemicals to Freshwater and Saltwater Organisms

Chemical	n	STD (log)	α	β	r^2	HC_5 (log µg/l)	HC_5 (µg/l)	Lower 95% CI (µg/l)	Ratio FW:SW HC_5
Dichloroaniline									
FW	14	0.959	3.497	0.539	0.960	2.398	250.04	244.71	
SW	11	0.587	3.717	0.269	0.941	2.672	469.89	467.11	0.53
Dieldrin									
FW	58	0.901	1.355	0.496	0.982	−0.230	0.589	0.376	
SW	33	0.951	1.458	0.523	0.966	−0.104	0.787	0.622	0.75
Endosulfan									
FW	76	1.362	1.231	0.749	0.952	−0.879	0.132	0.051	
SW	25	1.506	0.604	0.578	0.949	−2.142	0.007	0.006	18.85
Lead									
FW	11	1.117	3.261	0.614	0.886	1.334	21.58	19.41	
SW	6	0.182	3.880	0.100	0.971	3.510	3235.94	3234.96	0.006
Lindane									
FW	97	1.049	2.309	0.577	0.992	0.261	1.824	0.658	
SW	35	1.339	1.909	0.670	0.957	−0.296	0.506	0.495	3.60
Malathion									
FW	150	1.377	2.767	0.757	0.984	−0.036	0.92	nc	
SW	28	1.231	2.287	0.677	0.982	0.124	1.33	0.767	0.69
Mercury									
FW	15	1.150	2.242	0.632	0.983	−0.034	0.925	0.611	
SW	13	0.803	2.223	0.442	0.958	0.569	3.707	3.363	0.25

Nickel									
FW	11	1.0276	3.438	0.564	0.957	1.898	79.07	75.97	
SW	9	0.844	4.093	0.464	0.986	2.409	256.45	250.67	0.31
Pentachlorophenol									
FW	80	1.094	2.720	0.602	0.976	1.423	28.58	23.81	
SW	30	0.649	2.767	0.357	0.990	1.456	28.58	26.51	1
Phenol									
FW	144	0.967	4.857	0.532	0.987	3.347	2223.31	1577.69	
SW	28	0.654	4.549	0.359	0.980	3.387	2437.81	2415.65	0.91
Potassium dichromate									
FW	79	0.959	4.153	0.528	0.985	2.675	473.15	457.12	
SW	33	0.587	4.178	0.323	0.975	2.927	845.28	832.28	0.56
Thiobenocarb									
FW	28	0.520	3.379	0.296	0.991	2.423	264.85	261.15	
SW	6	0.259	2.642	0.142	0.938	2.149	140.93	140.47	1.88
Toluene									
FW	18	0.555	4.838	0.305	0.987	3.764	5807.64	5791.86	
SW	7	0.700	4.767	0.385	0.980	3.085	1216.19	1209.53	4.78
Trichloroethane									
FW	7	0.321	4.963	0.118	0.974	4.472	29648.34	29645.56	
SW	12	0.321	4.831	0.177	0.854	4.139	13772.10	13767.83	2.15
Zinc									
FW	28	1.092	3.284	0.600	0.983	1.351	22.44	15.08	
SW	12	0.538	3.513	0.296	0.980	2.458	287.08	284.86	0.078

FW = freshwater, SW = saltwater, nc = not calculable.

than List I compounds, and may be contained within a given area depending on the characteristics and location of that area.

The Dangerous Substances Directive required member states to draw up authorization limits for emissions on both lists. In the case of List I, the limit values were to be at least equivalent to those adopted by the Commission. So far, six daughter Directives have established emission limits for specific substances on List I, including mercury, cadmium, hexachlorocyclohexane, carbon tetrachloride, DDT, and pentachlorophenol. For List II substances, member states are required to set water quality objectives and prior authorization requirements on industry to reduce pollution from these substances.

Several other Directives have focused on protection of specific marine environments. For example, Council Directive 79/923/EEC (Quality of Shellfish Waters) was intended to protect and improve the quality of coastal and brackish waters designated by member states for shellfish growth. Council Decision 75/437/EEC approved the Paris Convention. Its aim was to prevent marine pollution from terrestrial sources, i.e., that emanating from water courses, underwater pipelines, and ports, to the northeastern Atlantic and Arctic Oceans, the North and Baltic Seas, and parts of the Mediterranean Sea. This was extended in 1986 to cover marine pollution by emissions into the atmosphere.

Dissatisfaction with the fragmentary nature of these pieces of legislation, and their apparent failure to reduce pollution as much as was hoped, has been the impetus behind legislation that seeks to draw all of these strands together: the Water Framework Directive.

2.5.2 THE WATER FRAMEWORK DIRECTIVE

2.5.2.1 Principles of the Water Framework Directive

The Water Framework Directive (2000/60/EC) will provide the basis for future EU water legislation.[30] The Directive aims to ensure the quality of EU waters and takes a holistic approach to water management. It will update existing water legislation through the introduction of a statutory system of analysis and planning based upon the river basin, the use of ecological as well as chemical standards and objectives, the integrated consideration of groundwater and surface water quality and quantity, the introduction of some new regulatory factors, and the phased repeal of several European Directives. The Directive will contain both environmental quality standards and emission limit values from point sources.

The main principles of the Water Framework Directive (WFD)[31] are as follows:

- Expansion of water protection measures. All European waters will be subject to protection under the Water Framework Directive. Unlike previous water legislation, the directive covers surface water, groundwater, estuaries (or "transitional" waters), and marine waters.
- "Good status" for all waters within the next few years, apart from limited exceptions. Good status for surface waters is measured in terms of ecological and chemical quality. Member states will need to establish programs for monitoring these criteria.

- Water management based on river basins, rather than administrative or political boundaries. For each river basin district, a River Basin Management Plan will have to be established and updated at regular intervals. Coastal waters will be assigned to the nearest or most appropriate river basin district.
- A program of emission limit values, water quality standards, and any other necessary measures. Within the framework of river basin management plans member states will need to establish a program of measures to ensure that all waters within a river basin achieve good water status. Foremost in this process is the application of local, national, and then EU regulation. Where necessary, this can be reinforced by implementing stricter controls for industry, agriculture, and urban wastewater. The Directive sets standards at two levels: at source by setting emission values and at the water body scale with water quality objectives.
- Getting citizens involved. The establishment of river basin management plans will require more involvement of, and consultation with, EU citizens, interested parties, and nongovernmental organizations.

2.5.2.2 What Is "Good Status" for Marine Waters?

Annex 5 of the WFD outlines technical specifications for the definition, classification, and monitoring of surface water ecological and chemical status. The annex does not attempt to define good ecological status for marine waters quantitatively, but states that a further proposal is required for this. Table 2.2 provides some examples of qualitative definitions of high, good, and moderate biological status for estuaries. These examples, although a starting point, will clearly be difficult to operationalize. The Commission rightly believes that monitoring of the marine environment on a consistent and systematic basis is important, so that it is possible to identify the information necessary to produce quantitative targets. It has proposed a basic set of monitoring obligations designed to be as consistent as possible with the current obligations of member states. The framework contains the following monitoring guidelines for estuarine and coastal waters.

- *Biological parameters:* Composition and abundance of aquatic flora (other than phytoplankton); composition, abundance and biomass of phytoplankton; composition and abundance of benthic invertebrate fauna; composition and abundance of fish fauna (in estuaries only).
- *Hydromorphological parameters to support the biological elements:* Tidal regime (freshwater flow and wave exposure in transitional waters, and direction of dominant currents and wave exposure in coastal waters); morphological elements (depth variation, quantity, structure and substrate of the bed, structure of the intertidal zone).
- *Chemical and physiochemical parameters to support the biological elements:* Transparency; thermal conditions; oxygen conditions; salinity; nutrient conditions; all priority polluting substances; other substances identified as being discharged in significant quantities into the body of water.

TABLE 2.2
Abbreviated Water Framework Directive Definitions of High, Good, and Moderate Biological Status for Estuaries (transitional waters)

Taxonomic Group	High Status	Good Status	Moderate Status
Phytoplankton	Taxonomic composition and abundance consistent with undisturbed conditions	Slight changes in composition and abundance	Moderate changes in composition and abundance
	The average biomass is at type-specific levels and does not alter transparency	Slight changes in biomass, but no indication of a potentially harmful acceleration in algal growth	Moderate changes in biomass, which may indicate a potentially harmful acceleration in algal growth
	Frequency and intensity of algal blooms consistent with type-specific conditions	A slight increase in the frequency and intensity of blooms may occur	A moderate increase in the frequency and intensity of blooms may occur; persistent summer blooms may occur
Macroalgae	Macroalgal taxa consistent with undisturbed conditions, and no detectable changes in cover due to anthropogenic activity	Slight changes in composition and abundance, but no indication of a potentially harmful acceleration in algal growth	Moderate changes in composition and abundance, which may indicate a potentially harmful acceleration in algal growth
Angiosperms	Angiosperm taxa consistent with undisturbed conditions, and no detectable changes in abundance due to anthropogenic activity	Slight changes in composition and abundance	Moderate changes in composition and abundance
Benthic invertebrates	Invertebrate diversity and abundance within range normally associated with undisturbed conditions, and sensitive taxa present	Diversity and abundance slightly outside type-specific range; most sensitive taxa present	Diversity and abundance moderately outside type-specific range; many sensitive taxa absent and pollution-tolerant taxa present
Fish	Species composition and abundance consistent with undisturbed conditions	Disturbance-sensitive species show slight signs of distortion from type-specific conditions	A moderate proportion of type-specific disturbance-sensitive species absent

Source: European Commission.[30]

Member states will have to monitor at a frequency chosen to achieve a stated level of confidence and precision. For biological monitoring, member states will be required to present the monitoring results for each site in terms of any deviation from "high status" reference conditions for that site. The establishment of what actually constitutes a high status reference condition should be completed soon after the Directive is implemented.

The Commission also wishes to ensure an exchange of information between member states leading to the identification across the community of a set of water bodies, representing a cross section of ecotypes and qualities. This group of sites will be collectively known as the "intercalibration network." A register of the sites comprising the intercalibration network will also be made available soon after the WFD has been implemented by member states.

As it stands, the WFD is an ambitious attempt to introduce explicit biological targets for water quality in the EU. However, establishing baselines, or "type-specific conditions" for different taxonomic groups, will be difficult, and may not be achievable within the time period set by the Directive. Because of this, a variety of complementary biological and chemical analyses may need to be used to establish whether a water body is of sufficiently high status.

To date, such studies have been rather limited in most member states. For example, the only regular, formal use of bioassays and ecological surveys to assess ambient saltwater quality in the United Kingdom is as part of the National Marine Monitoring Plan (NMMP), although even this use is only a nonstatutory national commitment.[32] The NMMP is an integrated bioassay, ecological survey, and chemical analytical program. Pacific oyster (*Crassostrea gigas*) embryo larval (OEL) bioassays and copepod (*Tisbe battagliai*) survival bioassays have been used to test the toxicity of water from estuaries and around the coast of the United Kingdom for several years as part of the NMMP. Sediment elutriates are also bioassayed with OEL bioassays, and whole sediments are tested in 10-day survival bioassays with the amphipod, *Corophium volutator*, and the lugworm, *Arenicola marina*. The feeding of lugworms is also assessed during the latter test by examining the quantity of casts deposited at the sediment surface. Surveys of benthic macroinvertebrate assemblage structure form part of the NMMP, as do population surveys of flatfish, which are selected for analysis of biomarkers or histological abnormalities because they live in close proximity to potentially contaminated sediment surfaces. Results from these bioassays have shown that toxicity is almost exclusively confined to industrialized estuaries in the United Kingdom, while sediments in coastal waters do not produce a toxic response.

Fleming et al.[33] provide a further example of how single-species bioassays could be used to assess water quality within the WFD, in lieu of direct ecological monitoring. Such alternatives may be important during the first few years after implementation of the WFD, at least until robust ecological monitoring schemes have been set up in member states, although it should be noted that the WFD does not explicitly mention the use of bioassays. It may even be the case that, under some environmental conditions, single-species bioassays provide more reliable and interpretable results than those obtained from ecological surveys.

2.5.2.3 Direct Biological Assessment of the Tees Estuary — A Case Study

The Tees estuary in the northeast of England (Figure 2.2) is heavily industrialized and flows into the North Sea. It has for many years been regarded as one of the most polluted estuaries in the United Kingdom, although levels of contaminants have declined and benthic diversity has increased in recent years as a result of improvements in effluent treatment and the closure of some industrial sites.[34–36]

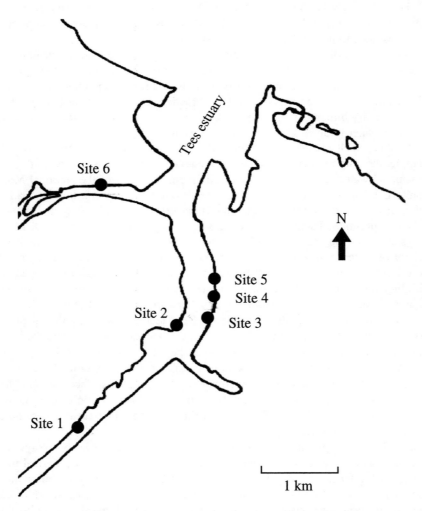

FIGURE 2.2 Diagram of the River Tees, showing six stations used to sample benthic macroinvertebrates, plus water and sediment for laboratory bioassays. *In situ* bioassays were also deployed at these stations.

Fleming et al.[33] took samples of benthic macroinvertebrates from several stations along the Tees, so that assemblage structure could be analyzed. They also performed laboratory bioassays with water and sediment samples from the same stations, using microbial bioluminescence, chemiluminescence, oyster embryo–larval development, copepod survival and reproduction, amphipod survival and growth, and polychaete cast formation. *In situ* studies were performed to examine mussel feeding rates and amphipod survival.

Nonmetric multidimensional scaling (MDS) of the benthic assemblage and bioassay data[37,38] (Figure 2.3) showed that station 4 in the Tees differed most from the others. This site, close to Dabholm Gut, has been identified in previous studies as particularly impacted by industrial pollutants.[34] When chemical data were compared with these biological data, using the BIOENV statistical procedure described by Clarke and Warwick,[37] nickel and zinc emerged as the measured chemical substances most highly correlated with both macroinvertebrate assemblage ($r = 0.91$) and bioassay ($r = 0.77$) data.

These types of direct biological assessment of coastal and estuarine quality, currently common in North America, will become much more common throughout Europe when the WFD is implemented. Their use should go some way toward addressing questions about contaminant bioavailability, mixture toxicity, and laboratory-to-field extrapolations that currently undermine monitoring approaches that rely solely on chemical analysis.

2.6 CONCLUSIONS

The EU represents a vast geographical area, which will become even larger when countries in Central Europe and the Near East join. Although originally formed in response to political and economic pressures, the EU is increasingly active in developing Directives and frameworks for both prospective and retrospective risk assessments of chemical pollution.

The environmental problems faced by EU member states are very similar to those faced in North America and in other parts of the industrially developed world. How do we prevent new dangerous chemicals from entering marine environments? How can we tell whether chemicals already in use are causing damage to coastal and estuarine systems? What are the best ways to assess risk, at what point should a decision be made under uncertainty, and how precautionary should we be in the face of international economic competition?

In the EU, prospective risk assessment of chemicals still depends largely on PEC/PNEC approaches, but there is increasing research activity and regulatory interest in the use of species sensitivity distributions and probabilistic risk assessment approaches. There is also a recognition that coastal and estuarine environments have been neglected during the development of EU prospective risk assessment strategies, and that this problem must be addressed.

A long history of urbanization and industrialization has left a legacy of polluted estuarine and coastal environments in the EU. Current retrospective risk assessment and monitoring strategies will largely be superseded by the Water Framework Directive, which provides ambitious environmental quality targets for all member states.

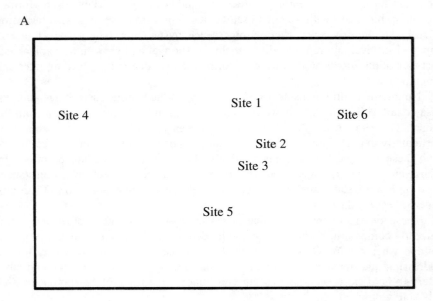

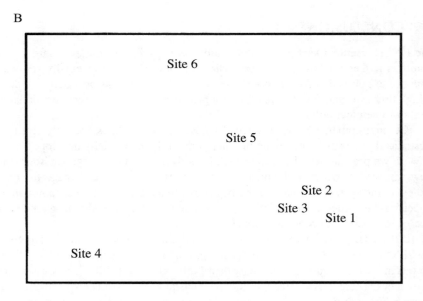

FIGURE 2.3 (A) Nonmetric multidimensional scaling ordination of benthic invertebrate data from the six stations in the Tees. (B) Nonmetric multidimensional scaling ordination of toxicity data from the six stations in the Tees. Stress for both representations was 0.00, suggesting excellent representation of these data.

It is likely that an appropriate combination of chemical analysis, ecological monitoring, and single-species bioassays will be necessary to monitor the achievement and maintenance of these targets.

At times, the diversity in language, culture, economy, and aspirations among the EU member states can make the differences between, for example, California and Louisiana appear trivial. These differences in the EU can only increase as more member states are accepted. Despite this, the marine and estuarine environments of the EU are generally improving in quality and look poised for further improvement if the Water Framework Directive delivers on its promise of more extensive and consistent biological monitoring.

ACKNOWLEDGMENTS

We thank the CEFIC Long Range Initiative and the Environment Agency of England and Wales for funding work described in this chapter, and Virginia Institute of Marine Sciences, Williamsburg, VA, for funding M.C. to deliver the presentation upon which this chapter is based. The comments of the editors and four referees on an earlier draft are also gratefully acknowledged.

REFERENCES

1. Blok, J. and Balk, F., Environmental regulation in the European Union, in *Fundamentals of Aquatic Toxicology,* Rand, G.M., Ed., Taylor & Francis, Washington, D.C., 1995, 775.
2. Shaw, I.C. and Chadwick, J., *Principles of Environmental Toxicology*, Taylor & Francis, London, 1998.
3. Abel, P.D., *Water Pollution Biology*, 2nd ed., Taylor & Francis, London, 1996.
4. Suter, G.W. II, *Ecological Risk Assessment*, Lewis Publishers, Boca Raton, FL, 1993.
5. European Union, Technical Guidance Document on Risk Assessment for New and Existing Substances, Part 2, Environmental Risk Assessment. Office for Official Publications of the European Community, Luxembourg, 1996.
6. Jager, T. and de Bruijn, J.H.M., The EU-TDG for new and existing substances: does it predict risk? in *Forecasting the Environmental Fate and Effects of Chemicals*, Rainbow, P., Hopkin, S., and Crane, M., Eds., John Wiley & Sons, Chichester, U.K., 2001, 71.
7. Swartz, S., Berding, V., and Matthies, M., Aquatic fate assessment of the polycyclic musk fragrance HHCB scenario and variability analysis in accordance with EU risk assessment guidelines. *Chemosphere,* 41, 671, 1999.
8. Eduljee, G.H., Trends in risk assessment and risk management, *Sci. Total Environ.,* 249, 13, 2000.
9. European Commission, Precautionary Principle communique of 02/02/2000. COM(2000)1, 2000.
10. Santillo, D. et al., The Precautionary Principle: protecting against failures of scientific method and risk assessment. *Mar. Pollut. Bull.,* 36, 939, 1998.
11. BUA Project Group. Assessment criteria for the marine environment. Final report, May 7, 1999.

12. Watts, C.D.W. et al., Risk management and assessment strategies for the marine environment. WRc report to the DETR 4607/2, 1999.
13. OECD, *Report of the OECD Workshop on the Extrapolation of Laboratory Aquatic Toxicity Data to the Real Environment,* OECD Monograph No. 60, Organisation for Economic Cooperation and Development, Paris, 1992.
14. Whitehouse, P. and Cartwright, N.G., Standards for Environmental Protection, in *Pollution Risk Assessment and Management,* Douben, P.E.T., Ed., John Wiley & Sons, Chichester, U.K., 1998, 235.
15. Solbé, J.F. et al., Developing hazard identification for the aquatic environment, *Sci. Total Environ.,* Supplement, 47, 1993.
16. Rainbow, P. S., Ecophysiology of trace metal uptake in crustaceans. *Estuarine Coastal Shelf Sci.,* 44, 169, 1997.
17. Ferguson, E.A. and Hogstrand, C., Acute silver toxicity to seawater-acclimated rainbow trout — influence of salinity on toxicity and silver speciation, *Environ. Toxicol. Chem.,* 17, 589, 1998.
18. Tachikawa, M. et al., Differences between freshwater and seawater killifish (*Oryzias latipes*) in the accumulation and elimination of pentachlorophenol, *Arch. Environ. Contam. Toxicol.,* 21, 146, 1991.
19. Tsuda, T. et al., Differences between freshwater and seawater-acclimated guppies in the accumulation and excretion of tri-n-butyltin chloride and triphenyltin chloride, *Water Res.,* 24, 1373, 1990.
20. Wong, C.K., Cheung, J.K.Y., and Chu, K.H., Effects of copper on survival, development and growth of *Metapenaeus ensis* larvae and postlarvae (Decapoda: Penaeidae), *Mar. Pollut. Bull.,* 31, 416, 1995.
21. Lee, R.F., O'Malley, K., and Oshima, Y., Effects of toxicants on developing oocytes and embryos of the blue crab, *Callinectes sapidus, Mar. Environ. Res.,* 42, 125, 1996.
22. Hutchinson, T.H., Scholz, N., and Guhl, W., Analysis of the ECETOC aquatic toxicity (EAT) database IV — comparative toxicity of chemical substances to freshwater versus saltwater organisms, *Chemosphere,* 36, 143, 1997.
23. Rainbow, P.S., Trace metal accumulation in marine invertebrates: marine biology or marine chemistry? *J. Mar. Biol. Assoc.,* 77, 195, 1997.
24. Dixon, E. and Gardner, M.J., Reactive aluminium in UK surface waters, *Chem. Speciation Bioavailability,* 101, 1, 1988.
25. Vaal, M. et al., Variation in the sensitivity of aquatic species in relation to the classification of environmental pollutants, *Chemosphere,* 35, 1311, 1997.
26. Dawson, G.W. et al., The acute toxicity of 47 industrial chemicals to fresh and saltwater fishes, *J. Hazardous Mater.,* 1, 303, 1975.
27. Robinson, P.W., The toxicity of pesticides and organics to mysid shrimps can be predicted from *Daphnia* spp. toxicity data, *Water Res.,* 33, 1545, 1999.
28. ECOFRAM (1999), available at http://www.epa.gov/oppefed1/ecorisk/index.htm.
29. Van Straalen, N. M. and Van Rijn, J.P., Ecotoxicological risk assessment of soil fauna recovery from pesticide application, *Rev. Environ. Contam. Toxicol.,* 154, 85, 1998.
30. European Commission, Directive 2000/60/EC of the European Parliament and of the Council of 23 October 2000 establishing a framework for community action in the field of water policy, *Off. J. Eur. Union,* L327, 1–72, 22 December 2000.
31. Bloch, H., The European Union water framework directive: taking European water policy into the next millennium, *Water Sci. Technol.,* 40, 67, 1999.
32. MPMMG, National Monitoring Programme Survey of the Quality of UK Coastal Waters. Marine Pollution Monitoring Management Group, Aberdeen, U.K., 1998.

33. Fleming, R. et al., Toxicity Based Criteria for Receiving Water Quality: Stage 2, R&D Technical Report P312. Environment Agency, Bristol, U.K., 1999.
34. Tapp, J.F., Shillabeer, N., and Ashman, C.M., Continued observations of the benthic fauna of the industrialised Tees estuary, 1979–1990, *J. Exp. Mar. Biol. Ecol.*, 172, 67, 1993.
35. Kirby, M.F. et al., Assessment of water quality in estuarine and coastal waters of England and Wales using a contaminant concentration technique, *Mar. Pollut. Bull.*, 36, 631, 1998.
36. Wedderburn, J. et al., The field application of cellular and physiological biomarkers, in the mussel *Mytilus edulis*, in conjunction with early life stage bioassays and adult histopathology, *Mar. Pollut. Bull.*, 40, 257, 2000.
37. Clarke, K.R. and Warwick, R.M., *Change in Marine Communities: An Approach to Statistical Analysis and Interpretation*, Plymouth Marine Laboratory, Plymouth, U.K., 1994.
38. McRae, G. et al., Relating benthic infaunal community structure to environmental variables in estuaries using nonmetric multidimensional scaling and similarity analysis, *Environ. Monit. Assessment* 51, 233, 1, 1998.

3 Emerging Contaminants of Concern in Coastal and Estuarine Environments

Robert C. Hale and Mark J. La Guardia

CONTENTS

3.1 Introduction 41
3.2 Brominated Fire Retardants 43
3.3 Polychlorinated Biphenyls 48
3.4 Natural and Synthetic Estrogens..49
3.5 Alkylphenol Ethoxylates and Associated Degradation Products 52
3.6 Other Pharmaceuticals 56
3.7 Nonpharmaceutical Antimicrobial Agents..58
3.8 Personal Care Products 58
3.9 Interaction of Multiple Stressors 59
    3.9.1 Multiple Xenobiotic Resistance 59
    3.9.2 STP Sludge..59
3.10 Conclusions 63
Acknowledgments..64
References..64

3.1 INTRODUCTION

Coastal and estuarine areas are strategically located, serving as focal points for commerce, as well as homes to a disproportionate share of the human population. As a consequence, they also receive a disproportionate share of the contaminants released. Because of their locations and their physical and chemical characteristics, they may also receive and trap additional contributions from upgradient watersheds and air sheds. Thus, these systems may be more vulnerable to degradation than less dynamic environments. Despite this, coastal and estuarine areas are very important wildlife habitats, serving as refuges and nurseries for a variety of organisms.

The initial and perhaps most important step in risk assessment, regardless of the system, is problem identification (see Chapter 1 for a discussion of the elements of risk assessment). Ideally, identification of environmentally problematic chemicals should be done prior to the occurrence of significant environmental damage. In practice, this process is often reactive, occurring after deleterious impacts of significant magnitude have already occurred. Chemicals that have emerged as problems in the past include organochlorine pesticides (e.g., effects on reproductive success of piscivorous birds and deformities in reptiles), mercury (e.g., accumulation in coastal marine life and resulting Minamata disease in Japanese residents), polybrominated biphenyls (e.g., PBB contamination of Michigan livestock and subsequent transfer to humans), tributytin (e.g., mortality and reproductive problems in European coastal shellfish), and Kepone (e.g., neurological disorders in Virginia chemical workers and contamination of estuarine biota of the tidal James River).[1-6] The time lapse between initial introduction of contaminants and assessment of impacts is critical, particularly when chemicals are resistant to degradation, are continuously introduced, or are widely dispersed. In some cases remediation is not possible or may be more destructive to the site than the contaminants themselves. Often chemical monitoring efforts, capable of detecting the presence of many contaminants prior to expression of widespread impacts, are retrospective and focus on so-called priority or historical pollutants.[7] Ironically, the justification offered in defense of this *modus operandi* is often that monitoring lists should not be expanded as current monitoring studies have failed to detect the compound in question. Analytical approaches also are increasingly specific, which is an asset when highly accurate results for selected chemicals at low environmental concentrations are required. However, this advantage may prevent recognition of the presence of new problem chemicals in the environment.[8] Deleterious effects are a culmination of all the chemicals (as well as other stressors) to which organisms are exposed, not just those chosen for study or regulation. We also are still learning what constitutes a significant effect. These effects may range from acute mortality to reallocation of valuable energy or other reserves.

Chemicals of concern are those for which the combination of toxicity and exposure exceeds a critical value, resulting in the expression of a deleterious effect. An emerging chemical of concern may be one that has been released into the environment for a considerable time, but for which effects have only now been recognized. The exact number of chemicals actually in commerce is uncertain, but estimates range as high as 100,000, with up to 1000 new compounds released each year.[9,10] The toxicological and environmental properties of only a fraction of these have been examined. An emerging contaminant of concern may also be a preexisting chemical whose production has increased, or for which a new use or mode of disposal has been found, increasing exposure. Existing chemicals for which important new modes of toxicity or environmentally important degradation intermediates have been discovered also may merit attention.

Persistent chemicals tend to accumulate in the environment, resulting in heightened ambient concentrations and exposure.[11] Bioaccumulative chemicals effectively increase the dosage within organisms themselves, although the location of these burdens may not coincide with the site of action. The impacts of so-called PBT (persistent, bioaccumulative, and toxic) chemicals have been recognized. The scientific literature is replete with studies on a few classes of these, notably polychlorinated biphenyls (PCBs), organochlorine pesticides, and polycyclic aromatic hydrocarbons. In fact, the U.S. EPA has recently established a PBT initiative in its Office of Pollution Prevention and Toxics. A significant portion of the emphasis has again been on organochlorine chemicals, banned in most developed countries. Mussel-watch data suggest that concentrations of these in U.S. coastal shellfish are decreasing.[12,13] Similar trends have been seen for organochlorines in other organisms such as Canadian seabirds.[14] While production of PCBs has stopped, large amounts remain in service and residues continue to be redistributed in the environment. Some organochlorine pesticides also remain in use, particularly in the tropics, on account of their effectiveness against disease-carrying insects and low cost. However, because of their physical properties, organochlorines continue to be transported to high latitudes, condense there, and accumulate in indigenous organisms. There they pose threats even to indigenous human populations that have never used the chemicals.[15,16] These "transboundary" contaminants have justifiably attracted the attention of the international scientific community, and efforts are expanding to elucidate their fate and consequences.

In contrast, the vast majority of chemicals released have received comparatively little attention from regulatory agencies and environmental scientists. A wider effort is needed to identify emerging contaminants of concern. While not exhaustive, several classes of these will be discussed here. Emphasis is on those that are bioaccumulative, have atypical degradation pathways, or interact with biological functions historically not fully considered by risk assessors.

3.2 BROMINATED FIRE RETARDANTS

Although we have learned much regarding designing chemicals with less deleterious environmental properties, some PBT chemicals are still being manufactured and used in large amounts. For example, a new generation of brominated fire retardants apparently has filled the niche formerly occupied by PBBs, largely deposited following the Michigan livestock feed incident. Fire retardants may be additive (present in, but not chemically bound to, the matrix) or reactive (covalently bound to the matrix). Tetrabromobisphenol A is one of the most widely used brominated fire retardants. It is reactive, limiting its dispersal somewhat. In addition, it has a log K_{ow} of 4.5; hence, its bioaccumulation potential is moderate.[17]

In contrast, brominated diphenyl ethers (BDEs) are additive fire retardants (see Figure 3.1A for a representative structure). BDEs are particularly important emerging contaminants as a result of their PBT properties. They are widely used in flammable polymers and textiles.[18] Their role there is critical, substantially decreasing the number of associated human fatalities. BDEs were first reported in soil and sediment in the United States in 1979 near manufacturing facilities and in a Swedish fish study in 1981.[19,20] However, their global distribution is only now becoming fully recognized.

Similar to PCBs and PBBs, BDEs are used commercially as mixtures, and 209 different congeners are theoretically possible, varying in their degree of halogenation. Three major commercial products are produced: Deca-, Octa-, and Penta-BDE (formulation designations will be capitalized to differentiate them from individual congener designations). Global BDE demand for the total of all three mixtures

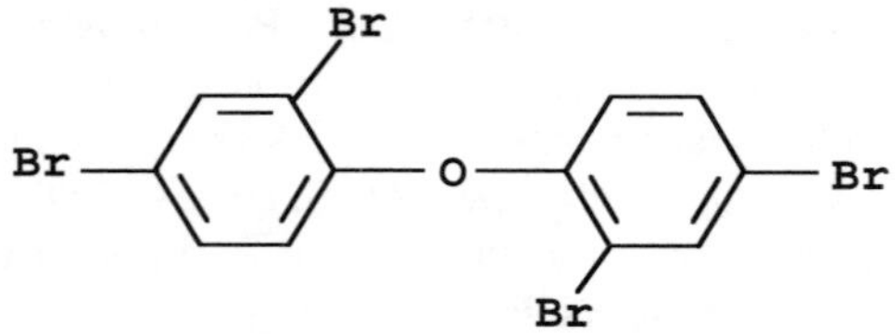

FIGURE 3.1 Structures of (A) a representative BDE (2,2′,4,4′-tetrabromodiphenylether, i.e., BDE-47); (B) a representative hydroxylated BDE; (C) the thyroid hormone thyroxin (also known as T4); and (D) triclosan. All have distinct similarities, i.e., a halogenated diphenyl or diphenyl ether backbone.

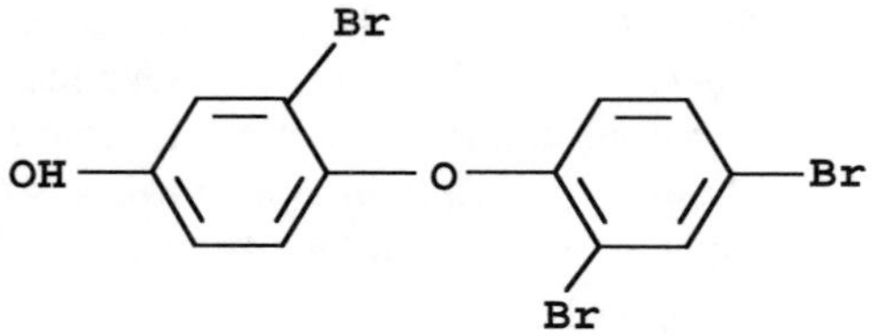

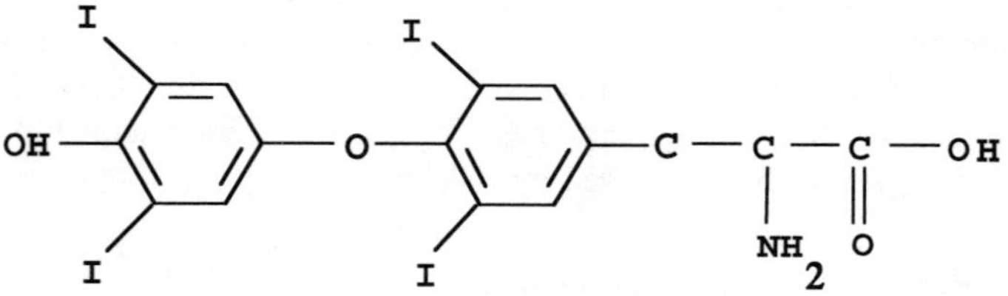

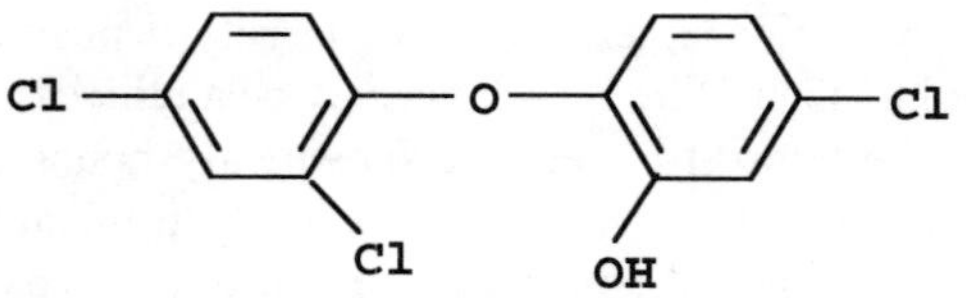

increased from 40,000 in 1992 to 67,125 metric tons in 1999.[18,21] The commercial Deca-BDE was reported to constitute about 82% of the reported total world BDE consumption in 1999.[21] It is used predominantly in plastics, such as high-impact styrenes, and on textiles. Commercial Deca-BDE consists mainly of a single fully brominated diphenyl ether (BDE-209, using the IUPAC PCB naming scheme), with contributions of <3% of the less-brominated diphenyl ethers.[22] Commercial Octa-BDE constituted less than 6% of total world BDE production in 1999, down from about 20% in 1992.[18,21] It also is used in plastics, including cabinets for computers, televisions, and other electronic devices. Hepta- and octa-congeners make up about

70 to 80% of this formulation, with hexa-, nona-, and deca-congeners constituting the remainder. The final widely used commercial mixture, the Penta-BDE formulation, is employed mostly in polyurethane foams, particularly in the United States. It also has been reported to be present in other products, e.g., in circuit boards.[18] The Penta-BDE formulation constituted 10% of the world market in 1992, increasing to approximately 13% in 1999.[18,21] It consists predominantly of congeners with 4 to 6 bromines, with the tetra- and penta-BDEs making up 74% or more of the total.[22] Sjodin et al.[23] characterized the congener composition of Bromkal 70-5DE, a widely used European-produced Penta-BDE formulation. They reported the major congeners 2,2',4,4'-tetra-BDE (BDE-47), 2,2'-4,4',5-penta-BDE (BDE-99), and 2,2',4,4',6-penta-BDE (BDE-100) contribute 37, 35, and 6.8% of the total, respectively. The major Penta-BDE mixture produced in the United States, DE-71, consists of similar proportions of these same congeners.

To date, no regulatory actions to restrict usage or releases of BDEs have been initiated in the United States. While the Deca-BDE mixture appears on the U.S. EPA Toxics Reduction Inventory (TRI), the Octa- and Penta-BDE formulations are not among the 600 chemicals designated by the 1986 Emergency Planning and Community Right-to-Know Act (EPCRA). Reporting of environmental releases of these chemicals is only required if facilities produce more than 25,000 lb or use more than 10,000 lb. Interestingly, a 10-lb threshold has now been proposed for a number of the banned organochlorines and 100 lb for tetrabromobisphenol A.[24] The latter compound is less bioaccumulative than several of the BDE congeners (notably the tetra- and penta-BDEs). As discussed above, most tetrabromobisphenol A in service is chemically bonded with its surrounding polymer matrix, decreasing its potential to migrate to the environment during product use. Although no current U.S. production figures for any of the three major BDE formulations are publicly available, all were listed as high-production-volume (HPV) chemicals on the 1990 Inventory Update Rule, required under the Toxic Substances Control Act, i.e., they were produced or imported to the United States in amounts exceeding 1 million lb annually.

Northern European nations have been quicker to take action to restrict usage of chemicals that may, by virtue of their properties, damage the environment under the so-called Precautionary Principle (see Chapter 2 for a discussion of this approach). Sweden and Denmark have called for a ban on BDE manufacture and the German Association of Chemical Industries voluntarily halted production in 1986.[25] The European Union completed a draft risk assessment in 2000 proposing an end to the use of the commercial Penta-BDE formulation.[25] Because of the global market for electronics, furniture, textiles, and automobiles and the use of BDEs in component parts, restrictions in selected countries may prove problematic.

Although toxicity studies on BDEs are limited, acute effects observed to date appear relatively modest.[18] In general, effects increase with decreasing bromination and most information available is on the commercial Deca-BDE formulation. The highly brominated BDEs are superhydrophobic (e.g., the BDE-209 log K_{ow} is 9.97).[22] They exhibit low bioaccumulation potentials, attributable to their large molecular sizes and tendency to remain associated with sedimentary organic matter in the environment. Reported effects of BDE exposure are similar to non-dioxin-type impacts of PCBs. Neurotoxic effects after neonatal exposure have been observed in mice after exposure to BDE-47 and BDE-99.[26] In addition, exposure to the commercial mixture Bromkal 70-5DE has been reported to decrease spawning success in sticklebacks.[27] Although shown to be a weaker inducer of the P-450 enzyme system than PCBs, BDEs may be hydroxylated *in vivo*. Asplund et al.[28] reported hydroxylated and methoxylated BDEs in concentrations similar to parent BDEs in blood of Baltic salmon. The former products possess an ether linkage analogous to the hormones thyroxine and triiodothyronine. Figure 3.1B and C contains structures of thyroxine and a hydroxylated BDE, respectively, for comparison. Meerts et al.[29] reported that some hydroxylated BDEs bind *in vitro* to the thyroid hormone transport protein transthyretin, with potencies similar to thyroxine.[29] Although hydroxylated PCBs have received considerable attention, BDEs may exhibit enhanced potency because bromine is intermediate between iodine and chlorine in the periodic chart. Under certain conditions, pyrolysis of BDE-containing products can result in the production of significant amounts of brominated dibenzo-p-dioxins and furans.[30–31] Although parent BDEs themselves appear to have low dioxin-like activity, some of these degradation products have been observed to be equal to, or more potent than, chlorinated analogues.[32]

The poor availability and high costs of authentic standards of individual BDE congeners have hampered research efforts. Nonetheless, BDEs are being detected in coastal and estuarine environments with increasing frequency. BDE-209 typically has been reported to be below quantitation limits in biota. In contrast, the less-brominated diphenyl ethers have been observed in wildlife and humans from several countries, including the United Kingdom, Germany, Sweden, Norway, the Netherlands, Japan, Canada, and the United States.[22,33,34] They recently were reported in remote arctic areas and in blubber of deepwater North Atlantic whales at about 100 µg/kg, suggesting entrance into the oceanic food web.[25,35] Marine mammals are prone to accumulate elevated concentrations of lipophilic contaminants, such as PCBs (see Chapter 9 for a further discussion of this subject). BDEs have also been observed in human adipose tissue, although typically at low microgram per kilogram levels.[36,37] Concentrations in breast milk in Swedish women, although still lower than PCBs, have been reported to be doubling at 5-year intervals since 1972.[38]

Figure 3.2 shows the BDE congener distribution observed in fish collected from south central Virginia. BDE-47 was the major congener detected in all species, in agreement with most published reports from other countries. BDE-99 and BDE-100, pentabrominated congeners, were also significant contributors in some species. As a consequence of the elevated burdens of tetra-BDE in edible fish tissue, assessments using data derived from exposure to the commercial formulations alone may underestimate risk because toxicity increases with decreasing bromination. The dominance of the less-brominated diphenyl ethers in fish appears at odds with production statistics. Industry has suggested these congeners are perhaps a legacy of historical usage of commercial penta-BDE formulations in offshore drilling, in European mining operations, or as a result of biosynthesis by marine invertebrates.[25] The Virginia fish data depicted in Figure 3.2 include freshwater and anadromous species. However, all samples were collected from freshwater systems with migration of the fish blocked by dams. Thus, the above explanations seem implausible here. BDEs were detected

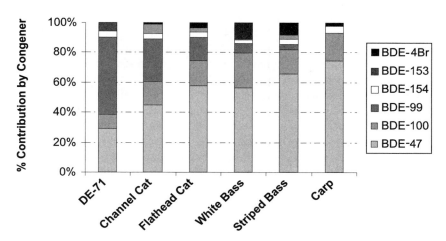

FIGURE 3.2 Percentage contributions of various congeners to the total BDEs detected in muscle tissue from five fish species, compared with DE-71, a commercial Penta-BDE formulation. BDE-47 was the dominant congener in fish, ranging from over 40% to over 70% of the total, as a function of species. BDE-99, the major pentabrominated congener in DE-71, was present at lower relative levels in the fish and was essentially absent in carp. BDE-4Br has been tentatively identified as BDE-49.

(quantitation limit ~5μg/kg on a lipid basis) in a surprisingly high 85% of the samples. In this case fillets, typically consumed by humans, were analyzed. Muscle generally contains lower burdens of lipophilic contaminants than more-fat-rich tissues such as liver. BDE-47 concentrations surpassed those of 4,4′-DDE and PCB-153 in 29 and 58% of the samples, respectively. The latter two compounds have been reported to be the most abundant organochlorine contaminants in U.S. fish.[39] The maximum BDE concentration (47.9 mg/kg lipid weight basis) in the Virginia fish was similar to the highest reported in Europe to date.[22] Yet, BDEs are not currently a U.S. EPA priority pollutant, nor have they typically been included in regulatory agency surveys. Lower concentrations of BDEs have recently been detected in fish from Chesapeake Bay tributaries, such as the Elizabeth and James Rivers.

Log K_{ow} values of 6.0 and 6.8 for BDE-47 and BDE-99, respectively, have been reported. These values are in the same range as the highly bioaccumulative tetra- to hexa-PCBs.[40] In contrast, BDE-209 has a log K_{ow} of 9.97.[33] Thus, it is not surprising that it is seldom detected in aquatic biota. High uptake efficiency of BDE-47, relative to both PCBs and more highly brominated diphenyl ethers has been observed for fish and bivalves.[40,41] Andersson et al.[42] recently compared uptake of selected BDE and chlorinated diphenyl ether (CDE) congeners in zebra fish (*Danio rerio*) from food. BDE-47 showed the greatest bioaccumulation factor, followed by BDE-28 (2,4,4′-tri-BDE), accumulating to a higher degree than the analogously substituted CDEs. BDEs with five or more bromines were accumulated to a lesser extent than either the analogous CDEs or the less-brominated BDEs. Interestingly, BDE congeners with adjacent bromines, i.e., BDE-85, BDE-99, and BDE-138, were not extensively concentrated. Differences in tissue burdens of individual BDE congeners in wild-caught fish by species have also been observed. In Virginia carp, disproportionately low amounts of BDE-99 (2,2′,4,4′,5-penta-BDE) were present relative to the total BDE burden (see Figure 3.2). This may be related to either differential uptake or elimination and could have important implications, particularly if degradation results in production of hydroxylated products.

BDE-99 contributions in sediments, unlike biota, typically rival those of BDE-47.[43–46] This suggests commercial Penta-BDE formulations might be sources. Deca-BDE (BDE-209), virtually absent in wild-caught biota, has been detected at high concentrations in some sediments.[22,43,44] However, BDEs with intermediate bromination (7 to 9 bromines) have seldom been reported in the environment. The long retention of BDE-209 on the gas chromatographic columns normally used for semivolatile compound determinations and its potential degradation during analysis may contribute to its underreporting.[47]

Photolytic debromination of deca-BDE in organic solvents, but not water, has been observed.[48] Tri- to octa-BDE and brominated furans were major products. Kierkegaard et al.[49] conducted a feeding study in rainbow trout using technical-grade, BDE-209 spiked into cod chips.[49] After 120 days of exposure, the trout contained predominantly hexa- through nona-BDE congeners. Interestingly, contributions from the tetra- and penta-congeners, the prominent congeners seen in wild fish, were not elevated compared with controls. The BDE-209 fed may have been debrominated in the fish or possibly trace amounts of the less-brominated BDEs, present as impurities in the dosing mixture, preferentially accumulated. Elevated levels of octa- and deca-BDE congeners have been seen in workers at a Swedish electronics-dismantling plant, confirming that these can be accumulated.[47] In that environment, employees likely were heavily exposed to the higher-brominated BDEs used in these products. Blood from hospital workers and office personnel using computers in the same study contained congener imprints dominated by the tetra- and penta-BDEs.

3.3 POLYCHLORINATED BIPHENYLS

PCBs were first reported in the environment in 1966[50] and since then have been widely detected in estuarine and coastal areas. Observation of alternative modes of toxicity, e.g., endocrine disruption, has resulted in their inclusion here as emerging contaminants of concern. PCBs were used commercially as complex mixtures of varying chlorination, e.g., Aroclors. It initially appeared logical to quantify environmental mixtures in terms of these formulations. However, the composition of the various commercial formulations released are modified as a function of the differing water solubilities, vapor pressures, vulnerability to degradation, and bioaccumulative potentials of the component congeners. Therefore, individual congener analysis was advanced to determine their true concentrations in environmental matrices more accurately.[51,52] As it is difficult to assess all the congeners potentially present, most monitoring efforts focus on a subset, e.g., those detectable in highest concentrations or diagnostic of the original commercial formulation. Alternatively, congeners are selected based on concerns over a specific mode of toxicity. This rationale is questionable, in light of the recognition of other toxicological end points as a function of chemical structure.

Establishment of the interaction of coplanar PCBs with the aryl hydrocarbon (AH) receptor leading to TCDD-like toxicity, has resulted in emphasis on these congeners.[53,54] As a result, only these PCBs, or their TCDD-equivalents, have been considered in many risk assessments. The coplanar PCBs typically are detectable in very low concentrations and require considerable sample manipulation to quantify accurately. Information on the remaining non-coplanar PCBs, which form the bulk of the total, may be overlooked in the process. Given our lack of knowledge regarding the possible modes of toxicity, this approach may not be as protective or cost-effective as often presumed. Recently, it has been determined that some non-coplanar PCBs, i.e., those with chlorines in the *ortho* position(s), may be hormonally active agents.[2,55] This substitution pattern results in barriers to ring rotation and a more rigid configuration. The high bioaccumulative potential of the PCBs themselves, as for the previously discussed BDEs, increases the opportunity for effects, compared with more polar compounds that possess lesser accumulative tendencies.

Studies have suggested neurotoxic effects and deficits in humans and other organisms as a function of PCB exposure during development.[56–58] Chauhan et al.[56] reported that *ortho*-substituted PCBs were capable of binding with transthyretin and thus interfering with normal thyroid hormone transport, as seen for BDEs. It also appears that the hydroxylated metabolites of these PCBs are hormonally active agents. Interestingly, this was suggested as early as 1970.[59] Metabolites with hydroxyl groups in the *para* position appear more potent.[60] Hydroxylation of PCBs does, however, increase their water solubility and potential clearance and thus is initially deemed beneficial (see Chapter 5 for a discussion of xenobiotic biotransformation and suborganismal effects). Hydroxylated PCBs have been reported to be more potent in *in vitro* assays than the alkylphenols (see below), but less than 17β-estradiol.[61] Bergman et al.[62] reported that certain hydroxylated PCBs appeared to be retained in the blood in seals and humans, perhaps associated with proteins. Concentrations there were in the same range as the most persistent PCBs. These and other organochlorines, e.g., DDT, have also been observed to interfere with endocrine function in birds and more recently in reptiles.[63]

3.4 NATURAL AND SYNTHETIC ESTROGENS

Naturally produced estrogen-related compounds have also been detected in the environment and may deleteriously impact organisms. These chemicals are being released to aquatic systems in elevated concentrations from humans, livestock, wildlife, and plants. Although very little research on these compounds in coastal and estuarine areas has been reported to date, their presence in other areas suggests they are relevant emerging contaminants. Sewage treatment plants (STPs) commonly discharge to these waters and appear to serve as conduits for toxicologically significant amounts of estrogens. These treatment facilities are preferentially sited in areas of greatest human populations; therefore, they are common in coastal areas. In terms of concentrations, 17β-estradiol and estrone have been reported as high as 12 and 47 ng/l; respectively, in treated Dutch wastewater.[64] Similar levels have been reported in Germany, Israel, and the United Kingdom.[64] Snyder[65] reported 17β-estradiol concentrations up to 3.7 ng/l in effluents from U.S. municipal STPs. Testosterone has also been detected in sewage and effluents at levels comparable to 17β-estradiol.[66] However, less work has been done on the toxicity and fate of the former chemical. Effects associated with STP discharges include intersexuality in wild populations of roach (*Rutilus retilus*) in the United Kingdom and elevated vitellogenin in carp in the United States.[67,68]

Determining the actual causative agents for the field effects seen can be difficult because of the effluent complexity and the myriad factors at work in receiving waters. In-laboratory exposures to 17β-estradiol have resulted in sex reversal, partial feminization and even death in aquatic organisms.[69] Desbrow et al.[70] chemically fractionated STP effluents observed to elicit estrogenic activity in a yeast-based assay. Extracts of particulates were inactive, suggesting the active elements were dissolved. Removal of activity by passage through a C_{18} cartridge and its presence in methylene chloride extracts suggested the responsible agents were organic. The authors were subsequently able to isolate the activity in a specific high-performance liquid chromatography (HPLC) fraction, which contained estrone and 17β-estradiol. Estriol was not detected and was not viewed as significant here because of its lower estrogenic potency. Rodgers-Gray et al.[71] recently reported that, although phenolic xenoestrogens are often present in effluents at higher concentrations, estrone and 17β-estradiol were likely responsible for the bulk of the effects due to their high potencies. Potencies of several natural estrogens have been obtained from *in vivo* and *in vitro* estrogenicity assays: typically 17β-estradiol ~ estrone > estriol.[72]

Routledge et al.[73] followed the Desbrow et al. work with *in vivo* experiments with roach and trout. They observed that estrone by itself was less potent at eliciting vitellogenin production in fish than 17β-estradiol. Interestingly, simultaneous exposure to both compounds produced a response greater than an equivalent concentration of the more potent 17β-estradiol. Intermittent exposures of male fish to 17β-estradiol have been found to be essentially as potent, in terms of vitellogenin induction, as continuous exposure. Levels remained high even after a 21-day depuration period.[74] Changes in biomarkers in fish have been reported after exposure to concentrations as low as 0.5 ng/l.[65]

Korner et al.,[75] using an *in vitro* breast cancer cell proliferation screen, determined the 17β-estradiol equivalent concentrations (EEC) of influents and effluents associated with a modern German STP. Water treatment reduced these by 90%, from >50 ng/l in the influent to 6 ng/l EEC in the effluent. They reported that less than 5% of the EEC in the effluent was attributable to phenolic xenoestrogens. Only 2.8% of the influent EEC was found in the sludge remaining after treatment. The authors suggested this activity was equivalent to 30 μg/g of 4-nonylphenol (NP), a detergent degradation product common in STPs (see discussion below), although the sludge was not chemically analyzed here. They concluded that biodegradation was a more important EEC removal process than sorption to particulates. Supporting this finding, Furhacker et al.[76] reported that over 90% of the total 17β-estradiol in wastewater remains in the dissolved phase. Ternes et al.[77] observed a 99.9% removal of 17β-estradiol during secondary degradation in a Brazilian STP. Degradation of estrone and 17α-ethinylestradiol (a synthetic estrogen, see discussion below) were less efficient, i.e., 83 and 78%, respectively. They reported lower removals in a German STP and suggested that colder *in situ* temperatures might be responsible. Estrone and 17β-estradiol concentrations were actually higher at this facility after preliminary clarification than in the raw influent, perhaps due to release of conjugated forms. Although the total EEC remained at a level expected to be sufficient to induce vitellogenesis in fish, Korner et al.[75] suggested that subsequent dilution in the receiving stream likely would be sufficient to negate this. However, adequate dilution may not occur in areas where receiving stream flow is low. Increased water flow may reduce residence time of wastewater in the STP, reducing degradation. Rodgers-Gray et al.[71] recently reported that concentrations of natural estrogens fluctuated up to an order of magnitude seasonally and that sensitivity to estrogenic effects in exposed organisms increased with the duration of exposure.

17β-Estradiol is normally produced and excreted by humans, particularly women of childbearing age. In addition, a number of conjugated and unconjugated estrogens (including equine derivatives and estrone), prescribed as hormone replacement drugs for postmenopausal symptoms and to prevent osteoporosis, can be released. In fact, these constitute some of the most prescribed drugs in current use.[78] Belfroid et al.[64] reported that hormones detected in surface water and effluents typically were in their unconjugated forms. Thus, some degradation and release of the free forms likely occur in STPs and receiving waters. Estrogens are also excreted by domestic livestock and other animals. It was estimated that growth-enhancing hormones were administered to 21.4 million feedlot cattle slaughtered in the United States in 1995.[78] Application of poultry litter on pastures is a common practice in areas such as the Eastern Shore of Virginia. Runoff from these applications has been reported to contain significant concentrations of 17β-estradiol and estrone.[69]

Although certainly not newly recognized contaminants, bleached pulpmill effluents have also been reported to affect reproductive systems and various physiological processes in wild fish.[79,80] Consideration of other pulp and wood-related compounds, such as phytosterols, may be in order with respect to endocrine system disruption as well. For example, β-sitosterol, derived from paper mill wastewater has been reported to alter the reproductive status of goldfish and to induce vitellogenin production in rainbow trout.[81,82] Interestingly, several structurally related compounds were found to be inactive in fish, but were estrogenic in an *in vitro* assay employing human breast cancer cells.[82] Consumption of phytoestrogens by livestock, quail, and mice have also been reported to elicit effects.[63] Greater resistance in general to deleterious effects of these natural estrogens may be related to multigenerational exposure.[63] Although estrogen supplements are often prescribed to menopausal women to relieve associated symptoms in the Western Hemisphere, these symptoms are rare in Asian countries where diets are rich in phytoestrogens, such as isoflavones. Several of these compounds are now being recommended as dietary supplements in the United States and other countries.

While natural hormones are prescribed for several health conditions, 17α-ethinylestradiol is widely used as a component of human birth control pills. Purdom et al.[83] reported that 17α-ethinylestradiol was more potent at stimulating vitellogenin production in exposed male trout than 17β-estradiol, eliciting a response at concentrations as low as 0.1 ng/l. It has recently been identified in effluents from a Swedish STP at levels well exceeding those shown to be estrogenic in fish.[84] Dutch STP effluents have been reported to contain up to 7.5 ng/l of 17α-ethinylestradiol.[64] In

Michigan, it was detected at concentrations up to 3.66 ng/l in STP effluent and 1.29 ng/l in surface waters.[65] It has been estimated that sufficient oral contraceptives are consumed annually in the United States, principally 17α-ethinylestradiol, to result in a potential average influent to STPs of 2.16 ng/l.[78] This concentration is consistent with some field measurements and suggests that degradation, sorption, and other removal processes may be relatively incomplete. Synthetic estrogens have been reported to be more persistent than natural steroids and occasionally occur in higher concentrations in raw and treated wastewaters.[78,84]

3.5 ALKYLPHENOL ETHOXYLATES AND ASSOCIATED DEGRADATION PRODUCTS

The presence of detergents in estuarine and coastal environments has historically been problematic, e.g., foaming in receiving waters, nutrient enrichment, and toxicity of oil spill dispersants. Recently, several alkylphenols have been found capable of interacting with the estrogen receptor.[85] Substitution with a hydrophobic group at the *para-* or 4-position of the phenol is typically required. Branching of the alkyl-substituent also increases environmental persistence.[86]

Of the alkylphenols, 4-nonylphenols (NPs) have been reported most frequently in the environment and are typically derived from the degradation of nonylphenol polyethoxylate (NPEO) surfactants. Nonylphenol polyethoxylate surfactants are interesting as some of their degradates, particularly the NPs, are more toxic, lipophilic, and persistent than the parent material. Nonylphenol polyethoxylate surfactants have been used since about 1950, primarily as surfactants in commercial applications. About 15% of production is employed in household cleaning products.[87] Minor uses include additives to pesticide formulations, plastics, cosmetics, and birth-control products.[61] Nonylphenol polyethoxylate surfactants contribute 80% of the total annual production of alkylphenol polyethoxylates (APEOs), with octylphenol polyethoxylates (OPEOs) contributing the majority of the remainder.[87,88] Nonylphenol polyethoxylate surfactants production in the United States was reported to be 227 million kg in 1999.[87]

Initial breakdown of APEOs occurs relatively rapidly in STPs. However, complete mineralization was estimated at only 53% for NPEOs in modern Canadian STPs.[89] A number of intermediates of varying stability are formed; most commonly reported are the NP mono- (NP1EO) and diethoxylates (NP2EO), NP carboxylates with one (NP1EC) or two ethoxylates (NP2EC), and the NPs themselves.[90–94] The extent of degradation is a function of environmental conditions, e.g., oxidation–reduction potential and presence of acclimated bacterial strains. The 4-nonylphenols themselves are generally the result of anaerobic pathways, whereas the NP carboxylates dominate under aerobic conditions (Figure 3.3).

Concentrations of NPEOs and related degradation products in STP effluents are typically in the low to sub microgram per liter range.[95] However, water volumes released from these facilities can be large. Thus, significant amounts of NP-related compounds may still be discharged. Effects will then be a function of the dilution capacity of the receiving waters. Reductions in NP, octylphenol (OP), and NP1EC downstream of a Canadian STP were attributed to dilution rather than further

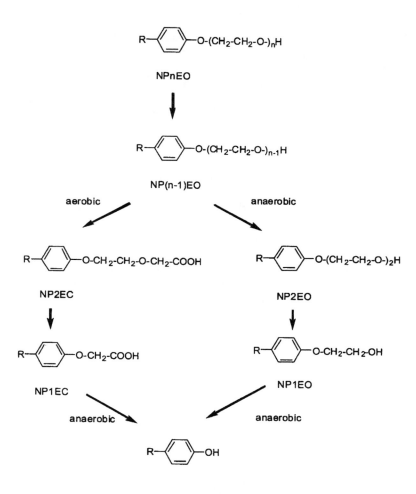

FIGURE 3.3 Degradation pathways of NPEOs (in NPnEO, "n" typically ranges from 4 to 20 ethoxylate groups) under aerobic and anaerobic conditions, such as those encountered in STPs. The principal intermediates under anaerobic conditions are NPs, while NP1EC and NP2EC are dominant under aerobic conditions. NPEOs may also be completely mineralized.[100]

degradation, as decreases in these paralleled those of chloride, a conservative water-soluble tracer of the effluent plume in the freshwater system studied.[96] As noted previously, STP discharges may contribute significantly to the total flow of some streams, particularly in arid regions. This may vary in coastal areas as a function of tides. Wastewater treatment efficiencies vary by facility type, as well as conditions such as residence time in the facility and season. Snyder et al.[65] reported NPs at 37 μg/l and total NPEOs an order of magnitude higher (332 μg/l) in an effluent from a STP using only primary treatment. Untreated wastewater may also directly enter receiving waters during stormwater runoff events. Some facilities routinely discharge to receiving waters without treatment, e.g., parking lot runoff and equipment-cleaning areas.[97] Other transport routes are also possible; e.g., NPs have recently been reported in atmospheric samples taken over the lower Hudson River estuary.[98]

The 4-nonylphenols are generally more acutely toxic than NPEOs and other associated degradation products.[99,100] Relative toxicities of these to several bioassay organisms are depicted in Figure 3.4. Exposure may also result in a variety of sublethal and chronic effects. Exposure of larval oysters (*Crassotrea gigas*) to NPs resulted in increased incidence of a convex hinge deformity at the lowest concentration tested, 0.1 µg/l.[101] This concentration also has been observed to delay larval development, which in turn can contribute to greater mortality. Reduction in larval settlement of the barnacle, *Balanus amphitrite*, has also been observed at nominal NP levels of 0.1 to 1.0 µg/l.[102] Additional research on NP interactions with aquatic plants appears to be in order. A surprisingly high BCF of 10,000 in algae[103] and an EC_{50} for vegetative growth in the diatom, *Skeletonema costatum*, of 27 µg/l have been reported.[104]

Laboratory exposure to NPs has been observed to stimulate egg yolk precursor protein production in trout hepatocytes[105] and to disrupt sexual differentiation in carp.[106] 4-Nonylphenol concentrations between 1.6 and 3.4 µg/l have been reported to cause alterations in the gonadal histology of male fathead minnows.[107] Increased vitellogenesis in males and frequency of intersex in wild U.K. fish populations

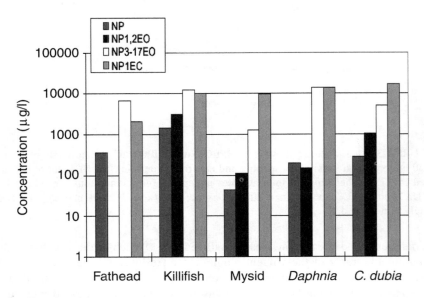

FIGURE 3.4 Acute toxicities of NPs, NP1EO + NP2EO, NPnEOs (where $n = 3$ to 17 ethoxylate groups), and NP1EC to fathead minnow (96-h LC_{50}), killifish (48-h LC_{50}), mysid shrimp (96-h LC_{50}), *Daphnia magna* (48-h LC_{50}), and *Ceriodaphnia dubia* (7-day LC_{50}).[100] NPs are generally the most toxic.

collected near sewage outfalls have also been detected.[67] Similarly, vitellogenin induction and altered serum testosterone levels have been observed in male carp collected near a U.S. STP.[68] However, field effects may be caused by other industrial chemicals, natural and synthetic hormones, or naturally occurring materials present. An interesting association between the presence of NPs in an aerially applied pesticide formulation was reported recently in eastern Canada.[108] Aminocarb was used there to control spruce budworm in forests. Spraying was coincident with salmon smolt development. Locations where the pesticide formulation used did not contain NPs had significantly higher salmon returns over a 17-year period. Declines in blueback herring populations were also observed in the NP-exposed areas. Preliminary laboratory work conducted in 1998, at the Canadian Department of Fisheries and Oceans St. Andrews Biological Station, suggested that NPs do indeed interfere with the smoltification process in salmon (unpublished data).

The 4-nonylphenols exhibit lower water solubilities and higher lipophilicities than their ethoxylate precursors.[89,109,110] Bioconcentration values provided in the literature vary, but typically are relatively modest (3400 or less) compared with chemicals such as PCBs.[100] Nonetheless, NPs and NP1EO were observed to accumulate to nearly 1 mg/kg (wet weight basis) in wild-caught flounder in the United Kingdom.[111] This accumulative tendency may enhance their potential to elicit deleterious effects in organisms compared with more polar chemicals.

A log K_{ow} of 4.48 for NPs has been reported and suggests it will partition preferentially from water to organic-rich sediments.[109,112] Concentrations in sewage sludge can reach gram per kilogram levels,[90] and milligram per kilogram concentrations have been previously reported in aquatic sediments near STPs.[91,93,94,97] The 4-nonylphenols are observed fairly frequently in areas receiving anthropogenic inputs. Detection in sediments is a function of the receiving water, distance from the discharge, and method quantitation limit. Several studies have observed that measurable NP concentrations were common (>50% of samples analyzed) in sediments taken near industrial and municipal outfalls in the United States and Canada.[94,97,113,114] One of the highest sediment concentrations of NPs reported to date, 54,400 µg/kg, was detected in 1998. The site was adjacent to a small Virginia STP, near the Chesapeake Bay, that had ceased operation in the 1970s.[97] Elements contributing to this high burden were the small dilution factor in the receiving stream and the low treatment efficiency of the STP during its operational lifetime. Sediment cores examined from the Fraser River Delta in the Strait of Georgia (British Columbia, Canada) also support the view that NPEO-related compounds may degrade slowly once deposited. While NP, NP1EO, and NP2EO were major contributors, in this case over 50% of the total NPEO-related material in the sediment consisted of NPEOs with greater than two ethoxylate groups.[115] This suggests that even these supposedly labile compounds can remain intact in sediments for long periods, particularly at low temperatures. In further support of this, Manzano et al.[116] reported primary biodegradation of long-chain NPEOs, i.e., loss of the polyethoxylate structure, in river water was only 68% at 7°C compared with 96% at 25°C after 30 days. Complete mineralization was only 30 and 70% at these respective temperatures. They also reported that low temperatures delayed the initiation of biodegradation.

Acidic metabolites constitute a significant portion of NPEO-related compounds in effluents from modern secondary treatment facilities. In Canadian STPs with secondary treatment, NPEOs with ethoxy chain lengths of 3 to 20 were reported to constitute 28% of the total NPEO-related material released. The NP1EC + NP2EC, NP1EO + NP2EO, and NPs contributed 46, 22, and 4%, respectively.[100] Ahel et al.[92] also reported that nearly half of the NP-related compounds in 11 Swiss secondary STP effluents were NP1EC or NP2EC. Concentrations of NP1EC + NP2EC have been reported as high as 1270 µg/l in effluents from paper mills, 270 µg/l from STPs, and 13.8 µg/l in receiving waters.[117]

Research on the toxicities of NPEO metabolites other than the NPs is limited. Estrogenicities of NP1EC and NP2EO were reported to be of the same order as the NPs in an *in vitro* assay based on vitellogenin production in trout hepatocytes.[105,118] In contrast, Metcalfe et al.[72] did not observe any endocrine-related effects while performing a battery of *in vivo* or *in vitro* tests for the acidic compounds at 100 µg/l. Routledge et al.[73] noted that OP was more potent during *in vivo* fish exposures than expected based on *in vitro* assays. This may be due to greater bioaccumulation of OP as a function of its higher lipophilicity and persistence. Apparently care must be taken when extrapolating *in vitro* values to *in vivo* scenarios.

Little information exists on the toxicity of sediment-associated NPs and related compounds. No observable effects concentrations (NOEC) greater than 20 mg/kg in sediments have been suggested.[100] A reproduction-based EC_{50} (21-day) for NP of only 3.4 and 13.7 mg/kg, respectively, was reported for the earthworm (*Apporectodea calignosa*).[100] Potential effects deserve additional examination as sewage sludges are applied to agricultural and other soils (see below).

As noted above, NPs are not the only alkylphenols apparently capable of exerting deleterious effects. Octylphenol and 4-butylphenol have been reported to be several times more estrogenic to fish hepatocytes than the NPs.[105] *In vivo* assays suggested OP inhibited testicular growth and enhanced vitellogenin synthesis in male fish.[119] In addition to metabolic alterations, sexual behavior of fish has also been reported altered by OP exposure.[120] Even though OP is usually observed at lower concentrations in the environment, its higher toxicological potency suggests it should not be dismissed. Octylphenol was reported coincident with NPs in surface sediments near the Virginia STP site described above at 8.22 mg/kg.[97] Bisphenol A, phenylphenol, cumylphenol, and various 4-pentylphenols[97,121–123] have also been detected in effluents, drinking water, fish, and sediments and may exert estrogenic and other effects via similar mechanisms.

3.6 OTHER PHARMACEUTICALS

Phytoestrogens, conjugated estrogens, and 17α-ethinylestradiol have already been discussed. Pharmaceuticals, like pesticides, have been engineered to have specific effects on the intended target organism. The exact mode of action is often unknown and "side effects" on other metabolic pathways are common. Potential effects on unintended recipients, such as wildlife, are even less well understood.[7] Many of these chemicals are metabolized or conjugated to new products. These, in turn, may be modified again when released to the environment or during wastewater treatment.

Pharmaceutical firms seldom investigate effects on nontarget organisms or the fate of a drug outside the intended recipient.

The fate of drugs is a function of their stability and partitioning properties. A number of pathways are possible, as shown in Figure 3.5. Unused, off-specification, or expired drugs may be disposed of in landfills and subsequently subject to leaching.[124] Most studies have examined the fate of drugs in STPs. Many pharmaceuticals are relatively water soluble and, as a consequence, are not removed by partitioning to solids in treatment works. In addition, their low relative concentrations in influents, typically sub microgram per liter, may limit the potential degradative capacity of resident bacterial populations. Removal efficiencies have also been observed to vary during periods of high and low STP residence time.[7] Ternes[125] examined effluents from German STPs for a suite of 32 drugs. About 80% of these drugs were detectable at nanogram per liter concentrations in releases from at least one of the 49 facilities examined. Sewage treatment plant removal efficiencies varied from 7% for the antiepileptic drug carbamasepine to 96% for the β-blocker metoprolol. The average removal for the drugs detected was about 60%. In these effluents, carbamasepine was observed at concentrations up to 6.3 μg/l. The lipid regulator benzafibrate was present as high as 3.1 μg/l and at a median concentration of 0.35 μg/l.

Approximately 23 million kg of antibiotics are produced in the United States annually.[126] Development of resistance to a number of antibiotics is well known and is viewed as an extremely serious problem. This is a result of not only overprescription of agents for treatment of humans, but also their widespread usage on animals. Agricultural applications include treatment of specific diseases and for the

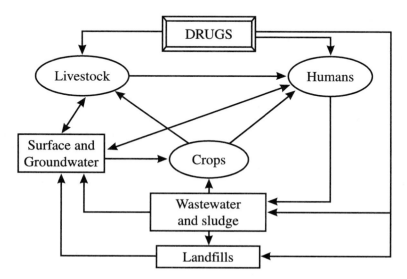

FIGURE 3.5 Flow diagram showing major potential pathways of drugs between major environmental compartments. Drugs may be present as parent compound, degradation intermediates, or be completely mineralized.

promotion of weight gain in livestock. Often these medicines are used in a purely preventative role on large numbers of animals. In some cases, the consumption of antibiotics in animal husbandry far outpaces that earmarked for humans. For example, in 1994 in Denmark 24 kg of vancomycin were used in humans, compared with 24,000 kg of the related avoparcin in animal feed.[127] In the United States, it has been estimated that 40% of all antibiotics produced are used on livestock.[126] Data on associated environmental distributions are limited. However, Meyer and Kolpin[126] detected a variety of antibiotics in groundwater near animal feeding operations and in streams downstream from STPs in the U.S. Midwest, typically at sub microgram per liter concentrations.

3.7 NONPHARMACEUTICAL ANTIMICROBIAL AGENTS

In addition to drugs, a variety of antimicrobial agents are used in household and industrial products. For example, triclosan (2,4,4'-trichloro-2'-hydroxydiphenylether) has been used for over 30 years as an antibacterial additive in shampoos, soaps, toothpaste, and medicinal agents.[128] Triclosan is structurally similar to the CDEs, hydroxylated PCB metabolites, and BDEs previously discussed (see Figure 3.1D). It has recently been suggested that widespread triclosan use at sublethal concentrations could foster biocide resistance in some bacterial populations.[129,130] Researchers have reported triclosan in U.S. wastewater, sediments, and river water associated with a specialty chemicals manufacturing facility as early as 1978.[131] We have also observed it at concentrations of up to 160 µg/kg in sediment near the previously mentioned decommissioned STP in the Chesapeake Bay watershed and in other sewage sludges.[97] Muller et al.[132] detected triclosan in effluents from seven STPs between 46 and 210 ng/l. Influent from one contained 520 ng/l. They estimated degradation in STPs at 80%, with 5% leaving in the effluent and 15% in sludge. From these data, it is apparent that microbial populations are exposed widely to this agent.

3.8 PERSONAL CARE PRODUCTS

Synthetic musks are extensively used fragrances in personal care products, including shampoos, detergents, cosmetics, and toiletries and thus are often discharged to the environment. Because of their lower costs, they are now used more frequently than natural musks. The synthetics include the nitro and the polycyclic musks, both of which appear to be persistent and bioaccumulative. Log K_{ow} values for musk ketone and xylene range from 4.1 to 5.2, whereas those for two polycyclic musks, galaxolide and tonalide, have been determined to be 5.9 and 5.8, respectively.[133] Bioaccumulation factors (wet weight basis) in trout have been reported as 4200 to 5100 for musk xylene.[134] The polycylic musks have begun to replace the nitro musks, particularly in Europe, and now constitute almost two thirds of worldwide production, estimated at 7000 tons annually in 1988.[7]

Yamagishi et al.[135] detected musk-related compounds in all finfish, shellfish, river water, and STP effluent samples they examined from Japan. Amino musks can apparently be generated from nitro musks during sewage treatment. Gatermann et al.[136] reported musk xylene and ketone concentrations in STP influent in Germany of 150 and 550 ng/l, respectively. Their concentrations in effluent were lower, 10 and 6 ng/l. However, levels of some amino derivatives increased from nondetectable in influent to 250 ng/l following treatment. Amino derivatives appear to be more toxic than their precursors (sub µg/l EC_{50} values for *Daphnia magna)*, and provide another example of degradation increasing toxicity.[137]

3.9 INTERACTION OF MULTIPLE STRESSORS

The typical approach taken in both research and regulation has been to consider chemicals individually, ignoring their additive, antagonistic, and possibly synergistic effects. This is obviously an oversimplification. Examples of interactions may be seen in everyday life, e.g., deleterious interactions of drugs and alcohol in humans. Effects are a function of the total exposure of an organism, not simply to those we have negotiated a "threshold" value. Therefore, increased research efforts are needed in this area. Despite this, experiments on interactive effects are relatively rare and results have sometimes been controversial, e.g., possible synergistic effects of exposure to multiple organochlorine pesticides.[138,139] The large number of potential endocrine-disrupting chemicals in the environment, which may elicit effects through similar common mechanisms, argue for the likelihood of additive effects.[2] In addition, chemical, physical, and biological agents may all contribute to deleterious outcomes. This may be seen in oyster (*Crassostrea virginica*) mortalities associated with interactions of infectious disease, toxic chemicals, temperature, salinity, and other factors.[140,141]

3.9.1 MULTIPLE XENOBIOTIC RESISTANCE

Another example of a complex interaction may be the inhibition of multiple xenobiotic resistance (MXR). This resistance is apparently conferred by the action of glycoproteins capable of ridding cells of accumulated toxic agents or inhibiting their entry. It is one mode for the development of drug resistance in bacteria. However, it has been detected in the cells of higher organisms as well, including aquatic species. Enhanced expression can be induced by exposure to these agents. Subsequent exposure to an inhibitor, or chemosensitizer, may impair this capability, resulting in novel expression of toxicity in the face of preexisting toxicant concentrations. Chemosensitizers include some drugs such as verapamil (cardiac drug), reserpine (antihypertensive), cyclosporins (immunosuppressants), and quinidine (antiarrythmic).[142] Klerks[143] recently observed lower resistance to contaminants in the grass shrimp (*Palaemonetes pugio)* after preexposure to mixtures compared with individual compounds.[143]

3.9.2 STP SLUDGE

As is apparent from the above discussions, a number of potentially interacting, emerging contaminants of concern might concentrate in STPs. These facilities face a daunting task, restoration of contaminated influents from diverse industrial and domestic discharges to acceptable standards. These standards change as our understanding of toxicological impacts advance. Ever-larger volumes of waste are directed to these STPs as human populations increase. Interest exists for reuse of resulting effluents or "gray water" for irrigation and other purposes, particularly where potable water shortages occur.[144] Based on the above discussions of contaminant content, additional research on the consequences of the direct usage of these effluents appears prudent.

The second product of STPs is sludge, consisting of particulates formed or collected during water treatment. Contaminants present in the influent may flocculate or preferentially partition to these high-organic particles. Chemicals of low persistence may degrade further, depending upon the conditions present within the sludge. While some chemicals may be completely removed; others with greater toxicity, e.g., NPs, may be formed. A major concern has been how to dispose of these sludges. Incineration may result in the production of toxic by-products, e.g., chlorinated and brominated dioxins and furans. Landfilling is becoming increasing difficult and expensive as a result of dwindling available space and increased "tipping fees" at landfills. Ocean dumping of sludges has been practiced in many countries. In the United States, releases initially were made in nearshore areas. However, dumping was subsequently moved farther offshore because of obvious coastal impacts, and the practice was banned completely in 1991.[144,145]

Sludge itself is rich in nutrients and organic matter, potential liabilities for ocean disposal, but attributes for a soil conditioner and fertilizer. It has been estimated that 7.5 million metric tons (dry) of sludge are currently produced in the United States annually and 54% of this is land-applied on agricultural, public, and residential lands.[146] Many human population centers and associated STPs are located in coastal areas and much of the biosolids produced are subsequently deployed in adjacent watersheds. Sewage sludge is typically processed by either composting or chemical treatment (e.g., heat or liming) before application. The term *biosolids* has been coined for the resulting material. Land application of sludges has been deemed feasible, even desirable, in part as a result of stricter industrial pretreatment standards for influents and limitations placed on PCB and organochlorine pesticide usage. Biosolids application is practiced in many countries and typically is regulated based on pathogen and heavy metal content. The metals are targeted because of their toxicity and inherent nondegradability. They thus have the potential to build up in soils after repeated land applications, eventually resulting in deleterious impacts. Concentrations of nine metals in sludge are currently regulated by the U.S. EPA (As, Cd, Cr, Cu, Hg, Ni, Pb, Se, and Zn).[146] In general, U.S. standards for land application are less restrictive than those in Europe.[147,148]

Burdens of organic contaminants in sewage sludges have been less thoroughly considered. A survey of U.S. STPs, concluded in 1988, suggested that burdens of regulated organic contaminants were low.[144] Since then, researchers have detected several additional organic pollutants in STP sludges, including surfactants, fire retardants, and pharmaceuticals. The 4-nonylphenols have been detected in sewage sludges in a number of countries, at concentrations up to 4000 mg/kg.[95] Although the U.S. EPA has further considered the question of organic pollutants in biosolids, to date it has only proposed new regulations for chlorinated dioxins and planar PCBs.[149] The typical toxicological end point considered in associated risk assessments for organic compounds has been cancer. As noted above, a number of emerging contaminants of concern act through other mechanisms.

We analyzed four different biosolids from U.S. STPs. All were derived from facilities processing predominantly domestic sewage. Treatments represented were composted, heat-treated, and lime-stabilized. As depicted in Figure 3.6, NPs were detected in all samples at concentrations from 5.38 to 820 mg/kg (dry weight basis). This compares well with the range of 8.4 to 850 mg/kg for Canadian sewage sludges reported by Bennie.[95] The bagged composted biosolid contained the lowest level and the lime-stabilized the highest. Octylphenol was detectable in all, ranging from 0.206 to 7.49 mg/kg. Both, NP1EO and NP2EO were much higher in the lime- and heat-treated samples. Other countries have begun to regulate NPs in sludge destined for land application. For example, Denmark has designated a 50-mg/kg limit (sum of NP, NP1EO, and NP2EO), with the expectation for a further reduction to 10 mg/kg in 2000.[95] Three of the four U.S. biosolids contained NP concentrations considerably greater than 50 mg/kg. The bagged compost was the exception, a so-called exceptional-quality (EQ) biosolid. EQ biosolids may be sold directly to the public and do not carry restrictions on application.

Halogenated organics were examined in these same four U.S. biosolids. Whereas PCBs and organochlorine pesticides were present as only minor constituents, BDEs were detected at relatively high levels. The sum of the concentrations of BDE-47, BDE-99, BDE-100, BDE-153, and BDE-154 in the four biosolids ranged from 1110 to 2290 μg/kg (Figure 3.7). These concentrations surpass those typically reported in sludges in Europe by over an order of magnitude.[46] It is also interesting how similar levels were among the sludges we examined. With the exception of the heat-treated biosolid, BDE-47 and BDE-99 were the dominant

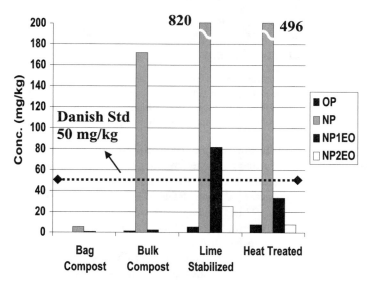

FIGURE 3.6 Concentrations (dry weight basis) of NP-related compounds in four different U.S. biosolids. Three of the four samples contained NP concentrations above the current 50 mg/kg Danish standard.

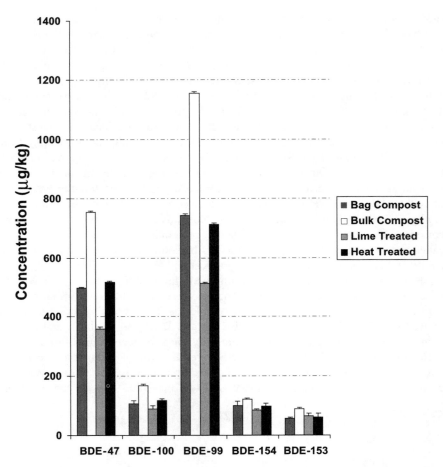

FIGURE 3.7 Concentrations (dry weight basis) of several BDE congeners in four different U.S. biosolids. BDE-47 and BDE-99 were the major congeners in three of four biosolids. In the heat-treated biosolid, BDE-209 was detected at high levels. It is not quantified here as definitive quantification methods remain under development.

congeners present in all. In this sample, BDE-209 was the most abundant. Efforts are under way to quantify this congener in sludges more accurately. The relative contributions of the five tetra- and pentabrominated congeners matched well with the proportions in the commercial Penta-BDE formulation, suggesting this mixture may be a major contributor. This product is used in the United States predominantly as a fire retardant in polyurethane foam and approximately 97% of the global demand for it now resides in North America.[21]

Polyurethane foam has a high surface area and degrades rapidly upon exposure to sunlight compared with denser plastics, e.g., high-impact polystyrenes. Commercial deca-BDE is the dominant formulation used in the latter polymers. The exposed surfaces of polyurethane foam may rapidly yellow, become brittle, and be eroded by weathering processes. Dementev et al.[150] observed a close relationship between weight loss and incidence of solar radiation in field-weathering trials of flexible foam. More rapid deterioration of polystyrene foam in seawater than in air, likely due to accelerated removal of the outer weathered layer, has also been noted.[151] The BDEs are added to some products at loadings as high as 30% by weight.[18] Thus, it seems plausible that eroded foam fragments could act as a source for the low-brominated congeners observed in sludge. To put this in context, if a single 1-kg foam seat cushion (containing a loading of 10% by weight of a commercial Penta-BDE mixture) completely disintegrated and was transported to a STP, 100 g of fire retardant would be contributed. As BDEs are quite resistant to microbial degradation and partition strongly to solids, these likely would accumulate in the resulting sludge. This scenario would result in a 1000 μg/kg burden, similar to that seen in the biosolid samples we examined, in 100,000 kg (dry weight) of sludge.

The presence of BDEs, NP-related chemicals, and other contaminants in STP effluents and sludges, as described above, and the increasing application of biosolids on soils in watersheds of coastal and estuarine areas argue for the consideration of these materials as emerging contaminants of concern. Additional research is needed to elucidate the concentrations of pollutants in sludge, especially the organic contaminants, their bioavailabilities, and ultimately their toxicological implications.

3.10 CONCLUSIONS

By definition, the list of emerging contaminants of concern must be viewed as constantly evolving. Obviously, all currently known examples could not be discussed here and additional ones will emerge over time. Analytical chemistry techniques are a powerful tool capable of revealing many of the xenobiotic chemicals present in the environment. However, the goals and priorities of any monitoring work must be carefully considered. For example, trade-offs often exist between the need for analytical specificity, accuracy, and sensitivity on the one hand and the need to detect environmentally significant concentrations of a wide range of contaminants on the other. Obviously, this analytical approach alone is insufficient. Clearly, considerable information is available from the manufacturing and regulatory arenas about compounds that may be potentially released to the environment and this is an additional starting point. We have learned much regarding specific chemical properties that may be problematic, e.g., hydrophobicity (and attendant bioaccumulation potential), modes of toxicity, and persistence. Some of the compounds discussed, e.g., low-brominated diphenyl ethers, are highly bioaccumulative and persistent, factors readily apparent from their structures. Monitoring efforts need to be augmented by other approaches, such as application of quantitative structure–activity relationships, to identify problematic chemicals. Ideally, this should be accomplished prior to their dissemination in the environment and the observation of deleterious effects or significant tissue burdens. Unfortunately, as modes of action may be unforeseen, it is likely that some toxic chemicals will escape scrutiny. These may not be identified as contaminants of concern until deleterious environmental effects are manifested. Some chemicals discussed here, e.g., natural and synthetic estrogens, have been investigated recently not because of their bioaccumulation potential or persistence, but rather because of the realization that they may interact with organisms in previously unappreciated ways, e.g., with the endocrine systems of exposed organisms.

The myriad drugs, antimicrobials, and personal care products produced and released to marine and estuarine environments may become the focus of concern in the future. In addition, while some chemicals exhibit low toxicity, bioaccumulation potential, and persistence, their breakdown products may behave considerably differently. The transformation of NPEOs to NPs is one example of this scenario. Finally, chemicals originating from different sources may come together and concentrate in the environment or a new matrix, e.g., sewage sludge. These chemicals may interact with each other, or other stressors, to produce unexpected biological effects. Sludge is of particular concern as it is being generated in increasing amounts and intentionally land-applied. This disposal practice thus disseminates any entrained contaminants. While some of these may be bound or subsequently degraded, the range of constituents in sludge is poorly known and their bioavailability and potential effects even less well studied.

ACKNOWLEDGMENTS

Ellen Harvey, Michael Gaylor, T. Matteson Mainor, Gregory Mears, Elizabeth Bush, and William Duff are acknowledged for their contributions in the laboratory. Some fish sample collection and financial support were provided by the Virginia Department of Environmental Quality. Partial support for BDE analytical standards was provided by the Brominated Chemical Manufacturers Association Fire Retardant Industry Panel.

REFERENCES

1. Fry, D.M., Reproductive effects in birds exposed to pesticides and industrial chemicals, *Environ. Health Perspect.*, 103 (Suppl. 7), 165, 1995.
2. DeRosa, C. et al., Environmental exposures that affect the endocrine system: public health implications, *J. Toxicol. Environ. Health*, Part B, 1, 3, 1998.
3. Igata, A., Epidemiological and clinical features of Minamata disease, *Environ. Res.*, 63, 157, 1993.
4. Di Carlo, F.J., Seifter, J., and De Carlo, V.J., Assessment of the hazards of polybrominated biphenyls, *Environ. Health Perspect.*, 23, 351, 1978.
5. Maguire, R.J., Environmental aspects of tributytin, *Appl. Organometallic Chem.*, 1, 475, 1987.
6. Huggett, R.J. and Bender, M.E., Kepone in the James River, *Environ. Sci. Technol.*, 14, 918, 1980.
7. Daughton, C.G. and Ternes, T.A., Pharmaceuticals and personal care products in the environment: agents of subtle change? *Environ. Health Perspect.*, 107 (Suppl. 6), 907, 1999.
8. Hale, R.C. and Smith, C.L., A multiresidue approach for trace organic pollutants: application to effluents and associated aquatic sediments and biota from the Southern Chesapeake Bay drainage basin 1985–1992, *Int. J. Environ. Anal. Chem.*, 64, 21, 1996.

9. Menditto, A. and Turrio-Baldassarri, L., Environmental and biological monitoring of endocrine disrupting chemicals, *Chemosphere*, 39, 1301, 1999.
10. Mumma, R.O., National survey of elements and other constituents in municipal sewage sludges, *Arch. Environ. Contam. Toxicol.*, 13, 75, 1984.
11. Jones, K.C. and de Voogt, P., Persistent organic pollutants (POPs): state of the science, *Environ. Pollut.*, 100, 209, 1999.
12. Lauenstein, G.G. and Daskalakis, K.D., U.S. long-term coastal contaminant temporal trends determined from mollusk monitoring programs, 1965–1993, *Mar. Environ. Res.*, 37, 6, 1998.
13. O'Connor, T.P., Mussel Watch results from 1986 to 1996, *Mar. Environ. Res.*, 37, 14, 1998.
14. Elliot, J.E. et al., Patterns and trends of organic contaminants in Canada seabird eggs, 1968–1990, in *Persistent Pollutants in Marine Ecosystems*, Walker, C.H. and Livingstone, D.R., Eds., Pergamon Press, Oxford, 1992, chap. 8.
15. MacDonald, R.W. et al., Contaminants in the Canadian Arctic: 5 years of progress in understanding sources, occurrence and pathways, *Sci. Total Environ.*, 254, 93, 2000.
16. Chan, H.M. et al., Evaluation of the population distribution of dietary contaminant exposure in an Arctic population using Monte Carlo statistics, *Environ. Health Perspect.*, 105, 316, 1997.
17. Sellstrom, U. and Jansson, B., Analysis of tetrabromobisphenol A in a product and environmental samples, *Chemosphere*, 31, 3085, 1995.
18. World Health Organization, Environmental Health Criteria 162: Brominated Diphenyl Ethers, WHO, Geneva, 1994.
19. DeCarlo, V.J., Studies on brominated compounds in the environment, *Ann. N.Y. Acad. Sci.*, 320, 678, 1979.
20. Andersson, O. and Blomkvist, G., Polybrominated aromatic pollutants found in fish in Sweden, *Chemosphere*, 10, 1051, 1981.
21. Renner, R., Increasing level of flame retardants found in North American environment, *Environ. Sci. Technol.*, 34, 452A, 2000.
22. de Boer, J., de Boer, K., and Boon, J.P., Polybrominated biphenyls and diphenyl ethers, in *The Handbook of Environmental Chemistry*, Paasivirta, J., Ed., Springer-Verlag, Berlin, 2000, Vol. 3, part K, chap 4.
23. Sjodin, A. et al., Gas chromatographic identification and quantitation of polybrominated diphenyl ethers in a commercial product, Bromkal 70-5DE, *J. Chromatogr. A*, 822, 83, 1998.
24. Hileman, B., Right to know: U.S. EPA proposes lower reporting thresholds for persistent bioaccumulative chemicals, *Chem. Eng. News*, 77, 4, 1999.
25. Renner, R., What fate for brominated fire retardants? *Environ. Sci. Technol.*, 34, 223A, 2000.
26. Eriksson, P., Jakobsson, E., and Fredriksson, A., Developmental neurotoxicity of brominated flame-retardant, polybrominated diphenyl ethers and tetrabromo-bis-phenol A, *Organohalogen Compd.*, 35, 375, 1998.
27. Holm, G.H. et al., Effects of exposure to food contaminated with BDDE, PCN or PCB on reproduction, liver morphology and cytochrome P450 in the three-spined stickleback, *Gasterosteus aculeatus*, *Aquat. Toxicol.*, 27, 33, 1993.
28. Asplund, L. et al., Organohalogen substances in muscle, egg, and blood from healthy Baltic salmon (*Salmo salar*) and Baltic salmon that produced offspring with the M74 syndrome, *Ambio*, 28, 67, 1999.
29. Meerts, I.A.T.M. et al., Intercation of polybrominated diphenyl ether metabolites (PBDE-OH) with human transthyretin *in vitro*, *Organohalogen Compd.*, 37, 309, 1998.

30. Zelinski, V., Lorenz, W., and Bahadir, M., Brominated flame retardants and resulting PBDD/F in accidental fire residues from private residences, *Chemosphere*, 27, 1519, 1993.
31. Buser, H., Polybrominated dibenzofurans and dibenzo-*p*-dioxins: thermal reaction products of polybrominated diphenyl ether flame retardants, *Environ. Sci. Technol.*, 20, 404, 1986.
32. Hornung, M.W., Zabel, E.W., and Peterson, R.E., Toxic equivalency factors of polybrominated dibenzo-*p*-dioxin, dibenzofuran, biphenyl, and polyhalogenated diphenyl ether congeners based on rainbow trout early life stage mortality, *Toxicol. Appl. Pharmacol.*, 140, 227, 1996.
33. Pijnenburg, A.M.C.M. et al., Polybrominated biphenyl and diphenyl ether flame retardants: analysis, toxicity, and environmental occurrence, *Rev. Environ. Contam. Toxicol.*, 141, 1, 1995.
34. Loganathan, B.G. et al., Isomer-specific determination and toxic evaluation of polychlorinated biphenyls, polychlorinated/brominated dibenzo-*p*-dioxins and dibenzofurans, polybrominated biphenyl ethers, and extractable organic halogen in carp from the Buffalo River, New York, *Environ. Sci. Technol.*, 29, 1832, 1995.
35. de Boer, J. et al., Do flame retardants threaten ocean life? *Nature*, 394, 28, 1998.
36. Meneses, M. et al., Polybrominated diphenyl ethers detected in human adipose tissue from Spain, *Chemosphere*, 39, 2271, 1999.
37. Linstrom, G. et al., Identification of the flame retardants polybrominated diphenyl ethers in adipose tissues from patients with non-Hodgkin's lymphoma in Sweden, *Oncol. Rep.*, 4, 999, 1997.
38. Noren, K. and Meironyte, D., Certain organochlorine and organobromine contaminants in Swedish human milk in perspective of past 20–30 years, *Chemosphere*, 40, 1111, 2000.
39. U.S. EPA, National Study of Chemical Residues in Fish, U.S. EPA 823-R-92-008a, Washington, D.C., 1992.
40. Gustafsson, K. et al., Bioaccumulation kinetics of brominated flame retardants (polybrominated diphenyl ethers) in blue mussels (*Mytilus edulis*), *Environ. Toxicol. Chem.*, 18, 1218, 1999.
41. Burreau, S. et al., Dietary uptake in pike (*Esox lucius*) of some polychlorinated biphenyls, polychlorinated naphthalenes and polybrominated diphenyl ethers administered in a natural diet, *Environ. Toxicol. Chem.*, 16, 2508, 1997.
42. Andersson, P.L. et al., Biomagnification of structurally matched polychlorinated and polybrominated diphenylethers (PCDE/PBDE) in zebrafish (*Danio rerio*), *Organohalogen Compd.*, 43, 9, 1999.
43. Watanabe, I., Kashimoto, T., and Tatsukawa, R., Polybrominated biphenyl ethers in marine fish, shellfish and river and marine sediments in Japan, *Chemosphere*, 16, 2389, 1987.
44. Allchin, C.R., Law, R.J., and Morris, S., Polybrominated diphenylethers in sediment and biota downstream of potential sources in the UK, *Environ. Pollut.*, 105, 197, 1999.
45. Sellstrom, U. et al., Polybrominated diphenyl ethers and hexabromocyclododecane in sediment and fish from a Swedish river, *Environ. Toxicol. Chem.*, 17, 1065, 1998.
46. Sellstrom, U. et al., Brominated flame retardants in sediments from European estuaries, the Baltic Sea and in sewage sludge, *Organohalogen Compd.*, 40, 383, 1999.
47. Sjodin, A. et al., Flame retardant exposure: polybrominated diphenyl ethers in blood from Swedish workers, *Environ. Health Perspect.*, 107, 643, 1999.

48. Watanabe, I. and Tatsukawa, R., Formation of brominated dibenzofurans from the photolysis of flame retardant decabromodiphenyl ether in hexane solution by UV and sunlight, *Bull. Environ. Contam. Toxicol.*, 39, 953, 1987.
49. Kierkegaard, A. et al., Dietary uptake and biological effects of decabromodiphenyl ether in rainbow trout (*Oncorhynchus mykiss*), *Environ. Sci. Technol.*, 33, 1612, 1999.
50. Jensen, S., Report of new chemical hazard, *N. Sci.*, 32, 612, 1966.
51. Eganhouse, R.P. and Gossett, R.W., Sources and magnitude of bias associated with determination of polychlorinated biphenyls in environmental samples, *Anal. Chem.*, 63, 2130, 1991.
52. Hale, R.C. and Greaves, J., Review: methods for the analysis of persistent chlorinated hydrocarbons in tissues, *J. Chromatogr.*, 580, 257, 1992.
53. Safe, S. et al., Halogenated biphenyls: molecular toxicology, *Can. J. Physiol. Pharmacol.*, 60, 1057, 1982.
54. Poland, A. and Knutson, J.C., 2,3,7,8-Tetrachlorodibenzo-*p*-dioxin and related halogenated aromatic hydrocarbons: examination of the mechanism of toxicity, *Annu. Rev. Pharmacol. Toxicol.*, 22, 517, 1982.
55. McKinney, J.D. and Waller, C.L., Polychlorinated biphenyls as hormonally active structural analogues, *Environ. Health Perspect.*, 102, 290, 1994.
56. Chauhan, K.R., Kodavanti, P.R.S., and McKinney, J.D., Assessing the role of ortho-substitution on polychlorinated biphenyl binding to transthyretin, a thyroxine transport protein, *Toxicol. Appl. Pharmacol.*, 162, 10, 2000.
57. Schantz, S.L., Moshtaghian, J., and Ness, D.K., Spatial learning deficits in adult rats exposed to ortho-substituted PCB congeners during gestation and lactation, *Fundam. Appl. Toxicol.*, 26, 117, 1995.
58. Jacobson, J.L. and Jacobson, S.W., Intellectual impairment in children exposed to polychlorinated biphenyls *in utero*, *N. Engl. J. Med.*, 335, 783, 1996.
59. Bitman, J. and Cecil, H.C., Estrogenic activity of DDT analogs and polychlorinated biphenyls, *J. Agric. Food Chem.*, 18, 1108, 1970.
60. Andersson, P.L. et al., Assessment of PCBs and hydroxylated PCBs as potential xenoestrogens: *in vitro* studies based on MCF-7 cell proliferation and induction of vitellogenin in primary culture of rainbow trout hepatocyes, *Arch. Environ. Contam. Toxicol.*, 37, 145, 1999.
61. Nimrod, A.C. and Benson, W.H., Environmental estrogenic effects of alkylphenol ethoxylates. *Crit. Rev. Toxicol.*, 26, 335, 1996.
62. Bergman, A., Klasson-Wehler, E., and Kuroki, H., Selective retention of hydoxylated PCB metabolites in blood, *Environ. Health Perspect.*, 102, 464, 1994.
63. Guillette, L.J. et al., Organization versus activation: the role of endocrine disrupting contaminants (EDCs) during embryonic development in wildlife, *Environ. Health Perspect.*, 103, 157, 1995.
64. Belfroid, A.C. et al., Analysis and occurrence of estrogenic hormones and their glucuronides in surface water and wastewater in the Netherlands, *Sci. Total Environ.*, 225, 101, 1999.
65. Snyder, S.A. et al., Analytical methods for the detection of selected estrogenic compounds in aqueous mixtures, *Environ. Sci. Technol.*, 33, 2814, 1999.
66. Shore, L.S., Gurevitz, M., and Shemesh, M., Estrogen as an environmental pollutant, *Bull. Environ. Contam. Toxicol.*, 51, 361, 1993.
67. Jobling, S. et al., Widespread sexual disruption in wild fish, *Environ. Sci. Technol.*, 32, 2498, 1998.

68. Folmar, L.C. et al., Vitellogenin induction and reduced serum testosterone concentrations in feral male carp (*Cyprinus carpio*) captured near a major metropolitan sewage treatment plant, *Environ. Health Perspect.*, 104, 1096, 1996.
69. Nichols, D.J. et al., Runoff of estrogen hormone 17β-estradiol from poultry litter applied to pasture, *J. Environ. Qual.*, 26, 1002, 1997.
70. Desbrow, C. et al., Identification of estrogenic chemicals in STW effluent. 1. Chemical fractionation and *in vitro* biological screening, *Environ. Sci. Technol.*, 32, 1549, 1998.
71. Rodgers-Gray, T.P. et al., Long-term temporal changes in the estrogenic composition of treated sewage effluent and its biological effects on fish, *Environ. Sci. Technol.*, 34, 1521, 2000.
72. Metcalfe, C.D. et al., Estrogenic compounds in sewage treatment plant effluents: their relative potency in the medaka assay, presented at 20th Annual Meeting of the Society of Environmental Toxicology and Chemistry, Philadelphia, 1999.
73. Routledge, E.J. et al., Identification of estrogenic chemicals in STW effluent. 2. *In vivo* responses in trout and roach, *Environ. Sci. Technol.*, 32, 1559, 1998.
74. Panter, G.H., Thompson, R.S., and Sumpter, J.P., Intermittent exposure of fish to estradiol, *Environ. Sci. Technol.*, 34, 2756, 2000.
75. Korner, W. et al., Input/output balance of estrogenic active compounds in a major municipal sewage plant in Germany, *Chemosphere*, 40, 1131, 2000.
76. Furhacker, M., Breithofer, A., and Jungbauer, A., 17β-Estradiol: behavior during wastewater analyses, *Chemosphere*, 39, 1903, 1999.
77. Ternes, T.A. et al., Behavior and occurrence of estrogens in municipal sewage treatment plants. I. Investigations in Germany, Canada and Brazil, *Sci. Total Environ.*, 225, 81, 1999.
78. Arcand-Hoy, L.D., Nimrod, A.C., and Benson, W.H., Endocrine-modulating substances in the environment: estrogenic effects of pharmaceutical products, *Int. J. Toxicol.*, 17, 139, 1998.
79. Andersson, T. et al., Physiological disturbances in fish living in coastal water polluted with bleached kraft pulp mill effluents, *Can. J. Fish. Aquat. Sci.*, 45, 1525, 1988.
80. Sandstrom, O., Neuman, E., and Karas, P., Effects of a bleached pulp mill effluent on growth and gonad function in Baltic coastal fish, *Water Sci. Technol.*, 20, 107, 1988.
81. Stahlschmidt-Allner, P. et al., Endocrine disruptors in the aquatic environment, *Environ. Sci. Pollut. Res.*, 4, 155, 1997.
82. Mellanen, P. et al., Wood-derived estrogens: studies *in vitro* with breast cancer cell lines and *in vivo* in trout, *Toxicol. Appl. Pharmacol.*, 136, 381, 1996.
83. Purdom, C.E. et al., Estrogenic effects from sewage treatment works, *Chem. Ecol.*, 8, 275, 1994.
84. Larsson, D.G.J. et al., Ethinylestradiol — an undesired fish contraceptive? *Aquat. Toxicol.*, 45, 91, 1999.
85. Mueller, G.C. and Kim, U., Displacement of estradiol from estrogen receptors by simple alkyl phenols, *Endocrinology*, 102, 1429, 1991.
86. Maguire, R.J., Review of the persistence of nonylphenol and nonylphenol ethoxylates in aquatic environments, *Water Qual. Res. J. Can.*, 34, 37, 1999.
87. APE Research Council, Alkylphenols and alkylphenol ethoxylates: an overview of safety issues, APE Research Council White Paper, Washington, D.C., 1999.
88. Renner, R., European bans on surfactant trigger transatlantic debate, *Environ. Sci. Technol.*, 31, 316A, 1997.
89. Lee, H.-B. and Peart, T.E., Determination of 4-nonylphenol in effluent and sludge from sewage treatment plants, *Anal. Chem.*, 67, 1976, 1995.

90. Giger, W., Brunner, P.H., and Schaffner, C., 4-Nonylphenol in sewage sludge: accumulation of toxic metabolites from nonionic surfactants, *Science*, 225, 623, 1984.
91. Marcomini, A. et al., Persistent metabolites of alkylphenol polyethoxylates in the marine environment, *Mar. Chem.*, 29, 307, 1990.
92. Ahel, M., Giger, W., and Koch, M., Behavior of alkylphenol polyethoxylate surfactants in the aquatic environment. I. Occurrence and transformation in sewage treatment, *Water Res.*, 28, 1131, 1994.
93. Ahel, M., Giger, W., and Schaffner, C., Behavior of alkyphenol polyethoxylate surfactants in the aquatic environment. II. Occurrence and transformation in rivers, *Water Res.*, 28, 1141, 1994.
94. Naylor, C.G., Environmental fate and safety of nonylphenol ethoxylates, *Textile Chem. Colorist*, 27, 29, 1995.
95. Bennie, D.T., Review of the environmental occurrence of alkylphenols and alkylphenol ethoxylates, *Water Qual. Res. J. Can.*, 34, 79, 1999.
96. Bennie, D.T. et al., The fate of STP-derived alkylphenol ethoxylate metabolites in an Ontario river, presented at 20th Annual Meeting of the Society of Environmental Toxicology and Chemistry, Philadelphia, 1999.
97. Hale, R.C. et al., Nonylphenols in sediments and effluents associated with diverse wastewater outfalls, *Environ. Toxicol. Chem.*, 19, 946, 2000.
98. Dachs, J., Van Ry, D.A., and Eisenreich, S.J., Occurrence of estrogenic nonylphenols in the urban and coastal atmosphere of the lower Hudson River estuary, *Environ. Sci. Technol.*, 33, 2676, 1999.
99. Servos, M.R., Review of the aquatic toxicity, estrogenic responses and bioaccumulation of alkyphenols and alkylphenol ethoxylates, *Water Qual. Res. J. Can.*, 34, 123, 1999.
100. Environment Canada, Nonylphenol and Ethoxylates, draft Canadian Environmental Protection Act Priority Substances List Assessment Report, 2000.
101. Nice, H.E. et al., Development of *Crassostrea gigas* larvae is affected by 4-nonylphenol, *Mar. Pollut. Bull.*, 491, 2000.
102. Billinghurst, Z. et al., Inhibition of barnacle settlement by the environmental oestrogen 4-nonylphenol and the natural oestrogen 17-β-estradiol, *Mar. Pollut. Bull.* 36, 883, 1998.
103. Ahel, M., McEvoy, J., and Giger, W., Bioaccumulation of the lipophilic metabolites of nonionic surfactants in freshwater organisms, *Environ. Pollut.*, 79, 243, 1993.
104. Brooke, L.T. and Thursby, G., U.S. EPA Draft Ambient Aquatic Life Water Quality Criteria for Nonylphenol, U.S. EPA Office of Water, Washington, D.C., 1998.
105. Jobling, S. and Sumpter, J.P., Detergent components in sewage effluent are weakly oestrogenic to fish: an *in vitro* study using rainbow trout (*Oncorhynchus mykiss*) hepatocytes, *Aquat. Toxicol.*, 27, 361, 1993.
106. Gimeno, S. et al., Disruption of sexual differentiation in genetic male common carp (*Cyprinus carpio*) exposed to an alkylphenol during different life stages, *Environ. Sci. Technol.*, 31, 2884, 1997.
107. Miles-Richardson, S.R. et al., Effects of waterborne exposure to 4-nonylphenol and nonylphenol ethoxylate on secondary sex characteristics and gonads of fathead minnows (*Pimephales promelas*), *Environ. Res. Section A*, 80, S122, 1999.
108. Fairchild, W.L. et al., Does an association between pesticide use and subsequent declines in catch of Atlantic salmon (*Salmo salar*) represent a case of endocrine disruption? *Environ. Health Perspect.*, 107, 349, 1999.
109. Ahel, M. and Giger, W., Aqueous solubility of alklyphenols and alkylphenol ethoxylates, *Chemosphere*, 26, 1461, 1993.

110. Ahel, M. and Giger, W., Partitioning of alkylphenols and alkylphenol ethoxylates between water and organic solvents, *Chemosphere,* 26, 1471, 1993.
111. Lye, C.M. et al., Estrogenic alkylphenols in fish tissues, sediments and waters from the U.K. Tyne and Tees estuaries, *Environ. Sci. Technol.,* 33, 1009, 1999.
112. Johnson, A.C. et al., The sorption potential of octylphenol, a xenobiotic oestrogen, to suspended and bed-sediments collected from industrial and rural reaches of three English rivers, *Sci. Total Environ.,* 210, 271, 1998.
113. Granmo, A. et al., Bioaccumulation of nonylphenol in caged mussels in an industrial coastal area on the Swedish west coast, in *Organic Micropollutants in the Aquatic Environment: Proceedings of the Sixth European Symposium*, Angeletti, G., Ed., Kluwer Academic Publishers, London, 1991, 71.
114. Bennie, D.T. et al., Occurrence of alkylphenols and alkylphenol mono- and diethoxylates in natural waters of the Laurentian Great Lakes basin and the upper St. Lawrence River, *Sci. Total Environ.,* 193, 263, 1997.
115. Shang, D.Y. et al., Persistence of nonylphenol ethoxylate surfactants and their primary degradation products in sediments from near a municipal outfall in the Strait of Georgia, British Columbia, Canada, *Environ. Sci. Technol.,* 33, 1366, 1999.
116. Manzano, M.A. et al., The effect of temperature on the biodegradation of a nonylphenol polyethoxylate in river water, *Water Res.,* 33, 2593, 1999.
117. Field, J.A. and Reed, R.L., Nonylphenol polyethoxy carboxylate metabolites of nonionic surfactants in US paper mill effluents, municipal sewage treatment plant effluents and river waters, *Environ. Sci. Technol.,* 30, 3544, 1996.
118. White, R. et al., Environmentally persistent alkylphenolic compounds are estrogenic, *Endocrinology,* 135, 175, 1994.
119. Jobling, S., et al., Inhibition of testicular growth in rainbow trout (*Oncorhynchus mykiss*) exposed to estrogenic alkylphenolic chemicals, *Environ. Toxicol. Chem.,* 15, 194, 1996.
120. Bayley, M., Nielsen, J.R., and Baatrup, E., Guppy sexual behavior as an effect biomarker of estrogen mimics, *Ecotoxicol. Environ. Saf.,* 43, 68, 1999.
121. Rudel, R.A. et al., Identification of alkylphenols and other estrogenic phenolic compounds in wastewater, septage, and groundwater on Cape Cod, Massachusetts, *Environ. Sci. Technol.,* 32, 861, 1998.
122. Shiraishi, H., Carter, D.S., and Hites, R.A., Identification and determination of tert-alkylphenols in carp from the Trenton Channel of the Detroit River, USA, *Biomed. Environ. Mass Spectrosc.,* 18, 478, 1989.
123. Gimeno, S. et al., Feminisation of young males of the common carp, *Cyprinus carpio*, exposed to 4-tert-pentylphenol during sexual differentiation, *Aquat. Toxicol.,* 43, 77, 1998.
124. Holm, J.V. et al., Occurrence and distribution of pharmaceutical organic compounds in the groundwater downgradient of a landfill (Grindsted, Denmark), *Environ. Sci. Technol.,* 29, 1415, 1995.
125. Ternes, T.A., Occurrence of drugs in German sewage treatment plants and rivers, *Water Res.,* 32, 3245, 1998.
126. Meyer, M.T. and Kolpin, D.W., Detection of antibiotics in surface and groundwater near confined animal feeding operations and wastewater-treatment plants by using radioimmunoassay and liquid chromatography/mass spectrometry, Issues in the Analysis of Environmental Endocrine Disruptors, American Chemical Society, Division of Environmental Chemistry, *Preprints Extended Abstr.,* 40, 106, 2000.
127. Witte, W., Medical consequences of antibiotic use in agriculture, *Science,* 279, 996, 1998.

128. Morse, P.M., Product report: antibacterial products grow despite controversy, *Chem. Eng. News*, 77, 39, 1999.
129. McMurray, L.M., Oethinger, M., and Levey, S.B., Triclosan targets lipid synthesis, *Nature*, 394, 531, 1998.
130. Routhi, M., Germ killers: could widely used biocide, found in common household products, pose a health risk? *Chem. Eng. News*, 76, 9, 1998.
131. Jungclaus, G.A., Lopez-Avila, V., and Hites, R.A., Organic compounds in an industrial wastewater: a case study of their environmental impact, *Environ. Sci. Technol.*, 12, 88, 1978.
132. Muller, S.R., Singer, H.P., and Canonica, S., Fate and behavior of the biocide triclosan in the aquatic environment, Specialty Chemicals in the Environment, American Chemical Society, Division of Environmental Chemistry, *Preprints Extended Abstr.*, 1, 166, 2000.
133. Winkler, M. et al., Fate of artificial musk fragrances associated with suspended particulate matter (SPM) from the river Elbe (Germany) in comparison to other organic contaminants, *Chemosphere*, 37, 1139, 1998.
134. Rimkus, G.G., Butte, W., and Geyer, H.J., Critical considerations on the analysis and bioaccumulation of musk xylene and other synthetic nitro musks in fish, *Chemosphere*, 35, 1497, 1997.
135. Yamagishi, T. et al., Identification of musk xylene and musk ketone in freshwater fish collected from the Tama River, Tokyo, *Bull. Environ. Contam. Toxicol.*, 26, 656, 1981.
136. Gatermann, R. et al., Occurrence of musk xylene and musk ketone metabolites in the aquatic environment, *Chemosphere*, 36, 2535, 1998.
137. Behechti, A. et al., Acute toxicities of four musk xylene derivatives on *Daphnia magna*, *Water Res.*, 32, 1704, 1998.
138. Arnold, S.F. et al., Synergistic activation of estrogen receptor with combinations of environmental chemicals, *Science*, 272, 1489, 1996.
139. Ramamoorthy, F. et al., Potency of combined estrogenic pesticides, *Science*, 275, 405, 1997.
140. Lenihan, H.S. et al., The influence of multiple environmental stressors on susceptibility to parasites: an experimental determination with oysters, *Limnol. Oceanogr.*, 44, 910, 1999.
141. Chu, F.L. and Hale, R.C., Relationship between pollution and susceptibility to infectious disease in the eastern oyster, *Crassostrea virginica*, *Mar. Environ. Res.*, 38, 243, 1994.
142. Kurelec, B., A new type of hazardous chemical: the chemosensitizers of multixenobiotic resistance, *Environ. Health Perspect.*, 105, 1997.
143. Klerks, P.L., Acclimation to contaminants by the grass shrimp *Palaemonetes pugio*: individual contaminants vs mixtures, *Ecotoxicology*, 8, 277, 1999.
144. National Research Council, Use of Reclaimed Water and Sludge in Food Crop Production, National Academy of Sciences, National Academy Press, Washington, D.C., 1996.
145. Bothner, M.H. et al., Sewage contamination in sediments beneath a deep-ocean dump site off New York, *Mar. Environ. Res.*, 38, 43, 1994.
146. Renner, R., Scientists debate fertilizing soils with sewage sludge, *Environ. Sci. Technol.*, 34, 242A, 2000.
147. U.S. EPA, A Plain English Guide to the U.S. EPA Part 503 Biosolids Rule, U.S. EPA/832/R-93/003, Washington, D.C., 1994.

148. McBride, M.B., Toxic metal accumulation from agricultural use of sludge: Are U.S. EPA regulations protective? *J. Environ. Qual.*, 24, 5, 1995.
149. 40 CFR Part 503, Federal Register 64, 72045, Standards for the Use or Disposal of Sewage Sludge, 1999.
150. Dementev, M.A., Demina, A.I., and Nevskii, LV., Ageing and salvaging of flexible polyurethane foam under atmospheric conditions, *Int. Polym. Sci. Technol.*, 26, 89, 1999.
151. Andrady, A.L. and Pegram, J.E., Weathering of polystyrene foam on exposure in air and in seawater, *J. Appl. Polym. Sci.*, 42, 1589, 1991.

4 Enhancing Belief during Causality Assessments: Cognitive Idols or Bayes's Theorem?

Michael C. Newman and David A. Evans

CONTENTS

4.1 Difficulty in Identifying Causality 73
4.2 Bacon's Idols of the Tribe 74
4.3 Idols of the Theater and Certainty 76
4.4 Assessing Causality in the Presence of Cognitive and Social Biases 77
4.5 Bayesian Methods Can Enhance Belief or Disbelief 80
4.6 A More Detailed Exploration of Bayes's Approach 82
    4.6.1 The Bayesian Context 82
    4.6.2. What Is Probability? 82
    4.6.3 A Closer Look at Bayes's Theorem 84
4.7 Two Applications of the Bayesian Method 86
    4.7.1 Successful Adjustment of Belief during Medical Diagnosis 86
    4.7.2 Applying Bayesian Methods to Estuarine Fish Kills
        and *Pfiesteria* 90
        4.7.2.1 Divergent Belief about *Pfiesteria piscicida*
                Causing Frequent Fish Kills 90
        4.7.2.2 A Bayesian Vantage for the *Pfiesteria*-Induced Fish
                Kill Hypothesis 91
4.8 Conclusion 93
Acknowledgments 94
References 94

4.1 DIFFICULTY IN IDENTIFYING CAUSALITY

At the center of every risk assessment is a causality assessment. Causality assessments identify the cause–effect relationship for which risk is to be estimated. Despite this, many ecological risk assessments pay less-than-warranted attention to carefully identifying causality, and concentrate more on risk quantification. The compulsion to quantify for quantification's sake (i.e., Medawar's *idola quantitatis*[1]) contributes to this imbalance. Also, those who use logical shortcuts for assigning plausible causality in their daily lives[2] are often unaware that they are applying shortcuts in their professions. A zeal for method transparency (e.g., U.S. EPA[3]) can also diminish soundness if sound methods require an unfamiliar vantage for assessing causality. Whatever the reasons, the imbalance between efforts employed in causality assessment and risk estimation is evident throughout the ecological risk assessment literature. Associated dangers are succinctly described by the quote, "The mathematical box is a beautiful way of wrapping up a problem, but it will not hold the phenomena unless they have been caught in a logical box to begin with."[4] In the absence of a solid causality assessment, the most thorough calculation of risk will be inadequate for identifying the actual danger associated with a contaminated site or exposure scenario. The intent of this chapter is to review methods for identifying causal relations and to recommend quantification of belief in causal relations using the Bayesian approach.

Most ecological risk assessors apply rules of thumb for establishing potential cause–effect relationships. Site-use history and hazard quotients are used to select chemicals of potential concern. Cause–effect models are then developed with basic rules of disease association.[3] This approach generates expert opinions or weight-of-evidence conjectures unsupported by rigor or a quantitative statement of the degree of belief warranted in conclusions. Expert opinion (also known as global introspection) relies on the informed, yet subjective, judgment of acknowledged experts; this process is subject to unavoidable cognitive errors as evidenced in analyses of failed risk assessments such as that associated with the *Challenger* space shuttle disaster.[5,6] The weight- or preponderance-of-evidence approach produces a qualitative judgment if information exists with which "a *reasonable* person reviewing the available information *could* agree that the conclusion was plausible."[7] Some assessments apply such an approach in a very logical and effective manner, e.g., the early assessments for tributyltin effects in coastal waters.[8,9] Although these and many other applications of such an approach have been very successful, the touchstone for the weight-of-evidence process remains indistinct plausibility.

4.2 BACON'S IDOLS OF THE TRIBE

How reliable are expert opinion and weight-of-evidence methods of causality assessment? It is a popular belief that, with experience or training, the human mind can apply simple rules of deduction to reach reliable conclusions. Sir Arthur Conan Doyle's caricature of this premise is Sherlock Holmes who, for example, could conclude after quick study of an abandoned hat that the owner "was highly intellectual ... fairly well-to-do within the last three years, although he has fallen upon evil days. He had foresight, but less now than formerly, pointing to a moral retrogression, which, when taken with the decline of his fortunes, seems to indicate some evil influence, probably drink, at work on him. This may account also for the obvious fact that his wife has ceased to love him."[10] As practiced readers of fiction, we are entertained by Holmes's shrewdness only after willingly forgetting that Doyle had complete control over the accuracy of Holmes's conclusions. In reality, including that surrounding ecological risk assessments, such conclusions and associated high confidence would be ridiculous. In the above fictional case, Doyle clearly generated the data that Holmes observed from the above set of conclusions the author had previously formulated; equally valid alternative conclusions that could be drawn from the observations were completely ignored. In the real world of scientific activity, the causes of the observations remain unknown. Reversal of the direction of causality to achieve an entertainingly high degree of belief is acceptable for fiction but should be replaced by more rigorous procedures for fostering belief.[11] Simple deductive (i.e., the hypotheticodeductive method of using observation to test a hypothesis) or inductive (i.e., methods producing a general theory such as a causal theory from a collection of observations) methods are sometimes insufficient for developing a rational foundation for a cause–effect relationship. Nevertheless, such informal conclusions are drawn daily in risk assessments.

Francis Bacon defined groupings of bad habits or "idols" causing individuals to err in their logic.[12] One, idols of the tribe, encompasses mistakes inherent in human cognition — errors arising from our limited abilities to determine causality and likelihood. Formal study of such errors lead Piattelli-Palmarini[2] to conclude that humans are inherently "very poor evaluators of probability and equally poor at choosing between alternative possibilities." As described below, expert opinion and weight-of-evidence approaches are subject to such errors. Key among these cognitive errors are anchoring, spontaneous generalization, the endowment effect, acquiescence, segregation, overconfidence, bias toward easy representation, familiarity, probability blindness, and framing.[2,13,14] Many of these general cognitive errors make their appearance in scientific thinking or problem solving as confirmation bias[15] or precipitate explanation,[16] belief enhancement through repetition,[17] theory immunization,[18] theory tenacity,[15] theory dependence,[18,19] low-risk testing,[4,13] and similar errors.

All of these cognitive errors are easily described. Two, anchoring and confirmation bias, are related. Anchoring is a dependency of belief on initial conditions: there is a tendency toward one option that appears in the initial steps of the process.[2] The flawed cognitive process results in a bias toward data or options presented at the beginning of an assessment. The general phenomenon of spontaneous generalization (the human tendency to favor popular deductions) is renamed "precipitate explanation" in the philosophy of science and can be described in the present context as the uncritical attribution of cause to some generally held mechanism of causality. Although formally denounced as unreliable in modern science, precipitate explanation emerges occasionally in environmental sciences. Other errors are less obvious than precipitate explanation. Confirmation bias emerges in the hypotheticodeductive or scientific method as the tendency toward tests or observations that bring support to a favored theory or hypothesis. It is linked to the practice of low-risk testing, which is the inclination to apply tests that do not place a favored theory in high jeopardy of rejection. In an ideal situation, tests with high capacity to negate a theory should be favored. Weak testing and the repeated invoking of a theory or casual structure to explain a phenomenon can lead to enhanced belief based on repetition alone, not on rigorous testing or scrutiny. Repetition is used to immunize a theory or favored causal structure from serious scrutiny or testing.[18] The endowment effect, recognized easily in the psychology of financial investing, is the tendency to believe in a failing investment's profitability or theory's validity despite the clear accumulation of evidence to the contrary. There is an irrational hesitancy in withdrawing belief from a failing theory. In scientific thinking, the endowment effect translates into theory tenacity, the resistance to abandon a theory despite clear evidence refuting it. Theory tenacity is prevalent throughout all sciences and science-based endeavors, and ecological risk assessment is no exception. Many of these biases remain poorly controlled because the human mind is poor at informally judging probabilities, i.e., subject to probability blindness. The theory dependence of all knowledge is an inherent confounding factor. In part, the context of a theory dictates the types of evidence that will be accumulated to enhance or reduce belief. For example, most ecological risk assessments for chemically contaminated sites develop casual structures based on toxicological theories. Alternative explanations based on habitat quality or loss, renewable resource-use patterns, infectious disease dynamics, and other candidate processes are too rarely given careful consideration. Toxicology-based theories dominate in formulating causality hypotheses or models. Other cognitive errors include acquiescence, bias toward easy representation, and framing. Acquiescence is the tendency to accept a problem as initially presented. Bias toward easy representation is the tendency to favor something that is easy to envision. For example, one might falsely believe that murders committed with handguns are a more serious problem than deaths due to a chronically bad diet. The image of the murder scenario is easier to visualize than the gradual and subtle effects of poor diet. Framing emerges from our limited ability to assess risk properly. For example, more individuals would elect to have a surgery if the physician stated that the success rate of the procedure was 95%, rather than that the failure rate was 5%. The situation is the same but the framing of the fact biases the perception of the situation.

4.3 IDOLS OF THE THEATER AND CERTAINTY

Bacon also described bad habits of logic associated with received systems of thought: idols of the theater. One example from traffic safety is the nearly universally accepted paradigm that seat belts save lives. To the contrary, Adams[20] suuggests that widespread use of seat belts does not reduce the number of traffic fatalities. Many people drive less carefully when they have the security of a fastened seatbelt, resulting in more fatalities outside of the car. The number of people falling victim to the incautious behavior of belted drivers has increased and negates the reduced number of fatalities to drivers.

Kuhn[19] describes many social behaviors specific to scientific disciplines including those easily identified as idols of the theater, e.g., maintaining belief in an obviously failing paradigm. Such a class of flawed methods also seems prevalent in ecological risk assessment. Some key theoretical and methodological approaches are maintained in the field by a collective willingness to ignore contradictory evidence or knowledge. (See Reference 21 for a more complete description of this general behavior.) Even when fundamental limitations are acknowledged, acknowledgment often comes in the form of an occultatio — a statement emphasizing something while appearing to pass it over. A common genre of ecotoxicological occultatio includes statements such as the following, "Although ecologically valid conclusions are not possible based solely on LC_{50} data, extrapolation from existing acute lethality data suggests that concentrations below X are likely to be protective of the community." Another example of our ability to ignore the obvious is that most ecological risk assessments are, in fact, hazard assessments. Insufficient data are generated to quantify the probability of the adverse consequence occurring. Instead, the term *likelihood* is used to soften the requirement for quantitative assessment of risk; and qualitative statements of likelihood become the accepted norm.[3] (This fact was briefly acknowledged in Chapter 2 for EU-related risk assessment.)

The application of short-term LC_{50} values to determine the hazard concentration below which a species population remains viable in a community is another example[7,22] already alluded to above. A quick review of population and community ecology reveals that such an assumption is not tenable because it does not account for pivotal demographic vital rates, e.g., birth or growth rates, and community interactions. Further assumptions associated with prediction of ecological consequences with short-term LC_{50}/EC_{50} data can be shown to be equally invalid. Two examples are the uncritical acceptance of the individual tolerance concept and trivialization of postexposure mortality.[23] The error of accepting such incorrect assumptions is hidden under accreted layers of regulatory language. This codification of error suggests what Sir Karl Popper[11] called the idol of certainty — the compulsion to create the illusion of scientific certainty where it does not exist. It grows from the general error of cognitive overconfidence. When rigorously examined, the confidence of most humans in their assessments of reality tends to be higher than warranted by facts.

4.4 ASSESSING CAUSALITY IN THE PRESENCE OF COGNITIVE AND SOCIAL BIASES

How is causality established in the presence of so many cognitive and knowledge-based biases? Ecological risk assessors follow qualitative rules of thumb to guide themselves through causality assessments. Commonly, one of two sets of rules are applied for noninfectious agents: Hill's rules of disease association[24] and Fox's rules of ecoepidemiology.[25] The first is the most widely applied, although the recently published U.S. EPA "Guidelines for Ecological Risk Assessment"[3] (Section 4.3.1.2) focuses on Fox's rules.

Hill[24] lists nine criteria for inferring causation or disease association with non-infectious agents: strength, consistency, specificity, temporality, biological gradient, plausibility, coherence, experiment, and analogy (Table 4.1). Fox[25] lists seven criteria: probability, time order, strength of association, specificity of association, consistency of association, predictive performance, and coherence (Table 4.2). Both authors follow explanations of their rules with a call for temperance. They emphasize that none of these rules allows causality to be definitively identified or rejected, but are aids for compiling information prior to rendering an expert opinion or a judgment from a preponderance of evidence. Therefore, these rules provide some degree of protection against the cognitive and social errors described above.

TABLE 4.1
Hill's Nine Aspects of Noninfectious Disease Association

Aspect	Description
Strength	Belief in an association increases if the strength of association is strong. An exposed target population with extremely high prevalence of the disease relative to an unexposed population suggests association and, perhaps, causality.
Consistency	Belief in an association increases with the consistency of association between the agent and the disease, regardless of differences in other factors.
Specificity	Belief is enhanced if the disease emerges under very specific conditions that indicate exposure to the suspected disease agent.
Temporality	To support belief, the exposure must occur before, or simultaneously with, the expressed effect or disease. Disbelief is fostered by the disease being present before any exposure to the agent was possible.
Biological gradient	Belief is enhanced if the prevalence or severity of the disease increases with increasing exposure to the agent. Of course, threshold effects can confound efforts to document a concentration- or exposure-dependent effect.
Plausibility	The existence of a plausible mechanism linking the agent to the expressed disease will enhance belief.
Coherence	Belief is enhanced if evidence for association between exposure to an agent and the disease is consistent with existing knowledge.
Experiment	Belief is enhanced by supporting evidence from experiments or quasi-experiments. Experiments and some quasi-experiments have very high inferential strength relative to uncontrolled observations.
Analogy	For some agents, belief can be enhanced if an analogy to a similar agent–disease association can be made. Belief in avian reproductive failure due to biomagnification of a lipophilic pesticide may be fostered by analogy to a similar scenario with DDT.

Hill's aspects of disease association are applied below in a causality assessment for putative polycyclic aromatic hydrocarbon (PAH)-linked cancers in English sole (*Pleuronectes vetulus*) of Puget Sound (condensed from Reference 22). Field surveys and laboratory studies were applied to assess causality for liver cancers in populations of this species endemic to contaminated sites.

1. *Strength of Association:* Horness et al.[26] measured lesion prevalence in English sole endemic to areas having sediment concentrations of <DL to 6,300 ng PAH/g dry weight of sediments. There was very low prevalence of lesions at low concentration sites and 60% prevalence at contaminated sites.
2. *Consistency of Association:* English sole from contaminated sites consistently had high prevalence of precancerous and cancerous lesions.[26–28] Myers et al.[27] found no evidence of viral infection so that alternate explanation was judged to be unlikely.
3. *Specificity of Association:* Prevalence of hepatic lesions in English sole at a variety of Pacific Coast locations was used to generate logistic regression models.[28] Included in these models were concentrations of a wide range of

TABLE 4.2
Fox's Rules of Practical Causal Inference

Aspect	Description
Probability	With sufficiently powerful testing, belief is enhanced by a statistically significant association.
Time order[a]	Belief is greatly diminished if cause does not precede effect.
Strength[b]	Belief is enhanced if the strength of the association between the presumptive cause and the effect (i.e., concordance of cause and disease, magnitude of effect, or relative risk) is strong.
Specificity	Given the difficulty of assigning causality when other competing disease agents exist, specificity of the agent–disease association enhances belief.
Consistency[a,b]	Belief is enhanced if the association between the agent and disease is consistent regardless of the circumstances surrounding the association, e.g., regardless of the victim's age, sex, or occupation.
Predictive performance[b]	Belief is enhanced if the association is seen upon repetition of the observational or experimental exercise.
Coherence	Belief is enhanced if a hypothesis of causal association is effective in predicting the presence or prevalence of disease.
Theoretical	Belief is enhanced if the proposed association is consistent with existing theory.
Factual[a]	Belief is enhanced if the proposed association is consistent with existing facts.
Biological	Belief is enhanced if the proposed association is consistent with our current body of biological knowledge.
Dose–response[b]	Belief is enhanced if the proposed association displays a dose– or exposure–response relationship. The dose– or exposure–response curve can be linear or curvilinear including thresholds.

[a] Strong inconsistency of these three rules can be used to reject causality.
[b] Strong adherence to these four rules can be used as clear evidence of causality.

pollutants in sediments. PAHs, polychlorinated biphenyls, DDT and its derivatives, chlordane, and dieldrin were all significant ($\alpha = 0.05$) risk factors, suggesting low specificity of association between PAHs and liver cancer.

4. *Temporal Sequence:* Temporal sequence is difficult to define clearly for cancers with long periods of latency. However, Myers et al.[27,29] produced lesions in the laboratory-exposed English sole that were indicative of early stages in a progression toward liver cancer.
5. *Biological Gradient:* A biological gradient with a threshold was indicated by the work of Myers et al.[29] and Horness et al.[26]
6. *Plausible Biological Mechanism:* General liver carcinogenesis following P-450-mediated production of free radicals and DNA adduct formation was the clear mechanism for production of precancerous and cancerous lesions. Myers et al.[29] documented the presence of DNA adducts in English sole and correlated these adducts with lesions leading to cancer.
7. *Coherence with General Knowledge:* The results with English sole are consistent with a wide literature on chemical carcinogenesis including that for rodent cancers due to PAH exposure.[27,30]

8. *Experimental Evidence:* Laboratory exposure to high PAH concentrations resulted in lesions characteristic of a progression to liver cancer.[29]
9. *Analogy:* The general causal structure of PAH exposure, P-450-mediated production of free radicals, DNA adduct formation, and the emergence of cancer are consistent with many examples in the cancer literature.

Applying Hill's criteria to this exemplary work, the conclusion would generally be drawn that high PAH concentrations in sediments were likely the causal agent for liver cancer lesions in English sole: high PAH concentrations in sediments will result in significant risk of liver cancer in this coastal species. Yet it would be difficult to aver that other carcinogens were not involved. It would also be difficult clearly to quantify one's belief in the relative dominance of PAHs vs. other carcinogens. Despite such ambiguity, a recommendation might emerge that PAH concentrations in sediments should be regulated to some concentration near or below the threshold of the logistic models described above. The weakness in the causal hypothesis, i.e., Points 3 and 4 above, might become the focus for a party with financial liability. In fact, this was the general strategy successfully taken by tobacco companies for many years relative to tobacco-induced lung cancer.[24]

4.5 BAYESIAN METHODS CAN ENHANCE BELIEF OR DISBELIEF

Sir Karl Popper[18] and numerous others concluded that scientific methods producing quantitative information are superior to qualitative methods. Relative to qualitative methods, quantitative measurement and model formulation permit more explicit statement of models (hypotheses), more rigorous testing (falsification), and clearer statements of statistical confidence. These obvious advantages motivate consideration of quantitative methods for enhancing belief during causality assessments. In fact, but not often in practice, the application of Hill's or Fox's rules within an expert opinion or weight-of-evidence process can be improved by a more explicit, mathematical method.

The expert opinion and weight-of-evidence approaches are qualitative applications of abductive inference. Simply put, abductive inference is inference to the most probable explanation. Josephson and Josephson[31] render abductive inference to the following thought pattern:

1. D is a collection of data about a phenomenon.
2. H explains D, the collection of data.
3. No other hypothesis (H_A) explains D as effectively as H does.
4. Therefore, H is probably true.

The logic used in applying Hill's aspects of disease association to liver cancers in English sole was clearly abductive inference.

An obvious shortcoming with such abductive inference as a means of enhancing belief is its qualitative nature. Quantification would allow a much clearer statement of belief in the conclusion that "*H* is *probably* true." Then, a hypothesis of causality could be judged as false if it were sufficiently improbable.[32] Conversely, a highly probable hypothesis of causality could be judged as conditionally true. The conceptual framework for such an approach would be the following.[32] Let E be a body of evidence and H be a hypothesis to be judged. Then $p(H)$ is the probability of H being true irrespective of the existence of E and $p(H|E)$ is the conditional probability of H being true given the presence of the evidence, E. [A conditional probability is the probability of something given another thing is true or present, i.e., p(Disease|Positive Test Result) is the probability of having a specific disease given that results of a diagnostic test were positive.]

1. E provides support for H if $p(H|E) > p(H)$
2. E draws support away from H if $p(H|E) < p(H)$
3. E provides no confirming nor undermining information regarding H if $p(H|E) = p(H)$.

The degree of belief in H given a body of information E would be a function of how different $p(H|E)$ and $p(H)$ are from one another. Abductive inference about causality can be quantified with Bayes's theorem (Equation 4.1) based on this context.

$$p(H|E) = \frac{p(H) \bullet p(E|H)}{p(E)} \qquad (4.1)$$

In Equation 4.1, H is the hypothesis and E is the new data or evidence obtained with the intent of assessing H. The posterior probability, $p(H|E)$, is the probability of H being true given the new information, E. The prior probability ($p(H)$) is the probability of the hypothesis being true as estimated prior to E being available. The $p(E|H)$ is the conditional probability of E given H, it is called the likelihood of E and is a function of H, and $p(E)$ is the probability of E regardless of H.

Bayes's theorem can be applied to determine the level of belief in the hypothesis after new information is acquired. The magnitude of the posterior probability suggests the level of belief warranted by the information in hand together with the prior belief in H. As more information is acquired, the posterior probability can be used as the new prior probability and the process repeated. The process can be repeated until the posterior probability is sufficient to decide whether the hypothesis is probable or improbable. This iterative application of Bayes's theorem is analogous to, but not equilvalent to, the hypotheticodeductive method in which a series of hypotheses are tested until only one explanation remains unfalsified. The dichotomous falsification process is replaced by one in which the probability or level of belief changes during sequential additions of information until the causality hypothesis becomes sufficiently plausible (probable) or implausible (improbable).

4.6 A MORE DETAILED EXPLORATION OF BAYES'S APPROACH

4.6.1 THE BAYESIAN CONTEXT

The Reverend Thomas Bayes died on 17 April 1761 in Tunbridge Wells, Kent, England. In 1763, a paper by Bayes was read to the Royal Society at the request of his friend, Richard Price. The paper[33] provided solution to the problem that was stated as follows:

> Given the number of times on which an unknown event has happened and failed [to happen]: Required the chance that the probability of its happening in a single trial lies somewhere between any two degrees of probability that can be named.

The 18th-century style is rather opaque to modern readers, but it can be seen that the problem addresses the advancement of the "state of knowledge or belief" by experimental results. The modern representation of Bayes's result is encapsulated in Equation 4.1. As this formulation may be similarly opaque to a reader unaccustomed to dealing with probability calculations, the purpose of this section is to clarify these statements.

4.6.2. WHAT IS PROBABILITY?

Bayesian methods are questioned by many statisticians, in large part because of the way the interpretation of probability is extended. Accordingly, we will review how probability can be defined. However, like pornography, while everyone knows what probability is when they encounter it, no one finds it easy to define.

Most courses in probability or statistics introduce probability by considering some kind of trial producing a result that is not predictable deterministically. A numerical value between 0 and 1 can be associated with each possible result or outcome. This value is the probability of that outcome. The classic example of such a trial is a coin toss with two possible outcomes, heads or tails. If a large number of trials were made, the ratio of the number of "heads" outcomes to the total number of trials almost always seems to approach a limiting value, or at least fluctuates within a range of values. The variability gets smaller as the number of trials increases. The probability of the "heads" outcome is then defined as the value that this ratio usually appears to stabilize around as the number of trials approaches infinity. It should be clear from this definition that the actual, or "true," value of the probability of an outcome cannot be determined experimentally. The definition suffers from the defect that it contains the words, "usually" and "almost always," that are themselves expressions of a probabilistic nature and is therefore circular. Probability is defined in terms of itself: the definition is not logically valid. However, it is a very helpful model in developing an understanding of stochastic events and dealing with them quantitatively.

The above is the frequentist approach to probability. It assists the prediction of what will happen "in the long run" or "on the average" for a finite series of trials. This is the sort of information that insurance companies or dedicated gamblers require to improve their chances of making money.

Enhancing Belief during Causality Assessments

While insurance companies depend upon what happens in the long run with many policies, the individual with a life insurance policy has only a single opportunity to die. A young person thinks little about obtaining life insurance, whereas the older a person becomes, the more concerned he or she is in obtaining protection. This is because the person's degree of belief in the hypothesis "I will die next year" increases as the years go by. Since the degree of belief is perceived as increasing, it is an ordinal quantity and can be assigned a numerical value. A sensible scale to choose is zero for absolute denial of the hypothesis and unity for certainty in the truth of the statement. As Benjamin Franklin might have written:

$$db(\text{death}) = db(\text{taxes}) = 1,$$

where $db(\ )$ stands for degree of belief in ( ).

But what shall we do about intermediate cases? How shall a value be assigned to a degree of belief? As noted above, one can accept that degrees of belief can be ordered or compared; for example, one's degree of belief in it raining today is lower on a day with no clouds in the sky than it is on a day with low gray clouds and a northeast wind. But, indeed, the weather forecast in the latter case could contain a numerical value of an 80% probability of rain. In fact, this quantity is the forecaster's degree of belief in the statement "it will rain today." How is it obtained?

If one examines closely the uses made of either probability or degrees of belief, they are intended to suggest decisions with regard to actions: to take an umbrella, to start a life insurance policy, to determine the premium of a policy, to publish results, or to market a drug. In all cases, one incurs an up-front cost of some kind that may or may not lead to a benefit greater than the cost. Whether we like it or not, it finally comes down to gambling — the very purpose for which probability studies were first made by Pascal and others. Accordingly, the interpretation of a degree of belief of 80%, for example, is that the forecaster is willing to pay 80¢ in the hope of receiving $1.00 if it rains (and losing the 80¢ if it does not). Fairly clearly, if there is 20% probability of rain, the forecaster is only willing to risk losing 20¢. In this example, it appears that the assignment of degrees of belief is very subjective. While there is some truth in this observation, probability considerations can be used to generate values. Consider the case of tossing a fair coin, that is, a perfectly symmetrical circular disk whose physical properties and appearance are exactly the same irrespective of which side of the disk is viewed. Without destroying the perfect physical symmetry, we mark one side of the disk "heads" and call the other side "tails." It is not unreasonable to assume that the degrees of belief are

$$db(\text{heads}) = db(\text{tails})$$

Denote this value by x. Thus, the amount of the bet on "heads" will be x. If one makes two bets, one on heads and the other on tails, the total outlay is $2x$. Because the two events are exclusive, the total winnings for the two bets is guaranteed to be $1. But this is betting on a certainty for which a fair outlay is $1 to win $1. Thus, x equals 0.5. In this argument, the determination of the degrees of belief follows directly from the knowledge of the symmetry of the disk. If one does not have this knowledge, one could initially hypothesize perfect symmetry giving *a priori* degrees of belief as above. Subsequent experiments on actual tosses of the coin are then needed to refine the degrees of belief in "heads" and "tails." This procedure is the essence of the Bayesian approach: a quantitative method for calculating how degrees of belief are altered by experiments.

In the previous paragraphs, probability and degrees of belief become apparently interchangeable terms. Not only do both take on values in the range 0 to 1, both also obey the same algebra or rules of combination. Bayesians effectively say that the frequentist and degree of belief contexts are just two interpretations of one underlying notion of probability. There continues to be an ongoing battle between statisticians who label themselves either Bayesians or frequentists. However, the recent resurgence of Bayesian methods shows that the approach gives useful results. The situation is somewhat analogous to the criticisms hurled by mathematicians at Newton's and Leibniz's introduction of the concept of infinitesimals used in calculus. It was rigorously unsupportable, but it worked perfectly in describing nature for the physicists and astronomers. Calculus had to wait two centuries for the mathematicians to put it on a sound footing.

4.6.3 A Closer Look at Bayes's Theorem

Central to Bayesian methods is the concept of *conditional* probability or degrees of belief. All probabilities are conditional because conditions of the system under consideration must be known or assumed, as was the case above where the symmetrical coin was described in some detail. We will present a simple example to demonstrate conditional probability.

Figure 4.1 shows a rectangle with two intersecting regions. Let this be a target on which small ball bearings are dropped. Assume that the landing places are randomly distributed throughout the rectangle. The following statements can be made based on intuition:

$$p(U) = 1;\ p(A) = a;\ p(B) = b;\ p(AB) = c$$

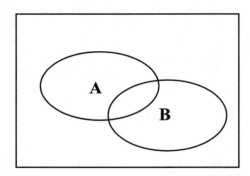

FIGURE 4.1 A rectangle with two intersecting regions, representing a "target" onto which small ball bearings can randomly drop.

where the events *U, A, B, AB* are the ball falls in the rectangle, region A, region B, the intersection of A and B, respectively. The rectangle has unit area and the areas of regions A and B and their intersection are *a, b,* and *c,* respectively. Consider the subset of cases where the ball falls in region A, i.e., the universe becomes region A. An outcome of the experiment is the event "the ball falls in B, conditional that it falls in A." The probability of this outcome is denoted by *p(B|A)*. Intuitively, this will be given by *c/a*. Thus,

$$p(B|A) = \frac{p(AB)}{p(A)} \text{ or } p(AB) = p(B|A) \cdot p(A).$$

If instead, the region B is taken as the universe one obtains:

$$p(A|B) = \frac{p(AB)}{p(B)} \text{ or } p(AB) = p(A|B) \cdot p(B).$$

The two expressions for *p(AB)* lead to the following relation:

$$p(B|A) = \frac{p(A|B) \cdot p(B)}{p(A)}$$

This is Bayes's theorem in its simplest form, i.e., Equation 4.1. Its importance is in relating the two conditional probabilities where the conditioning event and the "observed" event are interchanged. It shows clearly that, in general, $p(B|A) \neq p(A|B)$. As a homey example, this expression is just a symbolic way of stating: "All blackbirds are black birds, but not all black birds are blackbirds," or

p(black bird|blackbird) = 1
p(blackbird|black bird) < 1

A more serious case of the confusion of the two probabilities can be found in the use of racial or other profiling by law enforcement agencies. Suppose from arrest records that police determine *p*(bearded man|drugs in car) = 0.8, i.e., the driver was bearded in 80% of the cases where a traffic stop found drugs in the car. The result of the profiling procedure is that bearded drivers are more likely to be stopped. The assumption is that *p*(bearded man|drugs in car) is, if not 0.8, nonetheless large. However, Bayes's theorem gives:

$$p(\text{drugs in car}|\text{bearded}) = \frac{p(\text{bearded}|\text{drugs in car})}{p(\text{bearded})} \cdot p(\text{drugs in car})$$

Suppose that 0.1% of all traffic stops (without profiling) result in drugs being found and that 5% of all drivers are bearded. We obtain *p*(drugs in car|bearded) = (0.8/0.05) × 0.001 = 0.016. In traffic stops involving profiling, bearded drivers will have been unnecessarily inconvenienced and harassed in 100% − 1.6% or 98.4% of the time.

Bayes's theorem is primarily used for transforming *a priori* degrees of belief in a hypothesis to *a posteriori* degrees of belief as a result of experimental or observational data. Let $p(H)$ represent one's *a priori* degree of belief in a hypothesis, H. This will be based on the present state of knowledge. A body of data, E, is amassed as a result of experimentation or observation gathering. Bayes's theorem then becomes

$$p(H|E) = \frac{p(E|H) \cdot p(H)}{p(E)}$$

where $p(H|E)$ is the *a posteriori* degrees of belief in H, and $p(E|H)$ is called the likelihood of the data, E, given the hypothesis. The remaining expression, $p(E)$ is the probability of the observations irrespective of a particular hypothesis and is, in fact, the likelihoods summed over all possible hypotheses. This can be a complicated or even impossible operation. A simplification can be made if one considers the negation of H, usually written $\overline{H}$, meaning H is not true. Bayes gives

$$p(\overline{H}|E) = \frac{p(E|\overline{H})p(\overline{H})}{p(E)}$$

Dividing one equation by the other cancels out $p(E)$:

$$\frac{p(H|E)}{p(\overline{H}|E)} = \frac{p(E|H)}{p(E|\overline{H})} \cdot \frac{p(H)}{p(\overline{H})} \qquad (4.2)$$

The ratio of probabilities of an event to its negation or complement is called the odds of the event. For the toss of a fair coin, the odds of "heads" is 1 (usually called "evens"), for the roll of a fair die, the odds of a "6" is $1/5$, the odds of a throw less than "3" is $2/4 = 1/2$. The above relationship in words is

$$\text{Posterior Odds} = \text{Likelihood Ratio} \cdot \text{Prior Odds} \qquad (4.3)$$

4.7 TWO APPLICATIONS OF THE BAYESIAN METHOD

4.7.1 Successful Adjustment of Belief during Medical Diagnosis

The approach described above has been applied across many disciplines. An example is provided here from medical diagnostics, a field where global introspection is common but, on close study, has proved to be an inaccurate tool.[34] It illustrates the improvement in appropriate belief occurring if the expert opinion approach was replaced by a formal Bayesian analysis. The approach, formulations, and specific example are taken from work by Lane, Hutchinson, and co-workers.[34–37] The context is the application of likelihood ratios to modify prior odds for competing hypotheses of causality, i.e., application of Equation 4.3.

Lane[36] describes a case of a 38-year-old woman who lived in Gabon from 1981 to 1983. She took the antimalarial drug, chloroquine, during those years. Her prophylactic medication was switched from chloroquine to amodiaquine in mid-December 1983. She grew listless and began vomiting 36 days later. She became jaundiced 12 days after this but had no fever or joint pain. Testing showed no evidence of antibodies to the hepatitis B virus. Her bilirubin titer was fives times normal and she was immediately taken off the amodiaquine, that is, she was "dechallenged." Within 10 days of dechallenge, she felt better and her jaundice seemed to be diminishing. A week later and with no further testing, she was placed back on amodiaquine, i.e., she was "rechallenged." Jaundice returned 3 days after rechallenge and bilirubin titers were 18 times normal levels. After 12 more days, she was so ill that she was flown to a hospital in France. There she presented severe jaundice. Antibody testing for hepatitis A, B, and C were negative. The next day, she had bilirubin titers 20 times above normal levels. She slipped into a coma the next day, and died 3 days later. Her liver showed extensive necrolysis upon biopsy.

What was the cause of her death? The treating physician was clearly concerned about two potential causes, an adverse drug reaction to amodiaquine and viral hepatitis. Lane[36] presented this question to a panel of 40 physicians who overwhelmingly expressed the expert opinion that the drug caused her death. The presentation of symptoms upon initial challenge, improvement after dechallenge, and worsening with rechallenge weighed heavily in their conclusion.

Lane[36] moved beyond this informal expert opinion process to include a more formal Bayesian analysis. The same panel was asked to carefully apply Bayesian methods. They were asked to focus on the following: (1) establishing prior odds from information relevant to testing the alternate explanations, (2) establishing odds conditional on each explanation, (3) using this information to calculate the odds of one explanation vs. the other, and (4) producing a statement of the most probable cause based on this information. Production of some probabilities required the panel to use its shared experience and to search the literature. This shared information was used to estimate the various probabilities.

The following information suggested that, despite their first conclusion, an adverse reaction to amodiaquine might not have been the only plausible explanation:

- The viral hepatitis endemic in Africa puts Europeans at high risk. Risk increases during the first years of residence.
- Although tests suggest that hepatitis A and B were not the agents of disease, nonA-nonB hepatitis would not have been detected with the applied tests.
- NonA-nonB viral hepatitis displays symptom waxing and waning as noted for this patient.
- Amodiaquine has a half-life of approximately a week in the body. The patient appeared better 10 days after dechallenge. This seemed too rapid a recovery of normal liver function after an adverse reaction to a drug with such a long pharmacokinetic half-life.
- No liver function tests were done when the subjective judgment of improvement was made after dechallenge. The high bilirubin levels

measured after rechallenge suggest that liver function may not have been improving because the implied increase in bilirubin titers after rechallenge was improbably rapid.

The prior probability or odds for the adverse drug reaction hypothesis were those associated with a patient displaying symptoms who had not received the drug. The posterior probabilities or odds were calculated from all available information. Lane[36] defined the posterior odds as the probability of the drug causing the disease (p(Drug)) over the probability of the drug not causing the disease (p(Not Drug)). Both of these probabilities are conditional on the general background information (B) and specific clinical information on the patient (C).

$$\text{Posterior odds} = \frac{p(\text{Drug}|B, C)}{p(\text{Not Drug}|B, C)}$$

The same expert panel methodically organized information allowing posterior odds to be estimated for this case. First, they collectively estimated the probability of an acute amodiaquine adverse reaction to be approximately two orders of magnitude more likely than that for a long-term, adverse reaction to chloroquine. In coming to this conclusion, they assumed that onset of an adverse reaction to either drug was randomly and uniformly distributed within the interval of exposure, and that chloroquine exposure duration was approximately 36 months vs. the 36 days for amodiaquine. Also, symptoms reappeared quickly after rechallenge with amodiaquine. The adverse reaction to chlorodiaquine hypothesis was then rejected because it was two orders of magnitude less likely an explanation than acute reaction to amodiaquine. Only the acute amodiaquine reaction and nonA-nonB hepatitis hypotheses remained to be assessed.

The panel searched the literature, combining the members' collective knowledge to produce the following information:

- A survey of liver disease following amodiaquine administration estimated an odds of 1:15,000 but only 60% of the cases in the survey met the description of this particular case so the odds where modified to 4:100,000. The panel produced a 4: to 8:100,000 confidence interval for this estimate based on the probability of missing cases of adverse reaction to this drug. The high level of documentation of such adverse drug reaction cases was afforded by the seriousness of the reaction that usually resulted in hospitalization. The final odds estimated for calculations were 6:100,000.
- The odds of a middle-aged female contracting nonA-nonB hepatitis after living 3 years in Gabon were estimated from the odds published for American missionary females in Africa. American women in their third year of missionary work in Africa had a very high viral hepatitis attack rate of 2:100 per year. Of viral hepatitis cases in Africa, 20% were neither A nor B hepatitis; therefore, the odds of nonA-nonB hepatitis in the third year of residency for a middle-aged, European woman was estimated to

be approximately 4:1000 per year. This figure was adjusted downward to 1.5:1000 because of differences in behavior of an American missionary and a typical European resident. Missionary women were judged to be more likely to contract the disease because of their specific activities.
- Next, the panel determined that the fraction of nonA-nonB hepatitis cases conforming to the case at hand required that the odds be reduced to 2.5:10,000.

So, prior to considering the timing of events in the specific case, the odds of an adverse reaction to amodiaquine causing the fatality vs. a nonA-nonB virus was the following:

$$\text{Prior Odds} = \frac{6/100,000}{2.5/10,000} = \frac{6}{25} = 0.24$$

Because the odds were not sufficiently different to decide between the two causal hypotheses, the panel considered the timing of events in the case next. They considered events in the 16-week interval from first taking amodiaquine to death. The probability of nonA-nonB hepatitis presentation is uniform over that period. The odds of presentation of symptoms due to nonA-nonB viral infection were 1:16 (or 0.0625). Based on immunological and pharmacokinetic data, the odds of an adverse reaction to amodiaquine during the fifth week of drug treatment was 11:100. Therefore, the odds of a drug-related vs. a virus-related etiology was 0.11/0.0625 = 1.76. The posterior odds can be calculated again based on the prior odds and this new information regarding timing of symptom presentation.

$$\text{New Posterior Odds} = (0.24)(1.76) = 0.42$$

The ability to differentiate between the two causality hypotheses is still insufficient so the panel considered three factors judged to be particularly discerning: (1) death by liver necrosis after a fulminating hepatitis, (2) elevated bilirubin (and liver enzyme) titers at day 70, and (3) hepatic encephalopathy beginning on day 83. The daily rate of bilirubin increase implied by the drug reaction hypothesis was judged unlikely. The improved condition of the patient could have been a result of the waxing and waning characteristic of nonA-nonB viral hepatitis. They calculated from various reports a final factor of 3.5 that favored the drug explanation. Using the posterior odds just calculated above as the new prior odds, the odds of the drug explanation being correct was estimated.

$$\text{New Posterior Odds} = (0.42)(3.5) = 1.47$$

At this point, the odds of nonA-nonB hepatitis being the cause (2.5:100,000) can be used as the prior odds of nonA-nonB hepatitis etiology and 1.47 as the posterior odds after the addition of information about the specific fatality, i.e., facts relevant to the period of drug exposure. The posterior odds of the drug causing the event was 1.47/2.5 = 0.59.

All potential insights about the alternate hypotheses had been extracted with the available information so the panel stopped at this point. The panel's nearly unanimous initial conclusion of an adverse drug reaction was replaced by a conclusion that there was not enough information to select logically between the two explanations. Clearly, the formal application of a Bayesian context to this case reduced biases manifested in the initial judgment.

4.7.2 Applying Bayesian Methods to Estuarine Fish Kills and *Pfiesteria*

Men have been talking now for a week at the post-office about the age of the great elm, as a matter interesting but impossible to be determined. The very choppers and travellers have stood upon its prostrate trunk and speculated. ... I stooped and read its years to them (127 at nine and a half feet), but they heard me as the wind that once sighed through its branches. They still surmised that it might be two hundred years old. ... Truly they love darkness rather than light.

— **Henry David Thoreau**
quoted in Reference 38

4.7.2.1 Divergent Belief about *Pfiesteria piscicida* Causing Frequent Fish Kills

With notable exceptions (e.g., Reference 39), this Bayesian approach has also been ignored to the disadvantage of many disciplines. Stow[40] provides a particularly relevant example of assessing the causal relationship between the toxin-producing dinoflagellate, *Pfiesteria piscicida*, and frequent fish kills. Considerable debate has occurred in Maryland, Virginia, and North Carolina regarding the cause of recent coastal fish kills. Most of the debate emerges from contrasting expert opinions based on incomplete knowledge and a political imperative for a statement of risk.

In theory, the posterior probability of a fish kill given the presence of *Pfiesteria* can be calculated using Bayes's theorem (Equation 4.1),

$$p(\text{Fish Kill}|Pfiesteria) = \frac{p(\text{Fish Kill}) \cdot p(Pfiesteria|\text{Fish Kill})}{p(Pfiesteria)}$$

Like the problem of law enforcement profiling described above, the erroneous equating of p(Fish Kill|*Pfiesteria*) with p(*Pfiesteria*|Fish Kill) has led to confusion with this issue and has distracted risk assessors from the importance of generating the information needed to calculate p(*Pfiesteria*) and p(Fish kill). As an example of how easily these conditional probabilities can be confused, Burkholder et al.[41] found high densities of *P. piscicida* after fish kills (8 of 15 fish kills in 1991, 5 of 8 fish kills in 1992, and 4 of 10 fish kills in 1993) and stated, "*P. piscicida* was implicated as the causative agent of 52 ± 7% of the major fish kills (affecting 10^3 to 10^9 fish from May 1991 to November 1993) on an annual basis in North Carolina estuaries and coastal waters." Although *P. piscicida* certainly could have been the causative agent, implications are being made about p(Fish Kill|*Pfiesteria*) but the data strictly define p(*Pfiesteria*|Fish Kill).

Commercially and politically costly judgments are currently being made without reliable estimates of the crucial probabilities, p(Fish Kill), p(*Pfiesteria*), p(Fish Kill|*Pfiesteria*), and ultimately, p(Fish Kill|*Pfiesteria*). The result is a contentious decision-making process with arguments now focusing on questions of scientific ethics and regulatory stonewalling,[42] and risk exaggeration.[43] (See References 42 through 48 as examples.) This confused *Pfiesteria*–fish kill causality assessment is not an isolated instance of a suboptimal assessment process. Certainly, risk assessments for alar on apples[49,50] and climatic change[51] were at least as important and as garbled.

4.7.2.2 A Bayesian Vantage for the *Pfiesteria*-Induced Fish Kill Hypothesis

Bayes's theorem (Equation 4.1) will be applied directly to this problem. This approach intentionally contrasts with the medical diagnosis example described above which explored competing hypotheses with likelihood ratios and prior odds (Equation 4.3). The focus will be the Neuse and Pamlico River systems for which Burkholder et al.[41] formulated the above causal hypothesis regarding frequent fish kills.

Using North Carolina Department of Water Quality data (Table 4.3), p(fish kill) can be estimated as the number of days with fish kills divided by the total number

TABLE 4.3
Summary of North Carolina Department of Environmental Quality Fish Kill Data for 1997 to 2000 (~930 days)

Year	River System	Total No. of Fish Kills
1997[a]	Neuse River	12
	Pamlico/Tar	6
1998[b]	Neuse River	8
	Pamlico/Tar	5
1999[c]	Neuse River	16
	Pamlico/Tar	5
2000[d]	Neuse River	13
	Pamlico	10
Total	Neuse River	49
	Pamlico/Tar	26

[a] April through November 1997 (8 months).
[b] June through October 1998 (5 months).
[c] February 1999 through December 1999 (11 months).
[d] January 2000 through July 2000 (7 months).

Source: North Carolina Division of Water Quality Web site, http://www.esb.enr.stste.nc.us/Fishkill/Fishkill100.htm.

of observation days (31 months, or roughly 930 days): 75/930 or 0.081. It could be argued that only data for warm months when *Pfiesteria* blooms are likely should be used in these calculations. However, for illustrative purposes, all months for which data are available were used. The analysis can easily be redone based on warm months only.

Burkholder et al.[41] estimate $p(Pfiesteria|\text{fish kill})$ to be 0.52. However, there is an important caveat to this estimate. The occurrences involve presumptive PLO (*Pfiesteria*-like organisms) and definitive identification was not made.

The presence of PLOs in Virginia waters was explored by Marshall et al.[52] From these data, $p(Pfiesteria)$ = 496/1437 or 0.345. It is important to note that, again, $p(Pfiesteria)$ was estimated from PLO counts. Molecular techniques were applied by Rublee et al.[53] on East Coast sites to produce an estimate of *P. piscicida* presence in 35 out of 170 samples or 0.205.

The $p(\text{Fish kill}|Pfiesteria)$ can be calculated with these estimates of $p(Pfiesteria)$, $p(\text{Fish kill})$, and $p(Pfiesteria|\text{fish kill})$:

$$p(\text{Fish Kill}|Pfiesteria) = (0.52)(0.081)/0.345 = 0.122$$
or 12.2% based on $p(\text{PLO})$

$$p(\text{Fish Kill}|Pfiesteria) = (0.52)(0.081)/0.205 = 0.205$$
or 20.5% based on $p(Pfiesteria)$

Given the presence of *Pfiesteria* as defined above, the likelihood of a fish kill occurring is approximately 12 or 20%, not 52%. If one were to measure PLO in a water body, the likelihood or "belief" that a fish kill will occur is crudely estimated to be 12%. Belief increases to 20% if *Pfiesteria* detection was attempted rather than PLO detection.

Several important points should be made about these estimates. First, the initial misleading impression given by $p(Pfiesteria|\text{fish kill}) = 0.52$ is that there is a very high risk of a fish kill if *Pfiesteria* was present — roughly a 50:50 chance. As discussed earlier, this is a common and understandable error. Second, it cannot be overemphasized that the results of this type of analysis are only as good as the data used to calculate $p(Pfiesteria)$, $p(\text{Fishkill})$, and $p(Pfiesteria|\text{fish kill})$. Confidence in results will increase as more high-quality and explicit data are generated. The approach does not avoid the biases described early in this chapter: it only lessens their influence. Regardless, these results are an improvement over the qualitative conclusions drawn with criteria such as Hill's aspects of disease association or the inaccurate impression derived from $p(Pfiesteria|\text{fish kill})$ alone. Third, the Bayesian context allows one to identify the most important information required to estimate the likelihood of a causal relationship between the presence of *Pfiesteria* in a coastal water body and fish kills. For example, better definitions of the presence of "*Pfiesteria*" would be extremely helpful. Should simple presence/absence or cell density above a particular threshold be scored at each site? Should one monitor for PLO, PCO (*Pfiesteria* cluster organisms), *P. piscicida*, or only the toxin-producing stages of *P. piscicida*? An explicit definition of "fish kill" would be helpful because there might be characteristics of fish killed by *Pfiesteria* that would allow the exclusion from analysis of kills caused by other factors. Better means of defining the temporal sequence of *P. piscicida* bloom followed by a fish kill is needed because this dinoflagellate appears suddenly in its toxin-producing form and quickly disappears.[41,54,55]

Bayesian analysis of competing causes would also be helpful in this particular causality assessment. Low dissolved oxygen concentration can be used to make this point. Fish kills associated with episodes of low dissolved oxygen were also studied by Paerl et al.[56] in the Neuse River estuary. Workers in North Carolina[44] quickly responded to their conclusions,

> Paerl et al.'s central conclusion about finfish kills is not supported either by their data or by any statistical analysis. ... The paper contains numerous misinterpretations and misuse of literature citations. Paerl et al. also made serious errors of omission, germane from the perspective of science ethics, in failing to cite peer-reviewed, published information that attributed other causality to various fish kills that they described.

Bayesian analysis of the competing explanations, i.e., fish kills due to *Pfiesteria* toxin vs. fish kills due to low oxygen, in this estuary could be done as illustrated above for adverse drug reaction vs. viral hepatitis. The monetary and political costs associated with the current divergent states of belief among researchers would be lowered by such an analysis. It would provide an explicit statement of relative belief that could be used to make wise management decisions for marine resources.

4.8 CONCLUSION

At the core of each ecological risk assessment is an assessment of causality. Causality assessments identify the cause–effect relationship for which risk is to be estimated. Insufficient emphasis is placed on the quality of causality assessment relative to risk estimation. Expert opinion and weight-of-evidence methods are subject to cognitive and knowledge base biases. Rules of thumb such as Hill's aspects of disease association or Fox's rules of practical causal inference are often used to decrease such biases. To illustrate use of such rules, the exceptionally high quality evidence for PAH-induced liver cancers in English sole was assessed with Hill's aspects of disease association.

Abductive inference and its quantification by means of Bayes's theorem can further reduce biases and provide a framework for the efficient accumulation and use of evidence. Bayesian methods allow quantification of belief based on observational and experimental evidence. Belief in a causal hypothesis can be determined by simple or iterative application of Bayes's theorem (Equation 4.1) as illustrated here with *Pfiesteria*-linked fish kills in coastal waters. Likelihood ratios and prior odds (Equation 4.3) can be used to quantify relative belief in competing explanations, e.g., frequent fish kills in Neuse River due to *Pfiesteria* vs. low dissolved oxygen. Wider application of Bayesian methods would reduce problems associated with causality assessments, reduce conflicts emerging from less formal integration of available evidence during global introspection, and most effectively use limited resources needed for ecological risk assessments in coastal waters.

ACKNOWLEDGMENTS

The authors are grateful to Dr. J. Shields of the Virginia Institute of Marine Science who provided initial references to literature for and valuable advice about *Pfiesteria*-related fish kills.

REFERENCES

1. Medawar, P.B., *Pluto's Republic*, Oxford University Press, Oxford, 1982.
2. Piattelli-Palmarini, M., *Inevitable Illusions. How Mistakes of Reason Rule Our Minds*, John Wiley & Sons, New York, 1994.
3. U.S. EPA, Guidelines for Ecological Risk Assessment, U.S. EPA/630/R-95/002F, April 1998, U.S. Environmental Protection Agency, Washington, D.C., 1998.
4. Platt, J.R., Strong inference, *Science*, 146, 347, 1964.
5. McConnell, M., *Challenger. A Major Malfunction*, Doubleday & Co., Garden City, NY, 1987.
6. Dalal, S.R., Fowlkes, E.B., and Hoadley, B., Risk analysis of the space shuttle: pre-*Challenger* prediction of failure, *Am. Stat. Assoc.*, 84, 945, 1989.
7. Newman, M.C., *Fundamentals of Ecotoxicology*, Lewis Publishers, Boca Raton, FL, 1998.
8. Bryan, G.W. and Gibbs, P.E., Impact of low concentrations of tributyltin (TBT) on marine organisms: a review, in *Metal Ecotoxicology: Concepts & Applications*, Newman, M.C. and McIntosh, A.W., Eds., Lewis Publishers, Chelsea, MI, 1991, chap. 12.
9. Huggett, R.J. et al., The marine biocide tributyltin: assessing and managing the environmental risks, *Environ. Sci. Technol.*, 26, 231, 1992.
10. Doyle, A.C., The Adventure of the Blue Carbuncle, in *The Adventures of Sherlock Holmes* (1891), reprinted in *Sherlock Holmes: A Complete Novels and Stories*, Vol. 1, Bantam Books, New York, 1986.
11. Popper, K.R., *The Logic of Scientific Discovery*, Routledge, London, 1959.
12. Russell, B., *A History of Western Philosophy*, Simon & Schuster, New York, 1945.
13. Newman, M.C., *Quantitative Methods in Aquatic Ecotoxicology*, Lewis Publishers, Boca Raton, FL, 1995.
14. Newman, M.C., Ecotoxicology as a science, in *Ecotoxicology. A Hierarchical Treatment*, Newman, M.C. and Jagoe, C.H., Eds., Lewis Publishers, Boca Raton, FL, 1996, chap. 1.
15. Loehle, C., Hypothesis testing in ecology: psychological aspects and importance of theory maturation, *Q. Rev. Biol.*, 62, 397, 1987.
16. Chamberlin, T.C., The method of multiple working hypotheses, *J. Geol.*, 5, 837, 1897.
17. Popper, K.R., *Conjectures and Refutations. The Growth of Scientific Knowledge*, Harper & Row, New York, 1968.
18. Popper, K.R., *Objective Knowledge. An Evolutionary Approach*, Clarendon Press, Oxford, 1972.
19. Kuhn, T.S., *The Structure of Scientific Revolutions*, 2nd ed., University of Chicago Press, Chicago, IL, 1970.
20. Adams, J., *Risk*, UCL Press Limited, London, 1995.
21. Bailey, F.G., *The Prevalence of Deceit*, Cornell University Press, Ithaca, NY, 1991.
22. Newman, M.C., *Population Ecotoxicology*, John Wiley & Sons, Chichester, U.K., 2001.

23. Newman, M.C. and McCloskey, J.T., The individual tolerance concept is not the sole explanation for the probit dose-effect model, *Environ. Toxicol. Chem.*, 19, 520, 2000.
24. Hill, A.B., The environment and disease: association or causation? *Proc. R. Soc. Med.*, 58, 295, 1965.
25. Fox, G.A., Practical causal inference for ecoepidemiologists, *J. Toxicol. Environ. Health*, 33, 359, 1991.
26. Horness, B.H. et al., Sediment quality thresholds: estimates from hockey stick regression of liver lesion prevalence in English sole (*Pleuronectes vetulus*), *Environ. Toxicol. Chem.*, 17, 872, 1998.
27. Myers, M.S. et al., Overview of studies on liver carcinogenesis in English sole of Puget Sound; evidence for a xenobiotic chemical etiology. I: Pathology and epizootiology, *Sci. Total Environ.*, 94, 33, 1990.
28. Myers, M.S. et al., Relationships between toxicopathic hepatic lesions and exposure to chemical contaminants in English sole (*Pleuronectes vetulus*), starry flounder (*Platichthys stellatus*), and white croacker (*Genyonemus lineatus*) from selected marine sites on the Pacific Coast, USA, *Environ. Health Perspect.*, 102, 200, 1994.
29. Myers, M.S. et al., Toxicopathic hepatic lesions in subadult English sole (*Pleuronectes vetulus*) from Puget Sound, Washington, USA: relationships with other biomarkers of contaminant exposure, *Mar. Environ. Res.*, 45, 47, 1998.
30. Moore, M.J. and Myers, M.S., Pathobiology of chemical-associated neoplasia in fish, in *Aquatic Toxicology: Molecular, Biochemical and Cellular Perspectives*, Malins, D.C. and Ostrander, G.K., Eds., Lewis Publishers, Boca Raton, FL, 1994, chap. 8.
31. Josephson, J.R. and Josephson, S.G., *Abductive Inference. Computation, Philosophy, Technology*, Cambridge University Press, Cambridge, U.K., 1996.
32. Howson, C. and Urbach, P., *Scientific Reasoning. The Bayesian Approach*, Open Court, La Salle, IL, 1989.
33. Bayes, T., An essay toward solving a problem in the doctrine of chances, *Philos. Trans. R. Soc.*, 53, 370, 1763.
34. Lane, D.A. et al., The causality assessment of adverse drug reactions using a Bayesian approach, *Pharm. Med.*, 2, 265, 1987.
35. Hutchinson, T.A. and Lane, D.A., Assessing methods for causality assessment of suspected adverse drug reactions, *J. Clin. Epidemiol.*, 42, 5, 1989.
36. Lane, D.A., Subjective probability and causality assessment, *Appl. Stochastic Models Data Anal.*, 5, 53, 1989.
37. Cowell, R.G. et al., A Bayesian expert system for the analysis of an adverse drug reaction, *Artif. Intelligence Med.*, 3, 257, 1991.
38. Walls, L.D., *Thoreau on Science*, Houghton Mifflin, Boston, 1999.
39. Price, P.N., Nero, A.V., and Gelman, A., Bayesian prediction of mean indoor radon concentrations for Minnesota counties, *Health Phys.*, 71, 922, 1996.
40. Stow, C.A., Assessing the relationship between *Pfiesteria* and estuarine fishkills, *Ecosystems*, 2, 237, 1999.
41. Burkholder, J.M., Glasgow, H.B., Jr., and Hobbs, C.W., Fish kills linked to a toxic ambush-predator dinoflagellate: distribution and environmental conditions, *Mar. Ecol. Prog. Ser.*, 124, 43, 1995.
42. Burkholder, J.M. and Glasgow, H.B., Jr., Science ethics and its role in early suppression of the *Pfiesteria* issue, *Hum. Organ.*, 58, 443, 1999.
43. Griffith, D., Exaggerating environmental health risk: the case of the toxic dinoflagellate *Pfiesteria*, *Hum. Organ.*, 58, 119, 1999.

44. Burkholder, J.M., Mallin, M.A., and Glasgow, H.B., Jr., Fish kills, bottom-water hypoxia, and the toxic *Pfiesteria* complex in the Neuse River and Estuary, *Mar. Ecol. Prog. Ser.*, 179, 301, 1999.
45. Griffith, D., Placing risk in context, *Hum. Organ.*, 58, 460, 1999.
46. Lewitus, A.J. et al., Human health and environmental impacts of *Pfiesteria*: a science-based rebuttal to Griffith (1999), *Hum. Organ.*, 58, 455, 1999.
47. Oldach, D., Regarding *Pfiesteria, Hum. Organ.*, 58, 459, 1999.
48. Paolisso, M., Toxic algal blooms, nutrient runoff, and farming on Maryland's Eastern Shore, *Cult. Agric.*, 21, 53, 1999.
49. Ames, B.N. and Gold, L.S., Pesticides, risk, and applesauce, *Science*, 244, 755, 1989.
50. Groth, E., Alar in apples, *Science*, 244, 755, 1989.
51. Nordhaus, W.D., Expert opinion on climatic change, *Am. Sci.*, 82, 45, 1994.
52. Marshall, H.G., Seaborn, D., and Wolny, J., Monitoring results for *Pfiesteria piscicida* and *Pfiesteria*-like organisms from Virginia waters in 1998, *Vir. J. Sci.*, 50, 287, 1999.
53. Rublee, P.A. et al., PCR and FISH detection extends the range of *Pfiesteria piscicida* in estuarine waters, *Vir. J. Sci.*, 50, 325, 1999.
54. Burkholder, J.M. et al., New "phantom" dinoflagellate is the causative agent of major estuarine fish kills, *Nature*, 358, 407, 1992.
55. Culotta, E., Red menace in the world's oceans, *Science,* 257, 1476, 1992.
56. Paerl, H.W. et al., Ecosystem responses to internal and watershed organic matter loading: consequences for hypoxia in the eutrophying Neuse River Estuary, North Carolina, USA, *Mar. Ecol. Prog. Ser.*, 166, 17, 1998.

5 Bioavailability, Biotransformation, and Fate of Organic Contaminants in Estuarine Animals

Richard F. Lee

CONTENTS

5.1 Introduction 97
5.2 Bioavailability 98
5.3 Uptake 100
    5.3.1 Uptake from Water 100
    5.3.2 Uptake from Sediment 104
    5.3.3 Uptake from Food 105
5.4 Fate of Xenobiotics after Uptake by Estuarine Animals 106
    5.4.1 Biotransformation (Metabolism) 106
        5.4.1.1 Phase-One Reactions 106
        5.4.1.2 Phase-Two Reactions 108
    5.4.2 Fates and Metabolic Pathways for Xenobiotics and Metabolites within Tissues and Cells 111
    5.4.3 Binding of Xenobiotics to Cellular Macromolecules 115
5.5 Elimination 115
5.6 Summary 118
References 119

5.1 INTRODUCTION

An important component of ecological risk assessment studies in oceans and estuaries includes the characterization of the exposure of estuarine animals to contaminants. Data on the bioavailability, uptake, accumulation, and elimination of contaminants by animals are necessary to characterize contaminant exposure.[1] Contaminants found in estuarine and marine waters and sediments include aromatic hydrocarbons, organometallics, organohalogens, and pesticides, often referred to as organic xenobiotics. The high concentrations of various xenobiotics in aquatic animals from contaminated sites are indicative of the efficient uptake and accumulation of these xenobiotics.[2-14] As a result of the presence of these contaminants in tissues, many toxicological effects may be manifested including the following: growth, reproduction, and development.

The extent of uptake of xenobiotics by an estuarine animal depends on their bioavailability from various matrices, including water, sediment, or food. After entering from one of these matrices via the gill or digestive tract, the xenobiotic can be accumulated in the liver (fish) or hepatopancreas/digestive gland (annelid, crustacean, mollusk). Hemolymph or blood functions as an important avenue for transporting xenobiotics and xenobiotic metabolites (Figure 5.1). After entrance into an animal, the processes of accumulation, biotransformation, and elimination determine the fate of the xenobiotic. The relative importance of these different processes depends on a number of factors including the physicochemical properties of the xenobiotic, the ability of the animal's enzyme system to metabolize the compound, and the lipid content of the animal.

This chapter discusses bioavailability of contaminants in estuaries, followed by sections on the uptake, accumulation, metabolism, and elimination of xenobiotics. The focus is on fish and three groups of marine estuarine invertebrates, i.e., crustaceans, mollusks, and annelids. There are a number of reviews that have discussed the uptake, metabolism, and elimination of toxicants by aquatic animals.[5-22]

5.2 BIOAVAILABILITY

In this chapter, the bioavailable fraction is that fraction of a xenobiotic available for uptake by estuarine and marine animals. Matrices in the estuarine environment include water, sediment, and food. *Bioaccumulation* is a general term describing the processes by which bioavailable xenobiotics are taken up by estuarine animals from

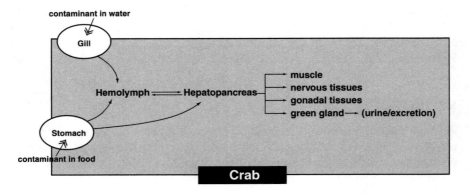

FIGURE 5.1 Uptake and bioaccumulation of organic contaminants by crabs.

the water, sediment, or food. To determine bioavailability, it is necessary to determine the relative partitioning between these matrices and the animal's gill or stomach (see Figure 5.1). The partitioning can be illustrated by the following expressions:

Water/gills of animal
Sediment/pore water or digestive juices/gills or stomach of animal
Food/stomach of animal

Xenobiotics in estuarine and marine waters are associated with both the dissolved and particulate phases. A xenobiotic in the dissolved phase can be freely dissolved, but in natural waters xenobiotics tend to bind to dissolved organic matter, primarily the humic fraction.[23-27] Landrum et al.[24] using the amphipod, *Pontoporeia hoyi*, found that the uptake rate constants for a series of xenobiotics increased as the dissolved organic carbon decreased. Thus, binding of xenobiotics to dissolved organic matter can reduce the amount that is bioavailable.

Particulates in estuarine water are often in high concentrations, ranging from 10 to 400 mg/l.[28,29] These particulates are mixtures of organic matter, living matter, and small clay particles. Scanning electron micrographs reveal rough surfaces on these detrital particles, with bacteria fastened by mucoid-like pads and fibrillar appendages[30-32] (Figure 5.2). Xenobiotics can bind to hydrophobic sites on the particulate surfaces. When radiolabeled benzo(*a*)pyrene was added to estuarine water, it was found by autoradiography that most of the benzo(*a*)pyrene was bound to detrital particles[33] (Figure 5.3). Particulates with associated xenobiotics are considered to be an important pathway by which contaminants enter estuarine food webs.

Bioavailability of xenobiotics in sediments is generally not related to the sediment concentration, but rather to organic carbon content and physicochemical properties of the sediment. Xenobiotics in sediments are partitioned among particles, pore water, and organisms. Estuarine sediments are composed of particles of various sizes with xenobiotics associated with particles in the 30 to 60 μm size range, which is in the silt-clay fraction.[34-36] In addition to the mineral phase, estuarine sediments can be high in organic carbon and xenobiotics bind to hydrophobic sites within the organic phase of the sediments. Three factors that are important in controlling the bioavailability of contaminants associated with sediment include the aqueous solubility of the xenobiotic, rate and extent of desorption from the solid phase into the pore water, and the ability of digestive juices of infaunal animals to solubilize the xenobiotic.[37-39] Some infaunal animals pass sediment particles through their digestive tract. Surfactants in their digestive juices solubilize a certain fraction of xenobiotics off the sediment particles.[38] Sediment organics can be labile or refractory. Xenobiotics bound to labile organics are more bioavailable because, during digestion, these xenobiotics are released within the animal.[40-42] There is some desorption of xenobiotics from sediment particles into pore water and xenobiotics in pore water are highly bioavailable.[36] Because of tight binding to humin-kerogen polymers in sediment, there are very low desorption rates of uncharged lipophilic xenobiotics, e.g., 5- and 6-ringed polycyclic aromatic hydrocarbons (PAHs) and polychlorinated hydrocarbons in organic-rich sediment.[38,43-46]

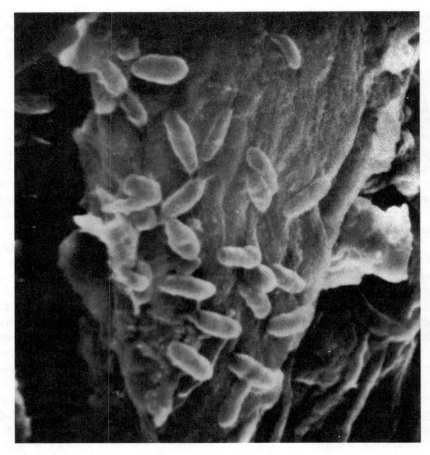

FIGURE 5.2 Scanning electron micrograph of detrital particle from Skidaway River, GA, showing attached bacteria (19,000×). (Courtesy of H. Paerl, University of North Carolina.)

5.3 UPTAKE

5.3.1 Uptake from Water

The simplest uptake occurs where the xenobiotic is in the dissolved phase of the water and uptake can be described by a first-order equation.[47] Other work described below elaborates on this basic equation:

$$C_A = K_U C_W T \tag{5.1}$$

where
- K_U = uptake rate constant (1/h)
- C_W = concentration of xenobiotic in water (ng/g)
- C_A = concentration of xenobiotic in animal (ng/g)
- T = time (h)

Bioavailability, Biotransformation, and Fate of Organic Contaminants   101

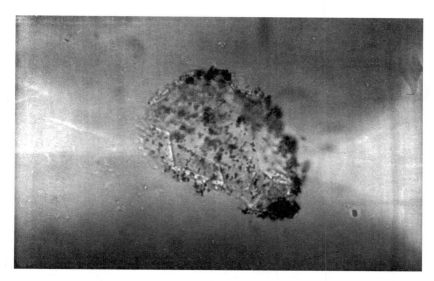

FIGURE 5.3 Autoradiography of detritus from estuarine river labeled with ^{3}H-benzo(a)pyrene. ^{3}H-Benzo(a)pyrene (25mci/mM) was added to 100 ml of Skidaway River, GA (final concentration: 0.1 μg/l). After 12 h of incubation, water was filtered onto a 0.2-μm filter followed by autoradiography using Kodak NTB-2 emulsion (H. Paerl and R. Lee, unpublished work). Note dark spots on detritus particle, which indicates binding of ^{3}H-benzo(a)pyrene.

Uptake of benzo(a)pyrene from seawater by the clam, *Mercenaria mercenaria*, fits this equation and has a rate constant of 5/day (Figure 5.4). Some of the factors that can affect K_U include water temperature, metabolic rate of the animal, and the efficiency of passage of xenobiotic across the gill.[47]

There is evidence that the rate of uptake into estuarine animals is determined by the hydrophobicity of the compound and the lipid content of the animal.[48] Gobas and Mackay[49] showed the importance of lipid content of tissues by using the following expression to describe the uptake of xenobiotics by fish where the xenobiotic is transferred from a water compartment to a lipid compartment in the fish.

$$V_F Z_F \, df_F/dt = V_L Z_L \, df_L/dt = D_F \, (f_W - f_L) \qquad (5.2)$$

where

V	= volume (m^3)
Z	= fugacity capacity (mol/m^3 · Pa)
f	= fugacity (Pa)
t	= time (s)
D	= transport parameter, including all resistances between the lipid compartment and the water (mol/Pa · s)

Subscripts W refer to water, F to fish, and L to lipid to which all the xenobiotic is assumed to partition.

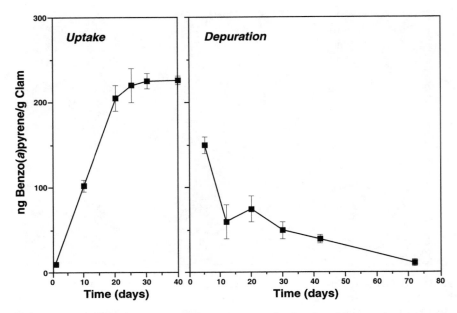

FIGURE 5.4 Uptake and depuration of benzo(*a*)pyrene by the clam, *Mercenaria mercenaria*: 66 clams were exposed in groups of three in 20-l aquaria containing benzo(*a*)pyrene (2 μg/l). Water was changed daily with new benzo(*a*)pyrene added. Three clams were extracted and separately analyzed for benzo(*a*)pyrene by high-performance liquid chromatography at each time interval. Results are mean ± standard deviation. After 40 days, clams were transferred to flowing seawater tanks for the depuration phase of the study.

Fugacity is the tendency of a chemical to escape from its existing phase into another phase. Fugacity has units of pressure and is to molecular diffusion what temperature is to heat diffusion. The fugacity capacity relates fugacity to chemical concentration and quantifies the capacity of a particular phase for fugacity. Fugacity and fugacity capacity are related by $C = Zf$, where C is the concentration, f is the fugacity, and Z is the fugacity capacity.[50]

Stegeman and Teal[51] noted a significant relationship between oyster lipid content and their accumulation of petroleum hydrocarbons. Oysters with high and low lipid contents accumulated 334 and 161 μg/g of petroleum hydrocarbons, respectively, after exposure to fuel oil in water. It has also been suggested that the lipid content of the gills is more important in controlling xenobiotic uptake than the lipid content of the whole animal.[48,49,51] An estimate of the partitioning of a xenobiotic between water and the gill is obtained from its K_{ow}, the octanol–water partition coefficient of the xenobiotic.

The uptake rate of different congeners of polychlorinated biphenyls (PCBs) by fish and polychaetes has been shown to be influenced primarily by the stereochemistry of the congeners.[52] Planar congeners were most efficiently taken up, whereas less planar congeners were less efficiently taken up. Thus, K_{ow} is not always the best estimator of uptake rate because steric factors can also be important in affecting uptake.

Bioavailability, Biotransformation, and Fate of Organic Contaminants

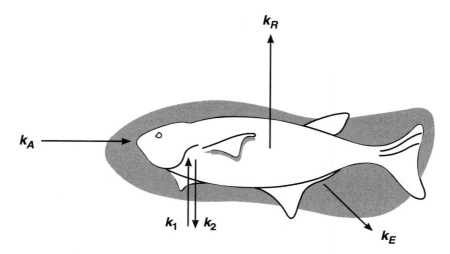

FIGURE 5.5 Uptake of contaminants by fish from water (k_1) and food (k_A) followed by metabolism (k_R) and elimination to the water (k_2) and feces (k_E). (Modified from Gobas et al.[146])

Bioconcentration takes place when the rate of uptake is greater than elimination. The bioconcentration factor is strongly related to the octanol–water partition coefficient of the xenobiotic.[48,53] Bioconcentration refers to the process by which, as a result of the uptake, there is a net accumulation of a xenobiotic from the water into an estuarine animal. The bioconcentration factor is a unitless value that describes the degree to which a xenobiotic is concentrated in the animal's tissues relative to the water concentration of the xenobiotic.[48–53] These relationships are defined by the following equations[54] (Figure 5.5).

$$C_a = (K_u/K_D)C_w\left(1 - \exp\left[f\left(\frac{1}{K_{D'}}\right)\right]\right) \quad (5.3)$$

where
- C_a = concentration in fish (ng/g)
- C_w = concentration in water (ng/g)
- K_U = uptake rate constant (1/h)
- K_D = depuration rate constant (1/h)

$$\text{Bioconcentration factor (BCF)} = C_a/C_w = K_U/K_D \quad (5.4)$$

$$\log_{10} \text{BCF} - 0.85 \log_{10} P - 0.70 \quad (5.5)$$

where
- P = octanol–water partition coefficient

5.3.2 Uptake from Sediment

A number of studies have found that estuarine and marine animals, including both fish and invertebrates, can take up xenobiotics from sediments or from food in the sediments.[36,55-63] Polychaetes and benthic copepods, which serve as food for many fish, can accumulate xenobiotics from sediment. In a series of experiments, fish were exposed to PCB-contaminated sediments (without polychaetes or benthic copepods) or to food (polychaetes or benthic copepods) previously exposed to the PCB-contaminated sediments.[61,63] The fish given the PCB-contaminated food accumulated more PCBs than fish exposed to the PCB-contaminated sediments. Infaunal animals can take up contaminants from the pore water or particles. Pore water concentrations of highly hydrophobic xenobiotics are quite low, but because uptake from water is quite rapid, pore water is an important pathway for uptake. Xenobiotic concentrations on sediment particles can be quite high, but significantly less bioavailable than xenobiotics in pore water. Infaunal animals can be selective feeders of food within the sediment, or they can be nonselective feeders and pass all sediment of particular size through their digestive tract. For example, the benthic amphipod, *Diporeia* spp., is a highly selective feeder, whereas the oligochaete, *Lumbriculus variegatus*, passes all fine-sized sediments through its intestinal tract.[64,66] As a result of these differences in feeding behavior, the assimilation efficiency of benzo(*a*)pyrene uptake from sediment was 45 to 57% for *Diporeia*, and 23 to 26% for *L. variegatus*.[32] One explanation for the differences between the two species could be that *Diporeia* selects very labile organic matter, so that much of the benzo(*a*)pyrene on these organics is bioavailable. In contrast, *L. variegatus* takes up particles of a certain size and proportionally less of the benzo(*a*)pyrene is bioavailable on these particles. The assimilation efficiency for hexachlorobenzene in sediment by the selective feeder, *Macoma nasuta*, an estuarine bivalve, was found to range from 38 to 56%.[67] For sediment ingesters, the amount of xenobiotic taken up depends on the amount of sediment ingested, so that high tissue concentrations are found when sediment ingestion is high.[34] Uptake of xenobiotics from ingestion of sediment particles depends on the feeding rate of the animal, assimilation efficiency, feeding selectivity and concentration of xenobiotics in ingested food particles.[34]

Kukkonen and Landrum[68] used the following first-order rate equation to describe the kinetics of benzo(*a*)pyrene accumulation from sediment by *Diporeia* spp.:

$$C_a = (K_s C_s (1 - e^{-K_e t}) / K_e) \tag{5.6}$$

where
K_s = uptake clearance coefficient (g dry sediment/g wet organism • h)
C_s = concentration of benzo(*a*)pyrene in sediment (mmol/g)
t = time (h)
K_e = elimination rate constant (1/h)
C_a = concentration of benzo(*a*)pyrene in *Diporeia* (mmol/g)

K_s used here is similar to K_U/K_D of Equation 5.3.

The bioaccumulation factor (concentration in animal/concentration in sediment), which takes into account both uptake and elimination, ranges from less than 0.1 to 20 for estuarine animals.[36] The lower bioaccumulation factors are associated with high-organic-content sediments, and higher factors are associated with low-organic-content sediments. In contrast, bioaccumulation factors for estuarine and marine animals exposed to contaminants in water is generally 1000 or more. It should be noted that because the sediment concentration is generally much higher than the water concentration, the sediment is still an important source for contaminant uptake. To allow comparisons with different compounds, different species, and different types of animals, the accumulation factors are often normalized with respect to lipid for animals and to total organic carbon for sediment, so the normalized bioaccumulation factor can be expressed as:[55,69]

$$BCF = \frac{\text{Conc. in animal/lipid of animal}}{\text{Conc. in sediment/total organic carbon of sediment}} \quad (5.7)$$

Normalized bioaccumulation factors for PCBs and dioxin accumulation by three estuarine animals (polychaetes — *Nereis virens*, clams — *Macoma nasuta*, grass shrimp — *Palaemonetes pugio*) ranged from 0.1 to 2.[55] The very low bioaccumulation factors for lower-chlorinated PCBs by *N. virens* were presumably due to metabolism of these congeners by this polychaete. The time to steady-state concentration for polychaetes exposed to PCB-contaminated sediment was between 70 and 120 days.[55]

5.3.3 Uptake from Food

Diet is a source of many of the highly hydrophobic contaminants found in fish. A number of studies have shown that diet was the major source of PCBs in various fish species.[63–65] For different xenobiotics, the relative importance of uptake from food and water can be quite different depending on the xenobiotic concentration in water and food, as well as the fluxes of food and feces, and bioconcentration factors.

A model for describing the uptake of a xenobiotic by estuarine animals that takes into account concentrations in the food, water, and sediment is the following:[70]

$$C_i = \{[k_1 C_w] + [(p_{ix}\ CAE\ I_{ix})\ C_x]\}/[k_2 + k_G + k_M + k_E] \quad (5.8)$$

where
- C_i = lipid-normalized xenobiotic concentration in animal (μg/kg lipid)
- k_1 = rate constant of xenobiotic uptake from water (l/day/g lipid)
- C_w = concentration of xenobiotic in water (μg/l)
- p_{ix} = feeding preference of animal on prey x
- CAE = chemical assimilation efficiency (g assimilated/g ingested)
- I_{ix} = ingestion rate of animal of prey x (g of x/g of i/day)
- C_x = lipid-normalized concentration of xenobiotic in prey x (μg/kg lipid)
- k_2 = depuration rate constant (1/day)
- k_M = rate constant of xenobiotic metabolism (1/day)
- k_E = excretion rate constant (1/day)
- k_G = growth rate constant (1/day)

The first bracketed term represents the uptake of xenobiotic from water. The second bracked term represents the uptake of xenobiotic from food or prey x. Uptake from food is determined by feeding preference (p), ingestion rate (I), and CAE, where CAE is the proportion of the total amount of xenobiotic that is ingested from food or sediment. The third bracketed term represents the loss of xenobiotic due to depuration (k_2), dilution from growth (k_G), xenobiotic metabolism (k_M), and excretion (k_E). For infaunal animals, biota-sediment accumulation factors (BSAFs) are incorporated into the model to estimate xenobiotic accumulation via sediment ingestion. The estimated C_i for polychaetes was equal to the organic carbon-normalized sediment concentrations and BSAF. The model was tested by comparing the estimated vs. measured concentration of some PCB congeners in members of a food web in a New Jersey estuary.[70] The model appeared to be accurate within an order of magnitude in estimating the bioaccumulation of PCBs in this food web.

5.4 FATE OF XENOBIOTICS AFTER UPTAKE BY ESTUARINE ANIMALS

5.4.1 Biotransformation (Metabolism)

The biotransformation of xenobiotics and the relationship of biotransformation to effects on fish and estuarine invertebrates are shown in Figure 5.6. Enzyme systems that add polar groups to hydrophobic xenobiotics increase their water solubility and thus facilitate elimination. However, the metabolites of some xenobiotics are more toxic than the parent compound. For example, the binding of certain reactive benzo(a)pyrene metabolites, i.e., arene oxides, to DNA in liver cells of mammals initiates carcinogenesis.[71-74] The reactions carried out by biotransformation enzyme systems can be broadly divided into two groups: phase-one reactions include oxidation, reduction, and hydrolysis: phase-two reactions involve conjugation of sulfate, sugars, and peptides to polar groups, such as –COOH, –OH, or –NH$_2$ groups, which in some cases, were added to the xenobiotic during phase-one reactions. Phase-two metabolites are highly water soluble and are rapidly eliminated from animals. Some xenobiotics already contain a polar group, e.g., phenols, and phase-two reactions would take place with these compounds.

5.4.1.1 Phase-One Reactions

One of the most investigated of phase-one enzyme systems is the cytochrome P-450–dependent monoxygenase (MO) system, which oxidizes xenobiotics by hydroxylation, O-dealkylation, N-dealkylation, or epoxidation. Examples of substrates metabolized by the MO system in estuarine animals are shown in Figure 5.7. Figure 5.8 diagrams the steps involved in the hydroxylation of the PAH, benzo(a)pyrene by the MO system. The steps shown here are based primarily on studies with the vertebrate MO system.[75-79] In summary, the benzo(a)pyrene binds to the oxidized cytochrome P-450 (Fe^{2+}), which then interacts with oxygen. A hydroxylated substrate, e.g., 3-hydroxybenzo(a)pyrene, and a molecule of water

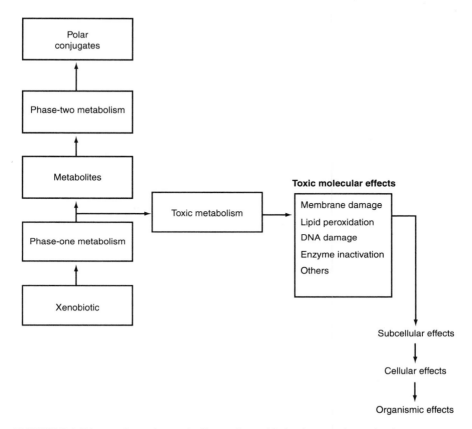

FIGURE 5.6 Biotransformation and effects of xenobiotics in estuarine animals.

leave the now-reoxidized cytochrome P-450. The substrate-oxidized P-450 complex is reduced by two electrons from NADPH carried by NADPH cytochrome P-450 reductase. The superoxide anion (O_2^-) is believed to be formed during the reaction and participates in the hydroxylation of the substrates. The MO system in estuarine animals, as in other animals, is a multicomponent system composed of phospholipid, cytochrome P-450, and NADPH cytochrome P-450 reductase.[80-84] Isozymes of cytochrome P-450 have been isolated and purified from fish and crustaceans, and partially purified from mollusks and annelids.[84-88]

Important intermediates in the oxidative metabolism of hydrocarbons and other xenobiotics are alkene and arene oxides, many of which are very reactive electrophiles capable of interactions with cellular macromolecules, such as DNA and proteins. Epoxide hydrase, another phase-one enzyme that metabolizes these epoxides to dihydrodiols, has been found in fish liver and crustacean hepatopancreas.[15]

Bivalves are often used in monitoring for contaminants, such as PAHs and PCBs. One reason bivalves are useful for this work is that bivalves accumulate such compounds because they have a very limited ability to metabolize PAHs[89,90] and seem to lack the ability to metabolize PCB congeners.[91]

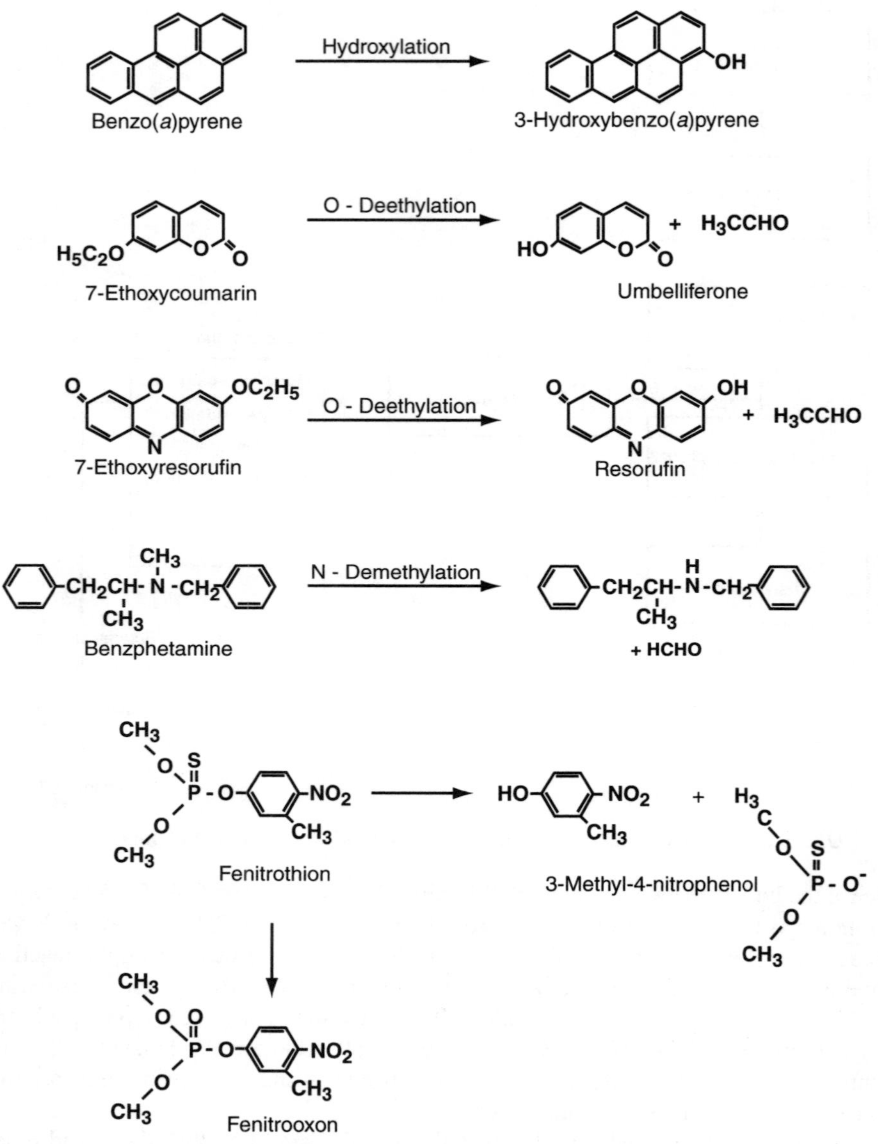

FIGURE 5.7 Mixed-function oxygenase reactions in estuarine animals.

5.4.1.2 Phase-Two Reactions

Phase-two reactions involve conjugation of phase-one products with a polar or ionic moiety (Figure 5.9). The most common polar or ionic moieties involved in these conjugation reactions are glucose, glucuronic acid, sulfate, and glutathione. In general, these conjugation products are quite water soluble so they are more easily eliminated from the animal than the parent compound. Many of the phase-one

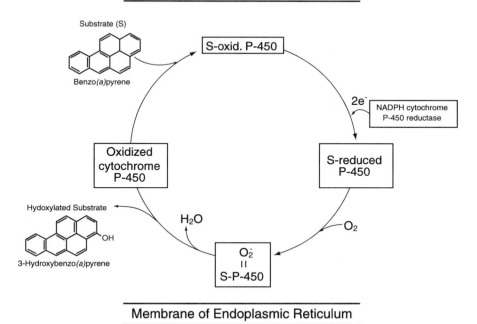

FIGURE 5.8 Reactions involved in the metabolism of benzo(*a*)pyrene by the mixed function oxygenase system.

products are electrophiles or nucleophiles. Electrophiles are molecules containing electron-deficient atoms with a partial or full positive charge that allows them to react by sharing electron pairs with electron-rich atoms in nucleophiles. Glutathione-*S*-transferase catalyzes the conjugation of the nucleophilic tripeptide glutathione (GSH, -γ-glutamylcysteinylglycine) to electrophiles that are produced by P-450 systems acting on various xenobiotics.[92] Electrophilic substrates shown to be conjugated to glutathione by glutathione-*S*-transferase of estuarine animals include 1-chloro-2,4-dinitrobenzene, 1,2-dichloro-4-nitrobenzene. 1,2-Epoxy-(*p*-nitrophenoxy)propane, styrene 7,8-oxide, *p*-nitrophenyl acetates, bromosulfophtalein, and benzo(*a*)pyrene-4,5-oxide.[15,93–98]

Nucleophiles formed by phase-one reactions are conjugated at the nucleophilic functional groups. For example, hydroxylated compounds are conjugated to sulfate or carbohydrates. *In vivo* studies with several estuarine fish and crustacean species have shown formation of sulfate and glycoside conjugates after exposure to various PAHs.[16,99–102] Sulfotransferases catalyze the transfer of the sulfuryl group, SO_3^-, from phosphoadenosyl phosphosulfate (PAPS) to a nucleophilic acceptor, either a hydroxyl or amino group. For example, pentachlorophenol is conjugated by sulfotransferases to form pentachlorophenol sulfate in estuarine animals (see Figure 5.9). Phase-two metabolites containing phenolic or carboxylic acid groups

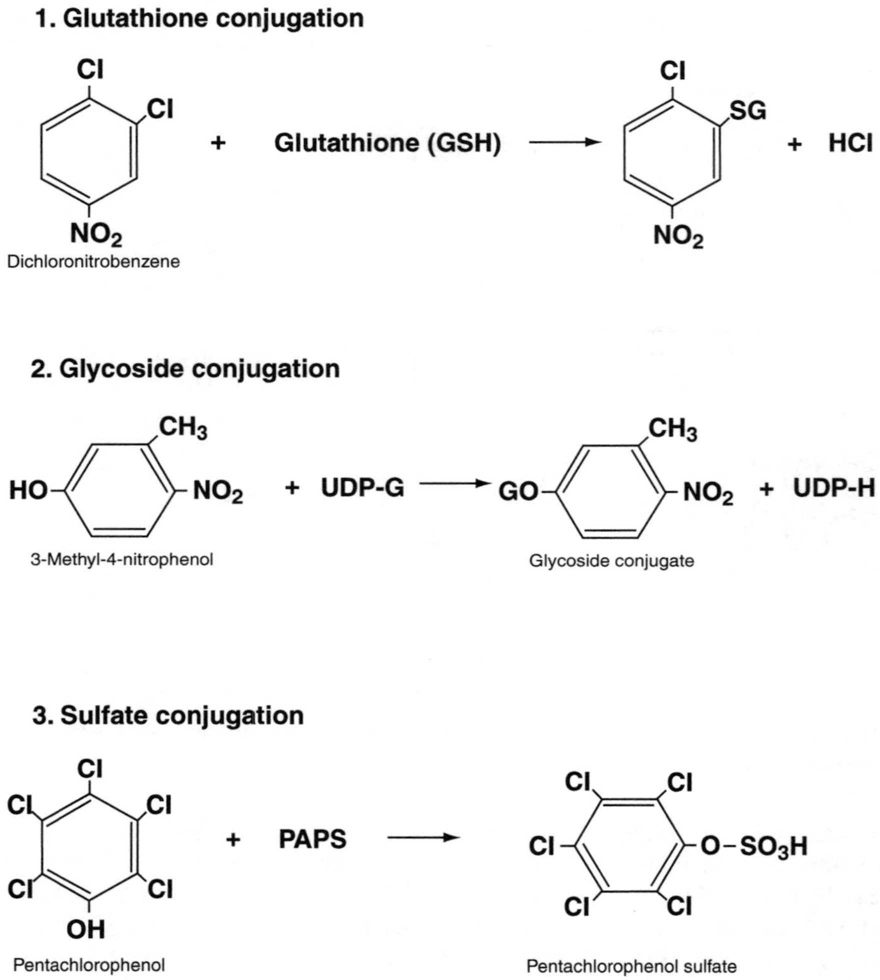

1. Glutathione conjugation

Dichloronitrobenzene (2,Cl; NO$_2$) + Glutathione (GSH) → (Cl, SG; NO$_2$) + HCl

2. Glycoside conjugation

3-Methyl-4-nitrophenol (HO—, CH$_3$, NO$_2$) + UDP-G → Glycoside conjugate (GO—, CH$_3$, NO$_2$) + UDP-H

3. Sulfate conjugation

Pentachlorophenol (C$_6$Cl$_5$OH) + PAPS → Pentachlorophenol sulfate (C$_6$Cl$_5$—O—SO$_3$H)

FIGURE 5.9 Phase-two conjugation reactions in estuarine animals. UDP-G = uridine diphospho-D-glucose or uridine diphospho-D-glucuronic acid; PAPS = 3′-phosphoadenosyl-5′-phosphosulfate.

or other nucleophilic centers can undergo glycosylation, as shown in Figure 5.9, where UDPG = uridine diphospho-D-glucose or its respective acid. The sugar moiety is often glucose in crustaceans while in fish the sugar moiety is often glucuronic acid.[16] An example is the glycoside conjugated to 3-methyl-4-nitrophenol, a metabolite of the organophosphate insecticide fenitrothion (see Figure 5.9).[103] *In vivo* studies have shown that estuarine animals form various conjugates of the metabolites of tributyltin oxide, benzo(*a*)pyrene, and chlorobenzoic acid. A significant amount of these metabolites are bound to macromolecules, including glutathione-S-transferase.[104]

5.4.2 Fates and Metabolic Pathways for Xenobiotics and Metabolites within Tissues and Cells

After entrance of a xenobiotic via the gill or stomach, the compound and its metabolites enter the blood or hemolymph, followed by entrance into various tissues as shown for a fish in Figure 5.10 and a crustacean in Figure 5.1. Within each tissue, there is partitioning of the chemical between the blood and the cells. There is partitioning after entering the cell between the outer membrane, organelles, lipid droplets, and the fluid cytosol. An example of this partitioning within the cells was shown in a series of the experiments with blue crabs using different ^{14}C-labeled xenobiotics in the food. The passage of the radiolabeled compound and metabolites were followed into the different cell types of the hepatopancreas.[104]

After entrance of a xenobiotic through the gill of fish, there is often accumulation of the compound in the liver with lesser amounts in other tissues, such as muscle, brain, and kidney.[22,65,105,114] If the xenobiotic was metabolized, metabolites can accumulate in the bile followed by elimination in the urine or feces (Figure 5.11). For example, monocyclic and polycyclic aromatic hydrocarbons that entered the fish via the water or food were rapidly metabolized, primarily by the liver, followed by buildup of conjugated metabolites in the bile and finally elimination via urine.[105]

The hepatopancreas of crustacea, digestive glands of mollusks and annelids, and livers of fish play important roles in the accumulation and metabolism of xenobiotics.[15,16,18,94,103–115] Cytochrome P-450, glutathione-S-transferase, and other

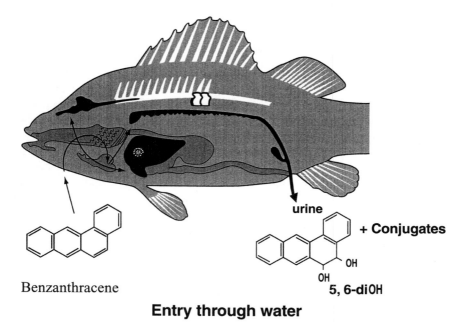

FIGURE 5.10 Schematic diagram of pathways for xenobiotics entering fish from water. (Modified from Newman.[19])

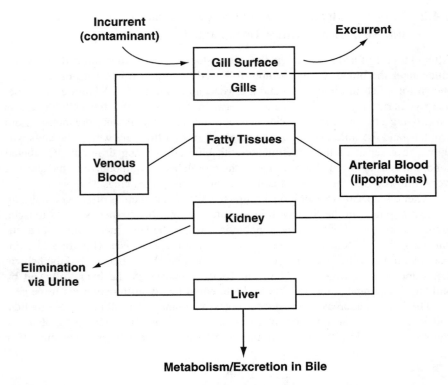

FIGURE 5.11 Xenobiotic and metabolite uptake, distribution, and elimination.

phase-two enzyme systems have been found in crustacean hepatopancreas and fish liver,[15,18,83,88,109,110,112,117–123] Mixed-function oxygenase (MO) activity was high in the blue crab stomach and green gland,[121] but low in blood, gill, reproductive tissues, eye stalk, cardiac muscle, and hepatopancreas.[122] Glutathione-S-transferase activity was high in blue crab hepatopancreas and gill.[123]

Different cell types found in crustacean hepatopancreas include E-, F-, R-, and B-cells (Figure 5.12).[124–126] The F-, R-, and B-cells are derived from embryonic or E-cells.[127,128] The R-cells are storage cells with large amounts of lipid (Figure 5.12), and the F- and B-cells are thought to be important in protein synthesis. The F-cells have a fibrillar nature due to the presence of abundant rough endoplasmic reticulum and Golgi network (Figure 5.12).[124,129,130] The cytochrome P-450 in blue crabs is associated with the endoplasmic reticulum of F-cells.[131] The highest activity of glutathione-S-transferase in blue crabs was in F-cell cytosol with significantly lower activity in B-cells and barely detectable activity in R-cells.[95] After the introduction of ^{14}C-xenobiotics into blue crab food, the distribution of the xenobiotics and their metabolites within hepatopancreas cells was determined.[104] Radioactivity was primarily in the storage lipid of the R-cells (Table 5.1, Figure 5.13B) for compounds not readily metabolized, i.e., hexachlorobiphenyl, Mirex, DDE. ^{14}C-chlorodinitrobenzene, which is a substrate for glutathione-S-transferase, accumulated in the cytosol with very little radioactivity in

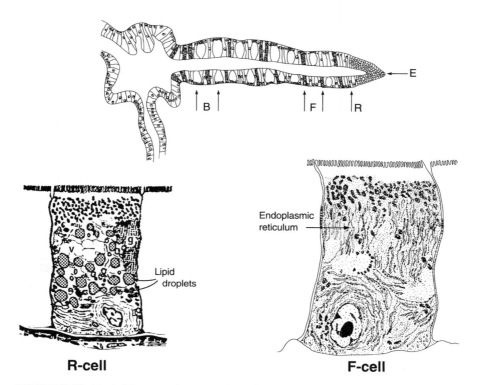

FIGURE 5.12 (Top): Diagram of cross section of crab hepatopancreas tubule showing location of E-, F-, R-, and B-cells. (Bottom): Diagrams of R- and F-cells.

TABLE 5.1
Distribution of Xenobiotics and Their Metabolites within Different Cells of Blue Crab Hepatopancreas after Being Fed on Food Containing ^{14}C-Xenobiotic

^{14}C-Xenobiotic	R-Cells (%)	F-Cells (%)	B-Cells (%)
Hexachlorobiphenyl	98	2	2
Benzo(a)pyrene	22	51	31
2,4-Dinitrochlorobenzene	12	68	27
Tributyltin chloride	10	72	14

Data taken from Lee.[104]

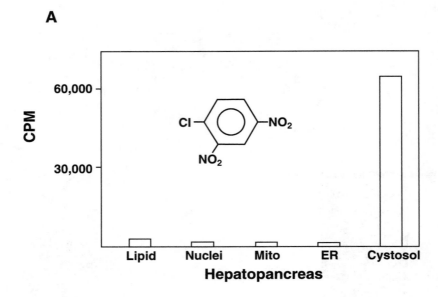

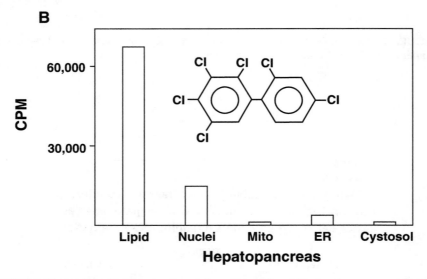

FIGURE 5.13 (A) Distribution of radioactivity within cells of the hepatopancreas at 3 days after feeding a male blue crab a diet containing ^{14}C-chlorodinitrobenzene (6×10^5 cpm). After homogenization of hepatopancreas, the different cell fractions were collected by centrifugation. Radioactivity determined with scintillation counter. ER = endoplasmic reticulum; mito = mitochrondria; lipid = lipid droplets that float to the surface after cell homogenization. (Data taken from Lee.[104]) (B) Distribution of radioactivity within cells of the hepatopancreas at 3 days after feeding a male blue crab a diet containing ^{14}C-hexachlorobiphenyl (4×10^5 cpm).

the lipid droplets (Figure 5.13A). Much of the radioactivity was in the cytosol of the F-cells for compounds that were more extensively metabolized, i.e., benzo(a)pyrene, tributyltin, pentachlorphenol, dinitrochlorobenzene, tetrachlrobenzene, dichlorobenzoic acid, chlorobenzene, trichlorophenol, dichloronitrobenzene, dichloroaniline, fluorene, and p-chlorobenzoic acid[104] (Table 5.1) (Figures 5.13 and 5.14). These compounds initially entered the lipid of R-cells followed by transfer and metabolism in F-cells with the metabolites being water-soluble conjugates that are eliminated from the crab. Most of the radioactivity from the experiment with ^{14}C-tetrachlorobenzene, which is slowly metabolized in crabs, was distributed among lipid droplets, nuclei, and cytosol (Figure 5.14A). It is assumed that metabolites of tetrachlorobenzene were conjugated by phase-two enzymes and entered the cytosol. The majority of the radioactivity from the experiment with ^{14}C-dichlorobenzoic acid, which can be quickly conjugated by phase-two enzymes, was in the cytosol (Figure 5.14B). In a study with ^{14}C-tributyltin (TBT), the radioactivity of the parent compound and its metabolites were followed in the hepatopancreas, stomach, and muscle tissues (Figure 5.15). Most of the radioactivity was in the hepatopancreas with more than half of the radioactivity in the form of metabolites after 1 day. After 3 days, approximately 60% of the radioactivity was lost from the hepatopancreas and other tissues with metabolites making up 80% of the radioactivity.

5.4.3 Binding of Xenobiotics to Cellular Macromolecules

After entrance of a xenobiotic into cells, the compound and its metabolites are distributed among the cytosol, outer membrane, lipid droplets, and different organelles. The compounds and their metabolites can be "dissolved" in the lipid droplets on in the hydrophobic part of the outer membrane or organelle, or covalently bound to cellular macromolecules such as DNA, RNA, or protein. For some compounds, cellular damage has been found only in organs where there was covalent binding of metabolites to macromolecules.[132,133] In mammals, the binding of xenobiotics to DNA has been used as a measure of their carcinogenesis potential.[134] Fish have been shown to form DNA adducts from PAH metabolites.[102,135–137] In blue crabs, introduction of radiolabeled benzo(a)pyrene, tributyltin, bromobenzene, and fluorene via the food lead to binding of their metabolites to macromolecules in hepatopancreas cells.[104] A portion of the metabolites were bound to cellular lipoproteins and glutathione-S-transferases, one of the major proteins in the cytosol of blue crab hepatopancreas.[95]

5.5 ELIMINATION

Elimination of xenobiotics from fish or invertebrates can be followed by transfer of the contaminated animals to clean seawater.

Xenobiotic elimination is often described by a decay function:[17]

$$C_t = C_0 e^{-bt} \tag{5.9}$$

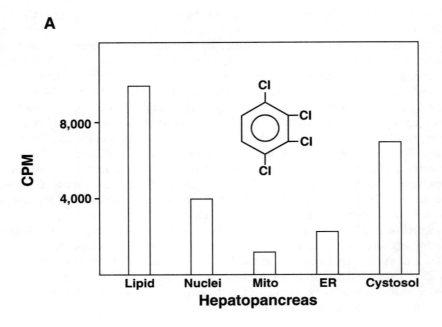
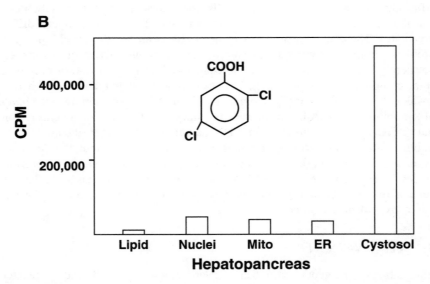

FIGURE 5.14 (A) Distribution of radioactivity within cells of the hepatopancreas at 3 days after feeding a male blue crab a diet containing ^{14}C-tetrachlorobezene (2×10^5 cpm). (B) Distribution of radioactivity within cells of the hepatopancreas at 3 days after feeding a male blue crab a diet containing ^{14}C-dichlorobenzoic acid (2×10^6 cpm). (Data taken from Lee.[104])

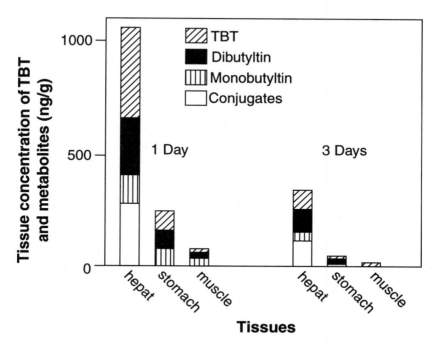

FIGURE 5.15 Distribution of radioactivity in different tissues after feeding a male crab a diet containing ^{14}C-tributyltin (12 µg/g food). Different tissues removed at 1 and 3 days, followed by extraction of tissues and thin-layer chromatography of extracts to separate metabolites. Separated metabolites were quantified with a scintillation counter. Conjugates in the cytosol were hydrolyzed by hydrolytic enzymes and quantified. (Data taken from Lee.[147])

where
C_t = concentration of xenobiotic at time t (µg/g)
C_0 = concentration of xenobiotic at time zero (µg/g)
b = elimination rate constant (1/h)
half-life = $(\ln 2)/b$

Elimination rates generally decrease as the K_{ow} of the xenobiotic increases.[17,138,139] Thus, highly hydrophobic compounds are eliminated at very slow rates. Elimination of anthracene taken up from the water by the amphipod *Pontoporeia hoyi* followed a first-order elimination curve.[140] Elimination of PAHs taken up from the sediment by *Pontopoireia* spp. also fit a first-order elimination model.[141] Since the metabolism of xenobiotics often produces highly polar metabolites, the elimination rate is greatly increased if a xenobiotic was metabolized by the animal.[17]

The elimination rate constant of benzo(*a*)pyrene was inversely proportional to the lipid content of *Pontopoireia*.[142] Lipid content did not affect the uptake rate constant. Other studies have shown that elimination rates of various contaminants in fish and invertebrates were inversely related to the lipid content of the animal.[49,114]

A xenobiotic can both enter and leave a fish by the gills. For example, Kobyashi[143] found that approximately half the pentachorophenol taken up by the goldfish *Carassius auratus* was eliminated via the gill, and the remainder was eliminated in the bile and urine.

Elimination of many xenobiotics by estuarine invertebrates is often biphasic with an initial rapid loss phase followed by a second much slower elimination phase.[17] Benzo(*a*)pyrene taken up from the water by the clam *Mercenaria mercenaria* for 40 days and then allowed to depurate in clean seawater for 70 days showed a biphasic elimination curve with rapid elimination for the first 10 days followed by a much slower elimination rate for the next 60 days (see Figure 5.4). When mussels, *Mytilus edulis*, that had accumulated petroleum hydrocarbons were transferred to clean seawater, between 80 and 90% of the hydrocarbons were eliminated within 10 to 15 days.[89,144] The elimination half-life of petroleum hydrocarbons by *M. edulis* ranged from 2.7 to 3.5 days.[144] However, the hydrocarbons remaining in the tissues were only slowly eliminated with 12% of the hydrocarbons retained after 57 days of depuration.[145]

5.6 SUMMARY

In an estuary or other marine system, bioavailability is that fraction of a xenobiotic available for uptake by a resident animal. These animals can take up xenobiotics from the water, sediment, and food. Physicochemical properties of a xenobiotic determines its bioavailability from these different matrices as well as the relative importance of uptake, accumulation, and elimination. Particulates are often in high concentrations in estuarine waters and constitute an important pathway by which xenobiotics enter estuarine food web. Three factors that are important in controlling the bioavailability of a sediment-associated xenobiotic include the aqueous solubility of the xenobiotic, rate and extent of desorption from the solid phase into the pore water, and the ability of digestive juices of infaunal animals to solubilize the xenobiotic in the sediment particles. Lipid content of the animal, particularly gill lipids, influences both uptake and elimination of hydrophobic xenobiotics.

The presence of phase-one and phase-two enzyme systems in fish and many invertebrate groups allows for rapid xenobiotic metabolism. Elimination rates of xenobiotics generally increase after metabolism, because the metabolites are generally more polar and hydrophilic. Estuarine bivalves accumulate many xenobiotics, e.g., PAHs, due in part to their limited ability to metabolize xenobiotics. Xenobiotic uptake, accumulation, metabolism, and elimination by estuarine fish and various invertebrates have been described by a series of equations, allowing modeling of the fate of xenobiotics in aquatic animals. The fate of a xenobiotic in estuarine animals can be related to its sublethal and toxic effects and can be used in ecological risk assessment to assess the importance of xenobiotics in affecting estuarine ecosystems.

REFERENCES

1. Norton, S.B. et al., The U.S. EPA's framework for ecological risk assessment, in *Handbook of Ecotoxicology*, Hoffman, D.J., Rattner, B.A., Burton, G.A., and Cairn, J., Eds., Lewis Publishers, Boca Raton, FL, 1995, 703.
2. Hale, R.C., Disposition of polycyclic aromatic compounds in blue crabs, *Callinectes sapidus*, from the southern Chesapeake Bay, *Estuaries*, 11, 255, 1988.
3. Marcus, J.M and Mathews, R.D., Polychlorinated biphenyls in blue crabs from South Carolina, *Bull. Environ. Contam. Toxicol.*, 39, 877, 1987.
4. Mothershead, R.F., Hale, R.C., and Greaves, J., Xenobiotic compounds in blue crabs from a highly contaminated urban subestuary, *Environ. Toxicol. Chem.*, 10, 1341, 1991.
5. Murray, H.E., Murphy, C.N., and Gaston, G.R., Concentration of HCB in *Callinectes sapidus* from the Calcasieu Estuary, Louisiana, *J. Environ. Sci. Health*, 27A, 1095, 1992.
6. Roberts, M.H., Jr., Kepone distribution in selected tissues of blue crabs, *Callinectes sapidus*, collected from the James River and the lower Chesapeake Bay, *Estuaries*, 4, 313, 1981.
7. Reimold, R.J. and Durant, C.J., Toxaphene content of estuarine fauna and flora before, during, and after dredging toxaphene-contaminated sediments, *Pestic. Monit. J.*, 8, 44, 1974.
8. O'Connor, T.P., Concentrations of organic contaminants in mollusks and sediments at NOAA National Status and Trend sites in coastal and estuarine United States, *Environ. Health Perspect.*, 90, 69, 1991.
9. Mearns, A.J., Contaminant trends in the U.S. coastal fish and shellfish: historical patterns, assessment methods and lessons, in *Proc. Bioaccumulation Workshop: Assessment of the Distribution, Impacts and Bioaccumulation of Contaminants in Aquatic Environments*, Miskiewicz, A.G., Ed., Water Board and Australian Marine Sciences Association, Sydney, 1992, 253.
10. Zdanowicz, V.S., Gadbois, D.F., and Newman, M.W., Levels of organic and inorganic contaminants in sediments and fish tissues and prevalences of pathological disorders in winter flounder from estuaries of the northeast United States, 1984, in *Proc. Oceans'86*, Vol. 3, Marine Technology Society, Washington, D.C., 1986, 578.
11. Sericano, J.L. et al., Status and Trends Mussel Watch Program: Chlorinated pesticides and PCBs in oysters (*Crassostrea virginica*) and sediments from the Gulf of Mexico, 1986–1987, *Mar. Environ. Res.*, 29, 161, 1990.
12. Butler, P.A., Organochlorine residues in estuarine mollusks, 1965–72. National Pesticide Monitoring Program, *Pestic. Monit. J.*, 6, 238, 1973.
13. Duke, T.W., Lowe, J.I., and Wilson, A.J., A polychlorinated biphenyl (Aroclor 1254) in the water, sediment, and biota of Escambia Bay, FL, *Bull. Environ. Contam. Toxicol.*, 5, 171, 1970.
14. Sanger, D.M., Holland, A.F., and Scott, G., Tidal creek and salt marsh sediments in South Carolina coastal estuaries: II. Distribution of organic contaminants, *Arch. Environ. Contam.*, 37, 458, 1999.
15. James, M.O. et al., Epoxide hydrase and glutathione S-transferase activities and selected alkene and arene oxides in several marine species, *Chem. Biol. Interact.*, 25, 321, 1979.
16. Kleinow, K.M., James, M.O., and Lech, J.J., Drug pharmacokinetics and metabolism in food producing fish and crustaceans, in *Xenobiotics and Food-Producing Animals*, Hutson, D.H., Hawkins, D.R., Paulson, G.D., and Struble, C.B., Eds., American Chemical Society, Washington, D.C., 1992, 98.

17. Spacie, A. and Hamelink, J.L., Bioaccumulation, in *Fundamentals of Aquatic Toxicology*, Rand, G.M. and Petrocelli, S.R., Eds., Chemisphere, New York, 1985, chap. 17.
18. Lech, J.J. and Vodicnik, M.J., Biotransformation, in *Fundamentals of Aquatic Toxicology*, Rand, G.M. and Petrocelli, S.R., Eds., Hemisphere, New York, 1985, chap. 18.
19. Newman, M.C., *Fundamentals of Ecotoxicology*, Ann Arbor Press, Chelsea, MI, 1998, chap. 3.
20. Livingstone, D.R., Organic xenobiotic metabolism in marine invertebrates, *Adv. Comp. Environ. Physiol.*, 7, 46, 1991.
21. Connell, D.W., *Bioaccumulation of Xenobiotic Compounds*, CRC Press, Boca Raton, FL, 1990, chaps. 6, 7.
22. Varanasi, U., Stein, J.E., and Nishimoto, M., Biotransformation and disposition of polycyclic aromatic hydrocarbons (PAH) in fish, in *Metabolism of Polycyclic Aromatic Hydrocarbons in the Aquatic Environment*, Varanasi, U., Ed., CRC Press, Boca Raton, FL, 1989, 93.
23. McCarthy, J.F. and Jimenez, B.D., Interactions between polycyclic aromatic hydrocarbons and dissolved humic material: binding and dissociation, *Environ. Sci. Technol.*, 19, 1072, 1985.
24. Landrum, P.F. et al., Predicting the bioavailability of organic xenobiotics to *Pontoporeia hoyi* in the presence of humic and fulvic materials and natural dissolved organic matter, *Environ. Toxicol. Chem.*, 4, 459, 1985.
25. Means, J.C. and Wijayaratne, R., Role of natural colloids in the transport of hydrophobic pollutants, *Science*, 215, 968, 1982.
26. Sigleo, A.C. and Means, J.C., Organic and inorganic components in estuarine colloids: implications for sorption and transport of pollutants, *Rev. Environ. Contam. Toxicol.*, 112, 123, 1990.
27. Melcer, M.E. et al., Evidence for a charge-transfer interaction between dissolved humic materials and organic molecules: I. Study of the binding interaction between humic materials and chloranil, *Chemosphere*, 16, 1115, 1987.
28. Boon, J.D., Suspended solids transport in a salt marsh creek — an analysis of errors, in *Estuarine Transport Processes*, Kjerfve, B., Ed., University of South Carolina Press, Columbia, 1978, 147.
29. Oertel, G.F. and Dunstan, W.M., Suspended-sediment distribution and certain aspects of phytoplankton production off Georgia, U.S.A., *Mar. Geol.*, 40, 171, 1981.
30. Paerl, H.W., Microbial attachment to particles in marine and freshwater ecosystems, *Microb. Ecol.*, 2, 73, 1975.
31. Paerl, H.W. and Shimp, S.L., Preparation of filtered plankton and detritus for study with scanning electron microscopy, *Limnol. Oceanogr.*, 18, 802, 1973.
32. Zabawa, C.F., Microstructure of agglomerated suspended sediments in northern Chesapeake Bay estuary, *Science*, 202, 49, 1978.
33. Lee, R.F., Fate of petroleum components in estuarine waters of the southeastern United States, in *Proc. 1977 Oil Spill Conference*, American Petroleum Institute, Washington, D.C., 1977, 611.
34. Kukkonen, J. and Landrum, P.F., Measuring assimilation efficiencies for sediment-bound PAH and PCB congeners by benthic organisms, *Aquat. Toxicol.*, 32, 75, 1995.
35. Karickhoff, S.W. and Brown, D.S., Paraquat sorption as a function of particle size in natural sediments, *J. Environ. Qual.*, 7, 246, 1978.
36. Neff, J.M., Bioaccumulation of organic micropollutants from sediments and suspended particulates by aquatic animals, *Fresenius Z. Anal. Chem.*, 319, 132, 1984.

37. Means, J.C. and McElroy, A.E., Bioaccumulation of tetrachlorobiphenyl and hexachlorobiphenyl congeners by *Yoldia limatula* and *Nephtys incisa* from bedded sediments: effects of sediment- and animal-related parameters, *Environ. Toxicol. Chem.*, 16, 1277, 1997.
38. Mayer, L.M. et al., Bioavailability of sedimentary contaminants subject to deposit-feeder digestion, *Environ. Sci. Technol.*, 30, 2641, 1996.
39. Swindoll, C.M. and Applehans, F.M., Factors influencing accumulation of sediment sorbed hexachlorobiphenyl by midge larvae, *Bull. Environ. Contam. Toxicol.*, 39, 1055, 1987.
40. Schlekat, C.E., Decho, A.W., and Chandler, G.T., Sorption of cadmium to bacterial extracellular polymeric sediment coatings under estuarine conditions, *Environ. Toxicol. Chem.*, 17, 1867, 1998.
41. Schlekat, C.E., Decho, A.W., and Chandler, G.T., Dietary assimilation of cadmium associated with bacterial exopolymeric sediment coatings by the estuarine amphipod *Leptocheirus plumulosus*: effects of Cd concentration and salinity, *Mar. Ecol. Prog. Ser.*, 183, 205, 1999.
42. Schlekat, C.E., Decho, A.W., and Chandler, G.T., Bioavailability of particle-associated silver, cadmium, and zinc to the estuarine amphipod *Leptocheirus plumulosus* through dietary ingestion, *Limnol. Oceanogr.*, 45, 11, 2000.
43. Freeman, D.H. and Cheung, L.S., A gel partition model for organic desorption from a pond sediment, *Science*, 214, 790, 1981.
44. Means, J.C. et al., Sorption properties of polynuclear aromatic hydrocarbons by sediments and soils, *Environ. Sci. Technol.*, 14, 1524, 1980.
45. Karickhoff, S.W., Brown, D.S., and Scott, T.A., Sorption of hydrophobic pollutants on natural sediments, *Water Res.*, 13, 241, 1979.
46. Borglin, S. et al., Parameters affecting the desorption of hydrophobic organic chemicals from suspended sediments, *Environ. Toxicol. Chem.*, 15, 2254, 1996.
47. Landrum, P.F., Uptake, depuration and biotransformation of anthracene by the scud *Pontoporeia hoyi*, *Chemosphere*, 10, 1049, 1982.
48. Barron, M.G., Bioconcentration, *Environ. Sci. Technol.*, 24, 1612, 1990.
49. Gobas, F.A.P.C. and Mackay, D., Dynamics of hydrophobic organic chemical bioconcentration in fish, *Environ. Toxicol. Chem.*, 6, 495, 1987.
50. Crosby, D.G., *Environmental Toxicology and Chemistry*, Oxford University Press, New York, 1998, 336 pp.
51. Stegeman, J.J. and Teal, J.M., Accumulation release and retention of petroleum hydrocarbons by the oyster *Crassostrea virginia*, *Mar. Biol.*, 22, 37, 1973.
52. Shaw, G.R. and Connell, D.W., Physicochemical properties controlling polychlorinated biphenyl (PCB) concentrations in aquatic organisms, *Environ. Sci. Technol.*, 18, 18, 1984.
53. Isnard, P. and Lambert, S., Estimating bioconcentration factors from octanol-water partition-coefficient and aqueous solubility, *Chemosphere*, 17, 21, 1988.
54. Spacie, A., Landrum, P.F., and Leversee, G.J., Uptake, depuration, and biotransformation of anthracene and benzo(*a*)pyrene in bluegill sunfish, *Ecotoxicol. Environ. Saf.*, 7, 330, 1983.
55. Pruell, R.J. et al., Accumulation of polychlorinated organic contaminants from sediment by three benthic marine species, *Arch. Environ. Contam. Toxicol.*, 24, 290, 1993.
56. Klump, J.V., Kaster, J.L., and Sierszen, M.E., *Mysis relicta* assimilation of hexachlorobiphenyl from sediments, *Can. J. Fish. Aquat. Sci.*, 48, 284, 1991.
57. Nimmo, D.R. et al., Polychlorinated biphenyl absorbed from sediments by fiddler crabs and pink shrimp, *Nature*, 231, 50, 1971.

58. Foster, G.D., Baksi, S.M., and Means, J.C., Bioaccumulation of trace organic contaminants from sediment by Baltic clams (*Macoma balthica*) and soft-shell clams (*Mya arenaria*), *Environ. Toxicol. Chem.*, 6, 969, 1987.
59. Clark, J.R. et al., Accumulation of sediment-bound PCBs by fiddler crabs, *Bull. Environ. Contam. Toxicol.*, 36, 571, 1986.
60. Elder, D.L., Fowler, S.W., and Polikarpov, G.G., Remobilization of sediment-associated PCB by the worm *Nereis diversicolor*, *Bull. Environ. Contam. Toxicol.*, 21, 448, 1979.
61. Rubinstein, N.I., Gilliam, W.T., and Gregory, N.R., Dietary accumulations of PCBs from a contaminated sediment source by a demersal fish (*Leiostomus xanthurus*), *Aquat. Toxicol.*, 5, 331, 1984.
62. Thomann, R.V., Equilibrium model of fate of microcontaminants in diverse aquatic food chains, *Can. J. Fish. Aquat. Sci.*, 38, 280, 1981.
63. DiPinto, L.M. and Coull, B.C., Trophic transfer of sediment-associated polychlorinated biphenyls from meiobenthos to bottom-feeding fish, *Environ. Toxicol. Chem.*, 16, 2568, 1997.
64. McMurtry, M.J., Rapport, D.J., and Chau, K.E., Substrate selection by tubificid oligochaetes, *Can. J. Aquat. Sci.*, 40, 1639, 1983.
65. Pizza, J.C. and O'Connor, J.M., PCB dynamics in Hudson River striped bass. II. Accumulation from dietary sources, *Aquat. Toxicol.*, 3, 313, 1983.
66. Quigley, M.A., Gut fullness of the deposit-feeding amphipod, *Pontoporeia hoyi*, in southeastern Lake Michigan, *J. Great Lakes Res.*, 14, 178, 1988.
67. Lee, H. et al., A method for determining gut uptake efficiencies of hydrophobic pollutants in a deposit-feeding clam, *Environ. Toxicol. Chem.*, 9, 215, 1990.
68. Kukkonen, J.V.K. and Landrum, P.F., Effect of particle-xenobiotic contact time on bioavailability of sediment-associated benzo(a)pyrene to benthic amphipod, *Diporeia* spp., *Aquat. Toxicol.*, 42, 229, 1998.
69. Lake, J.L., Rubinstein, N., and Pavignano, S., Predicting bioaccumulation: development of a simple partitioning model for use as a screening tool for regulating ocean disposal of wastes, in *Fate and Effects of Sediment-Bound Chemicals in Aquatic Systems*, Dickson, K.L., Maki, A.W., and Brungs, W.A., Eds., Pergamon, Oxford, 1984, 151.
70. Iannuzzi, T.J. et al., Distributions of key exposure factors controlling the uptake of xenobiotic chemicals in an estuarine food web, *Environ. Toxicol. Chem.*, 15, 1979, 1996.
71. Ames, B., Sims, N.P., and Grover, P.L., Epoxides of carcinogenic polycyclic aromatic hydrocarbons are frame shift mutagens, *Science*, 176, 47, 1972.
72. Weisburger, E.K., Mechanisms of chemical carcinogenesis, *Annu. Rev. Pharmacol. Toxicol.*, 18, 395, 1978.
73. Brookes, P., Newbold, R.F., and Osborne, M.R., Mechanism of the carcinogenicity of polycyclic hydrocarbons, in *Environmental Carcinogenesis*, Emmelot, P. and Kriek, E., Eds., Elsevier, Amsterdam, 1979, 123.
74. Miller, J.A. and Miller, E.C., Perspectives on the metabolism of chemical carcinogens, in *Environmental Carcinogenesis*, Emmelot, P. and Kriek, E., Eds., Elsevier, Amsterdam, 1979, 25.
75. Ahbad, S., The functional roles of cytochochrome P-450 mediated systems: present knowledge and future areas of investigations, *Drug Metab. Rev.*, 10, 1, 1979.
76. Hodgson, E., Comparative aspects of the distribution of cytochrome P-450 dependent monooxygenase systems: an overview, *Drug Metab. Rev.*, 10, 15, 1979.

77. Neber, D.W. and Jensen, N.M., The Ah locus: genetic regulation of the metabolism of carcinogens, drugs, and other environmental chemicals by cytochrome P-450-mediated monooxygenases, *Crit. Rev. Biochem.*, 6, 401, 1979.
78. Guengerich, F.P., Enzymatic oxidation of xenobiotic chemicals, *Crit. Rev. Biochem. Mol. Biol.*, 25, 97, 1990.
79. Guengerich, F.P., Cytochrome P450 enzymes, 81, 440, 1993.
80. Lu, A.Y.H., Liver microsomal drug-metabolizing enzyme system: functional components and their properties, *Fed. Proc.*, 35, 2460, 1976.
81. Philpot, R.M., James, M.O., and Bend, J.R., Metabolism of benzo(*a*)pyrene and other xenobiotics by microsomal mixed-function oxidases in marine species, in *Sources, Effects and Sinks of Hydrocarbons in the Aquatic Environment*, American Institute of Biological Sciences, Washington, D.C., 1976, 187.
82. Porter, T.D. and Coon, M.J., Cytochrome P-450 multiplicity of isoforms, substrates, and catalytic and regulatory mechanisms, *J. Biol. Chem.*, 266, 13469, 1991.
83. Stegeman, J.J., Polynuclear aromatic hydrocarbons and their metabolism in the marine environment, in *Polycyclic Hydrocarbons and Cancer*, Vol. 3, Gelboin, H.V. and Ts'o, P.O.P., Eds., Academic Press, New York, 1981, 1.
84. Lee, R.F., Annelid cytochrome P-450, *Comp. Biochem. Physiol.*, 121C, 173, 1998.
85. Peters, L.D., Nasci, C., and Livingstone, D.R., Immunochemical investigations of cytochrome P450 forms/epitopes (CYP1A, 2B, 2E, 3A and 4A) in digestive gland of *Mytilus* sp., *Comp. Biochem. Physiol.*, 121C, 361, 1998.
86. Stegeman, J.J., Cytochome P-450 forms in fish: catalytic, immunological and sequence similarities, *Xenobiotica*, 19, 1093, 1989.
87. Buhler, D.R. and Wang-Buhler, J.-L., Rainbow trout cytochrome P450s: purification, molecular aspect, metabolic activity, induction and role in environmental monitoring, *Comp. Biochem. Physiol.*, 121C, 107, 1998.
88. James, M.O. and Boyle, S.M., Cytochrome P450 in crustacea, *Comp. Biochem. Physiol.*, 121C, 157, 1998.
89. Lee, R.F., Sauerheber, R., and Benson, A.A., Petroleum hydrocarbons: uptake and discharge by the marine mussel, *Mytilus edulis*, *Science*, 177, 344, 1972.
90. Carlson, G.P., Detoxification of foreign compound by the quahaug, *Mercenaria mercenaria*, *Comp. Biochem. Physiol.*, 43B, 295, 1972.
91. Boon, J.P. et al., A structure-activity relationship (SAR) approach toward metabolism of PCBs in marine animals from different trophic levels, *Mar. Environ. Res.*, 27, 159, 1989.
92. Armstrong, R.N., Glutathione S-transferases; reaction mechanisms, structures, and function, *Chem. Res. Toxicol.*, 4, 131, 1991.
93. Tate, L.G. and Herf, D.A., Characterization of glutathione S-transferase activity in tissues of the blue crab, *Callinectes sapidus*, *Comp. Biochem. Physiol.*, 61C, 165, 1978.
94. Johnston, J.J. and Corbett, M.D., The effects of salinity and temperature on the *in vitro* metabolism of the organophosphorus insecticide fenitrothion by the blue crab, *Callinectes sapidus*, *Pestic. Biochem. Physiol.*, 26, 193, 1986.
95. Keeran, W.S. and Lee, R.F., The purification and characterization of gluthatione S-transferase from the hepatopancreas of the blue crab, *Callinectes sapidus*, *Arch. Biochem. Biophys.*, 255, 233, 1987.
96. Van Veld, P.A. et al., Glutathione S-transferase in intestine, liver and hepatic lesions of mummichog (*Fundulus heteroclitus*) from a creosote-contaminated environment, *Fish Physiol. Biochem.*, 9, 369, 1991.

97. Stenersen, J. et al., Glutathione transferases in aquatic and terrestrial animals from nine phyla, *Comp. Biochem. Physiol.*, 86C, 73, 1987.
98. Nimmo, I.A., The glutathione S-transferases of fish, *Fish Physiol. Biochem.*, 3, 163, 1987.
99. Corner, E.D.S., Kilvington, C.C., and O'Hara, S.C.M., Qualitative studies on the metabolism of naphthalene in *Maia squinada* (Herbst), *J. Mar. Biol. Assoc.*, 9, 167, 1973.
100. Sanborn, H.R. and Malins, D.C., The disposition of aromatic hydrocarbons in adult spot shrimp (*Pandalus platyceros*) and the formation of metabolites of naphthalene in adult and larval spot shrimp, *Xenobiotica*, 10, 192, 1980.
101. Little, P.J. et al., Temperature-dependent disposition of ^{14}C-benzo(*a*)pyrene in the spiny lobster, *Panulirus argus*, *Toxicol. Appl. Pharmacol.*, 77, 325, 1985.
102. Reichert, W.L., LeEberhart, A.-T., and Varanasi, U., Exposure of two species of deposit-feeding amphipods to sediment associated ^{3}H-benzo-(a)pyrene: uptake, metabolism and covalent binding to tissue macromolecules, *Aquat. Toxicol.*, 6, 45, 1985.
103. Johnston, J.J. and Corbett, M.D., The uptake and *in vivo* metabolism of the organophosphate insecticide fenitrothion by the blue crab, *Callinectes sapidus*, *Toxicol. Appl. Pharmacol.*, 85, 181, 1986.
104. Lee, R.F., Metabolism and accumulation of xenobiotics within hepato-pancreas cells of the blue crab, *Callinectes sapidus*, *Mar. Environ. Res.*, 28, 93, 1989.
105. Lee, R.F., Sauerheber, R., and Dobbs, G.H., Uptake, metabolism and discharge of polycyclic aromatic hydrocarbons by marine fish, *Mar. Biol.*, 37, 363, 1972.
106. Bend, J.R., James, M.O., and Dansette, P.M., *In vitro* metabolism of xenobiotics in some marine animals, *Ann. N.Y. Acad. Sci.*, 298, 505, 1977.
107. Sheridan, P.F., Uptake, metabolism, and distribution of DDT in organs of the blue crab, *Callinectes sapidus*, *Chesapeake Sci.*, 16, 20, 1975.
108. Lee, R.F., Ryans, C., and Heuhauser, M.L., Fate of petroleum hydrocarbons taken up from food and water by the blue crab *Callinectes sapidus*, *Mar. Biol.*, 37, 363, 1976.
109. Lee, R.F., Mixed function oxygenases (MFO) in marine invertebrates, *Mar. Biol. Lett.*, 2, 87, 1981.
110. James, M.O., Biotransformation and disposition of PAH in aquatic invertebrates, in *Metabolism of Polycyclic Aromatic Hydrocarbons in the Aquatic Environment*, Varanasi, U., Ed., CRC Press, Boca Raton, FL, 1989, 70.
111. Rice, S.D., Short, J.W., and Stickle, W.B., Uptake and catabolism of tributyltin by blue crabs fed TBT-contaminated prey, *Mar. Environ. Res.*, 27, 137, 1989.
112. Oberdorster, E., Rittschof, D., and McClellan-Green, P., Induction of cytochrome P450 3A and heat shock protein by tributyltin in blue crab, *Callinectes sapidus*, *Aquat. Toxicol.*, 41, 83, 1998.
113. Lee, R.F., Metabolism of tributyltin by aquatic organisms, in *Organotin — Environmental Fate and Effects*, Champ, M.A. and Seligman, P.F., Eds., Chapman & Hall, London, 1996, 369.
114. Boon, J.P., Oudejans, R.C.H.M., and Duinker, J.C., Kinetics of individual polychlorinated biphenyl (PCB) components in juvenile sole (*Solea solea*) in relation to their concentrations in food and to lipid metabolism, *Comp. Biochem. Physiol.*, 79C, 131, 1984.
115. James, M.O., Khan, M.A.Q., and Bend, J.R., Hepatic microsomal mixed-function oxidase activities in several marine species common to coastal Florida, *Comp. Biochem. Physiol.*, 62C, 155, 1979.

116. Lindstron-Seppa, P. and Hanninen, O., Induction of cytochrome P-450 mediated monooxygenase reactions and conjunction activities in freshwater crayfish (*Astacus astacus*), *Arch. Toxicol. Suppl.*, 9, 374, 1986.
117. James, M.O., Xenobiotic conjugation in fish and other aquatic species, in *Xenobiotic Conjugation Chemistry*, Paulson, G.D., Caldwell, J., Hutson, D.H., and Menn, J.J., Eds., American Chemical Society, Washington, D.C., 1986, 29.
118. James, M.O., Isolation of cytochrome P450 in crustaceans, *Comp. Biochem. Physiol.*, 121C, 157, 1998.
119. James, M.O., Isolation of cytochrome P450 from hepatopancreas microsomes of the spiny lobster, *Panulirus argus*, and determination of catalytic activity and NADPH cytochrome P450 reductase from vertebrate liver, *Arch. Biochem. Biophys.*, 282, 8, 1990.
120. Almar, M.M. et al., Organ distribution of glutathione and some glutathione-related enzymatic activities in *Procambarus clarkii*., Effect of sex, size and nutritional state, *Comp. Biochem. Physiol.*, 89B, 471, 1988.
121. Singer, S.C. et al., Mixed function oxygenase activity in the blue crab, *Callinectes sapidus*: characterization of enzyme activity from stomach tissue, *Comp. Biochem. Physiol.*, 65C, 129, 1980.
122. Singer, S.C. and Lee. R.F., Mixed function oxygenase activity in blue crab, *Callinectes sapidus*: tissue distribution and correlation with changes during molting and development, *Biol. Bull.*, 153, 377, 1977.
123. Lee, R.F., Keeran, W.S., and Pickwell, G.V., Marine invertebrate glutathione-S-transferases: purification, characterization and induction, *Mar. Environ. Res.*, 24, 97, 1988.
124. Robinson, A.G. and Dillaman, R.M., The effects of naphthalene on the ultrastructure of the hepatopancreas of the fiddler crab, *Uca minax*, *J. Invert. Pathol.*, 45, 311, 1985.
125. Al-Mohanna, S.Y. and Nott. J.A., B-cells and digestion in the hepatopancreas of *Penaeus semisulcatus* (Crustacea: Decapoda), *J. Mar. Biol. Assoc. U.K.*, 66, 403, 1986.
126. Wright, S.H. and Ahearn, G.A., Nutrient absorption in invertebrates, in *Handbook of Physiology*, Vol. II, Dantzler, W.H., Ed., Oxford University Press, New York, 1997, 1137.
127. Biesiot, P.M. and McDowell, J.E., Mudgut-gland development during early life-history stages of the American lobster *Homarus americanus*, *J. Crustacean Biol.*, 15, 679, 1995.
128. Vogt, G., Life-cycle and functional cytology of the hepatopancreatic cells of *Astacus astacus* (Crustacea, Decapoda), *Zoomorphology*, 114, 83, 1994.
129. Al-Mohanna, S.Y., Nott, J.A., and Lane, D.J.W., Mitotic E- and secretory F-cells in hepatopancreas of the shrimp *Penaeus semisulcatus* (Crustacea: Decapods), *J. Mar. Biol. Assoc. U.K.*, 65, 901, 1985.
130. Ahern, G.A., Nutrient transport by the invertebrate gut, *Adv. Comp. Environ. Physiol.*, 2, 91, 1988.
131. Lee, R.F., Metabolism of benzo(*a*)pyrene by a mixed function oxygenase system in blue crabs, *Callinectes sapidus*, in *Aquatic Toxicology and Environmental Fate*, Vol. 9, ASTM STP 921, Poston, T.M. and Purdy, R., Eds., American Society for Testing and Materials, Philadelphia, 1986, 233.
132. Bartolone, J.B. et al., Immunochemical detection of acetaminophen-bound proteins, *Biochem. Pharmacol.*, 36, 1193, 1987.
133. Hodgson, E. and Levi, P.E., *A Textbook of Modern Toxicology*, Elsevier, Amsterdam, 1987.
134. Lutz, W.K., *In vivo* covalent binding of organic chemicals to DNA as a quantitative indicator in the process of chemical carcinogenesis, *Mutat. Res.*, 65, 289, 1979.

135. Varanasi, U. et al., Comparative metabolism of benzo(a)pyrene and covalent binding to hepatic DNA in English sole, starry flounder, and rat, *Cancer Res.*, 46, 3817, 1986.
136. Von Hofe, E. and Puffer, H.W., *In vitro* metabolism and *in vivo* binding of benzo(a)pyrene in the California killifish (*Fundulus parvipinnis*) and speckled sanddab (*Citharicthys stigmaeous*), *Arch. Environ. Contam. Toxicol.*, 15, 251, 1986.
137. Sikka, H.C. et al., Metabolism of benzo(a)pyrene and persistence of DNA adducts in the brown bullhead (*Ictalurus nebulosus*), *Comp. Biochem. Physiol.*, 100C, 25, 1991.
138. Hawker, D.W. and Connell, D.W., Predicting the distribution of persistent organic chemicals in the environment, *Chem. Aust.*, 53, 428, 1986.
139. Neely, W.B., Estimating rate constants for the uptake and clearance of chemicals by fish, *Environ. Sci. Technol.*, 13, 1506, 1979.
140. Landrum, P.F., Toxicokinetics of organic xenobiotics in the amphipod, *Pontoporeia hoyi*: role of physiological and environmental variables, *Aquat. Toxicol.*, 12, 245, 1988.
141. Harkey, G.A., Van Hoff, P.L., and Landrum, P., Bioavailability of polycyclic aromatic hydrocarbons from a historically contaminated sediment core, *Environ. Toxicol. Chem.*, 14, 1551, 1995.
142. Kukkonen, J.V.K. and Landrum, P.F., Effect of particle-xenobiotic contact time on bioavailability of sediment-associated benzo(a)pyrene to benthic amphipod, *Diporeia* spp., *Aquat. Toxicol.*, 42, 229, 1998.
143. Kobayashi, K., Metabolism of pentachlorophenol in fish, in *Pesticides and Xenobiotic Metabolism in Aquatic Organisms*, Khan, M.A.Q., Lech, J.J., and Menn, J.J., Eds., American Chemical Society, Washington, D.C., 1979, 131.
144. Fossato, V.U. and Canzonier, W.J., Hydrocarbon uptake and loss by the mussel *Mytilus edulis*, *Mar. Biol.*, 36, 243, 1976.
145. Fossato, V.U., Elimination of hydrocarbons by mussels, *Mar. Pollut. Bull.*, 6, 7, 1975.
146. Gobas, F.A.P.C., Muir, D.C.G., and Mackay, D., Dynamics of dietary bioaccumulation and faecal elimination of hydrophobic organic chemicals in fish, *Chemosphere*, 17, 943, 1988.
147. Lee, R.F., Metabolism of tributyltin by marine animals and possible linkages to effects, *Mar. Environ. Res.*, 32, 29, 1991.

6 The Bioaccumulation of Mercury, Methylmercury, and Other Toxic Elements into Pelagic and Benthic Organisms

Robert P. Mason

CONTENTS

6.1 Introduction 127
6.2 Bioaccumulation in Pelagic Food Webs..129
6.3 Bioaccumulation in Benthic Organisms..134
6.4 Membrane Transport Processes 140
6.5 Summary 143
Acknowledgments..145
References..145

6.1 INTRODUCTION

Many elements are toxic to organisms, but often only in a specific chemical form. For example, although inorganic mercury (Hg) is toxic to organisms at low concentrations, it is the organic form of Hg, monomethylmercury (MMHg), that is highly bioaccumulative and accounts for the wildlife and health concerns resulting from the consumption of fish with elevated MMHg burdens.[1] For other metals and metalloids, such as cadmium (Cd), lead (Pb), arsenic (As), and selenium (Se), it is also often specific chemical forms, such as the free ion, e.g., Cd^{2+}, or the methylated or reduced species, e.g., mono- and dimethylAs or As(III), that are the most toxic.[2] Thus, knowledge of the total concentration (i.e., the sum of all chemical forms) of a potentially toxic element in the environment is insufficient to assess its toxicity accurately. Furthermore, it is accepted that contaminants must be in solution to be taken up directly from water[3,4] and, as a result, it is the competitive binding of contaminants to dissolved organic and inorganic ligands, colloids, and to particulate phases that ultimately controls the availability of an element in an aquatic system.[4]

In sediments, it is also the specific composition of the solid matrix, such as the amount of organic carbon or sulfide (acid volatile sulfide, or AVS; or pyrite), that determines the amount in solution in the sediment pore water,[4,5] as well as the bioavailability of the contaminants in the solid phase. For example, Lawrence and Mason[6] showed that the MMHg bioaccumulation factor (BAF) for amphipods living in sediment was a function of the sediment particulate organic content (POC). Additionally, a number of studies have shown that the particulate–water distribution coefficient (K_d) for Hg and MMHg is a function of POC.[5,7,8] Metal concentrations in sediments away from point source inputs are often strongly correlated with sedimentary parameters such as POC, AVS, or Fe,[4,5] and, as these parameters are often co-correlated, it is difficult to determine the controlling phase. Nonetheless, the binding strength of a metal or metalloid to the sediment influences its bioavailability and bioaccumulation in benthic organisms. This has been shown for copper (Cu), as well as other metals and organic contaminants.[9,10] Recently, Lawrence et al.[11] also showed that the bioavailability of Hg and MMHg to benthic organisms during digestion depended on the organic content of the sediment and further studies have extended these ideas to other metals.[12]

Understanding the sources, fate, and bioaccumulation of Hg and MMHg in the environment has received heightened attention primarily as a result of human and wildlife concerns resulting from the consumption of fish with elevated Hg.[1,13,14] In the United States, the U.S. EPA has targeted anthropogenic sources of Hg for regulation[1,15] to reduce Hg inputs to the atmosphere. It is apparent that future regulatory policies will focus on other metals and metalloids, such as As, Se, and Cd, that are volatilized to the atmosphere during high temperature combustion processes.[16] Each element has a particular anthropogenic source inventory, e.g., coal combustion and waste incineration (both medical and municipal) for Hg; coal combustion for Se; waste incineration for Cd; and smelting and other industrial activities for As.[17] It has been estimated that the input of metals to the atmosphere as a result of human activities has increased emissions by a factor of 5 for Cd, 1.6 for As, and 3 for Hg. Selenium anthropogenic inputs are about 60% of natural inputs.[18–20]

In addition to anthropogenic inputs to the atmosphere, metal and metalloids are also introduced directly into the aquatic environment as a result of activities such as mining and smelting and other industrial processes. These elements are typically retained within watersheds,[21,22] and postindustrialization activities have likely resulted in a general increase in their burden in surface soils, lake sediments, and other aquatic systems. For example, studies in contaminated environments such as the Clark Fork Superfund Site in Montana[23] have documented the bioaccumulation of metals in stream biota and have documented the environmental perturbation resulting from elevated metals in sediments and water.

The knowledge that chemical speciation controls bioavailability has become the guiding principle for research into the toxicity and bioaccumulation of inorganic contaminants in the environment.[4,5,24] However, while recent research has advanced the knowledge of the important differences in toxicity and fate of inorganic species, corresponding environmental regulations, especially for coastal waters, are typically still based on total dissolved concentrations of the contaminant in water or on total concentrations in sediments. To some degree, the lack of change in the regulatory framework is the result of the fact that, while much has been learned about the impact of chemical speciation on bioaccumulation and fate, knowledge is incomplete. This chapter provides a review of what is currently known about the factors controlling the bioaccumulation and fate of inorganic Hg and MMHg and other toxic metals and metalloids, such as As, Se, Cd, and Pb, in estuarine and coastal environments.

6.2 BIOACCUMULATION IN PELAGIC FOOD WEBS

For most trace metals, the largest bioconcentration occurs between water and phytoplankton/microorganisms,[13,24–27] and it is uptake at the base of the food chain that likely exerts the primary control on the amount of contaminant reaching higher trophic levels.[24,25,28] A comparison between the bioaccumulation of inorganic Hg, MMHg, As, Se, Cd, zinc (Zn), and silver (Ag) (Table 6.1) shows that although all these elements are concentrated in fish above their concentration in water, it is only MMHg that bioaccumulates at each stage of the food chain.[25–28]

The mechanism of accumulation plays a significant role in determining the magnitude of the accumulated concentration and the fate of the elements during trophic transfer.[25,29,30] For many metals, it is thought that the accumulation into phytoplankton and microbes is controlled by the free metal ion concentration in solution,[24,31] and a large body of research has documented this for both essential and potentially toxic metals, such as Cu, Zn, Cd, and Fe. In most cases, it is the free metal ion that is the form taken up in an active process through specific ion channels in the membrane. Metals are either specifically taken up, because they are essential for growth (e.g., Fe, Zn, and Cu), or inadvertently (e.g., Cd, Cu), as the transport sites are not entirely element specific. For example, Cd^{2+} and Pb^{2+} have been shown to be taken up through the Ca^{2+} channels in membranes,[4,30] whereas As, which exists

TABLE 6.1
Representative BAFs for a Variety of Elements for Both Phytoplankton and Piscivorous Fish

Element	BAF Algae[a]	BAF Fish[a]
Zn	4.7	3.2
Cd	3.7	3.0
Ag	5.0	2.7
Hg	4.5	3.8
MMHg	5.0	6.3
As	—	3.0
Se	—	3.8

[a] Values estimated from a variety of sources including Mason et al.,[25] Watras and Bloom,[26] and Reinfelder and Fisher.[29,30]

as an oxyanion in water, is thought to be an analogue for phosphate, and therefore, is transported into the cell. Metals such as Cu are required at low concentration by phytoplankton but can be toxic at high concentration.[24] There are a number of excellent reviews of trace metal uptake by microorganisms,[4,24,30] and the topic will not be dealt with in detail here. All these mechanisms involve energy and are therefore considered active processes.[24]

For Hg, Mason et al.[25,32] demonstrated passive uptake, likely by diffusion through the lipid bilayer, of neutral inorganic complexes of both inorganic Hg and MMHg by the estuarine diatom, *Thalassiosira weisflogii*. In these experiments, uptake was most efficient for the neutral chloride complexes, $HgCl_2$ and CH_3HgCl, and it was shown that these complexes have octanol–water partition coefficients (K_{ow}) that were one to two orders of magnitude higher than those of the neutral hydroxide complexes ($Hg(OH)_2$, CH_3HgOH), which were not taken up as efficiently (Table 6.2). Overall, at a given chloride concentration, the fraction of Hg or MMHg present as chlorocomplexes decreases with increasing pH. Further studies with both diatoms[33] and sulfate-reducing bacteria[34] have similarly shown that neutral complexes with sulfide — HgS and CH_3HgSH — and with organic thiols, such as cysteine, are also efficiently taken up and have K_{ow} values that are higher than those of the chloride complexes (Table 6.2). These results suggest that in the presence of neutral inorganic or simple organic complexes, passive accumulation of Hg and MMHg occurs by partitioning of these complexes into the cell membrane.

Passive accumulation of neutral inorganic complexes has also been demonstrated for other metals. For Ag, it has been shown that the complex AgCl has a higher K_{ow} than the free metal, and that it is taken up by phytoplankton more rapidly

TABLE 6.2
Estimated Octanol–Water Partition Coefficients for Neutrally Charged Inorganic and Organic Complexes of Metals

Metal	Inorganic Complexes	K_{ow}	Organic Complexes	K_{ow}
Hg	$HgCl_2$	3.3	$Hg(cysteine)_2$	3.7
	HgOHCl	1.2	$Hg(thiourea)_2$	4.6
	$Hg(OH)_2$	0.05		
	HgS	26		
MMHg	CH_3HgCl	1.7	$CH_3Hg(cysteine)$	50
	CH_3HgOH	0.07	$CH_3Hg(thiourea)$	630
	CH_3HgSH	28		
Cd	$CdCl_2$	0.002	$Cd(dithiocarbamate)_2$	1000
Ag	AgCl	0.09	—	
Cu	—		$Cu(oxine)_2$	400
	—		$Cu(dithiocarbamate)_2$	630
Pb	—		$Pb(dithiocarbamate)_2$	10,000

Adapted from Mason et al.,[25] Lawson and Mason,[33] Benoit et al.,[34] Phinney and Bruland,[38] and Reinfelder and Chang.[35]

than the free metal ion.[35] The K_{ow} is, however, much lower than that of $HgCl_2$, and is of similar magnitude to that of $Hg(OH)_2$ (Table 6.2). Additionally, uptake rates of AgCl, when normalized to exposure concentration, are similar to those for $Hg(OH)_2$, and both are less than that for $HgCl_2$. The differences in K_{ow} between the complexes are expected based on the relative "ionic" character of the complexes. Similarly, $CdCl_2$ has a low K_{ow} compared with $HgCl_2$.[35] Uptake of $CdCl_2$ by artificial membranes was examined by Gutknecht,[36] who showed that $CdCl_2$ was taken up much more rapidly than Cd^{2+}, but the rate for $CdCl_2$ was many orders of magnitude less than that for $HgCl_2$ under the same conditions,[37] in accordance with the measured differences in K_{ow}. These results confirm that there is the potential for uptake of neutral inorganic metal complexes by passive diffusion into phytoplankton. However, for complexes with substantial "ionic" character, other mechanisms likely dominate as the uptake rates are relatively slow. It is only for relatively "covalent" complexes such as $HgCl_2$ and CH_3HgCl, and the corresponding sulfide complexes, which have significant K_{ow} values, that passive diffusion rates are significant compared with other accumulation mechanisms.

Overall, except for Hg and MMHg, the neutral inorganic complexes of metals do not appear to be rapidly taken up by passive processes. This is not true for neutral organic complexes. The accumulation by diatoms of Hg and MMHg as thiol complexes (e.g., with cysteine and thiourea) has been demonstrated.[33] These complexes have relatively high K_{ow} values (Table 6.2). Additionally, Phinney and Bruland[38] showed that, although Cu, Cd, and Pb were not accumulated as charged complexes of these metals with EDTA, all were taken up if complexed to other organic ligands (e.g., oxine, dithiocarbamate) that formed neutral complexes with substantial K_{ow} values (Table 2). In all cases, initial accumulation rates were much higher for the neutrally complexed metal than they were in the absence of the ligand. The studies with Cu-oxine have been repeated with a variety of algae.[39] The observed permeability of the complex varied little across species as expected for a passive accumulation process. However, the observed permeability of the Cu-oxine complex was similar to that of $HgCl_2$ even though the K_{ow} of the Cu complex is two orders of magnitude higher.[38] As demonstrated by others, the size of the molecule is an important consideration as diffusion through the cell membrane likely limits the accumulation rate by passive processes for large compounds, even if they are highly lipophilic.[40]

Bioaccumulation is further complicated by the presence of dissolved organic carbon (DOC) in most natural systems. Dissolved concentrations of Hg and MMHg in natural waters are often positively correlated with DOC, but are negatively correlated with the BAF for phytoplankton, invertebrates, and fish[6,41–45] in the same systems. In this chapter, the BAF is defined as the concentration of the contaminant in the organism relative to the concentration of the medium in which it resides or upon which it feeds (e.g., water or sediment). These Hg–DOC relationships suggest that organic matter complexation makes Hg and MMHg much less bioavailable, so that positive relationships between lake DOC and MMHg in fish[43,44] cannot be solely explained by enhanced uptake of MMHg at the base of the food chain. Recent measurements of octanol–water partitioning of Hg in the presence of DOC extracted from the Florida Everglades confirm that the Hg–DOC complexes do not partition into octanol to any significant degree[46] and, consequently, will not passively diffuse across the cell membrane. However, when complexed to DOC, Hg and MMHg are still taken up by phytoplankton, albeit less efficiently than the neutral complexes.[33] Additionally, as discussed below, complexation to organic matter does not hinder accumulation across the gut lining of higher organisms. These observations suggest that Hg and MMHg–DOC complexes are taken up across membranes by other processes as well. Overall, in the aquatic environment, DOC affects MMHg bioaccumulation into fish via a number of conflicting interactions as it increases dissolved concentrations while decreasing bioavailability of Hg to methylating bacteria, phytoplankton, and to benthic invertebrates.

For other metals and metalloids, the influence of DOC is likely less marked than it is for Hg. For Cd, which binds relatively weakly to DOC,[47] the influence of DOC on water column speciation is small. This is not so for Cu, which is almost entirely bound to organic matter in seawater.[24] Demonstration of the influence of DOC on metal accumulation in phytoplankton is limited to synthetic ligands, as discussed above. It has recently been suggested that zebra mussels can take up DOC directly and that metals could be similarly assimilated.[48] Such a pathway has not been considered previously, and its general applicability needs to be demonstrated. For fish, studies of uptake across gills[49] have shown that addition of DOC can reduce Cu accumulation, but that realistic DOC additions have little influence on Cd uptake. These studies also confirm the interaction between Cd and Ca, and demonstrate that Cd is taken up by higher organisms through Ca ion channels but Cu is not.

Bioavailability of metalloids that exist as oxyanions in aqueous solution is not directly influenced by DOC. Both As and Se are present in surface waters in two oxidation states primarily because the reduced state, As(III) or Se(IV), is kinetically relatively stable to oxidation. As discussed above, As(V) is taken up as a phosphate analogue and is either incorporated into organic compounds within phytoplankton or released, either as As(III) or as methylated As compounds.[50,51] For Se, preferential uptake of Se(IV) over Se(VI) has been shown[52] and phytoplankton also excrete organo-Se compounds. Selenium is required at low levels by most organisms and is incorporated into protein, but it is toxic at higher levels.

Chemical speciation modeling allows estimation of the impact of DOC on metal bioavailability and toxicity if the binding constants for the metal to DOC are known. In these estimations it is assumed that the metal–DOC complex is not taken up and this assumption is for the most part true although there is the potential for absorption and/or competitive exchange reactions occurring between the metal complex and surface active sites on the membrane.[4,47] There is increasing evidence[6,28,33] that adsorption may be an important mechanism for the accumulation of Hg and MMHg as accumulation into both algae and higher organisms occurs under conditions where equilibrium speciation modeling indicates that all the metal should be bound to DOC. There is the potential for such interactions with other metals as well.

Although all metals enter phytoplankton cells, more efficient trophic transfer of some constituents leads to their enhanced bioaccumulation upon grazing by primary consumers.[25,29,30] Of all the toxic metals, MMHg transfer from diatoms to copepods is the greatest and this coincides with the relatively greater sequestration of MMHg in the diatom cytoplasm compared with binding to cellular membranes. Similarly, more MMHg is associated with the soft tissue of copepods, and this correlates with the higher assimilation of MMHg over inorganic Hg by fish feeding on these organisms.[33] A similar trend is found for other metals and elements.[53] Overall, Ag, Cd, and Hg behave similarly with relatively low assimilation efficiencies (<30%) during trophic transfer between phytoplankton and zooplankton. Transfer efficiency is somewhat higher for bivalves feeding on algae.[30] For metals such as Zn, regulation of concentration occurs in higher organisms. Assimilation of Se by copepods feeding on diatoms is high as most of the Se is found in the cytoplasm of the algae.[29] Similarly, Luoma et al.[54] showed that uptake from food (algae or sediment) for a deposit-feeding bivalve was much more important than direct uptake from water. Recent studies with insects confirm that elements like As and inorganic Hg are stored in the carapace after ingestion from food or absorption directly from water, and are not readily available for biotransfer to consumers.[55]

Therefore, because of the more efficient trophic transfer of MMHg,[55-57] the percent of the total Hg as MMHg tends to increase with increasing trophic levels[58-61] until the bulk of Hg (>90%) in predatory fish tissues is MMHg. Thus, the fraction of body burden Hg as MMHg in fish is a function of trophic position. This is also true for invertebrates. Riisgård and Famme[62] observed 73% MMHg in carnivorous shrimp compared with 17% in suspension-feeding mussels collected from the same area, and Mason et al.[55] found 60 to 100% MMHg in predatory insect larvae compared with <50% MMHg in herbivorous insects. It should be noted that many measurements of Hg in fish tissue focus only on muscle because muscle is the tissue fraction consumed by humans. In general, the fraction of the total Hg as MMHg in muscle tissue is much greater than that of other tissues,[55] and thus measurements of muscle tissue alone do not reflect the overall fraction of the Hg as MMHg for the whole organism.

Bioenergetic models[63] and feeding experiments[64] indicate that diet accounts for greater than 85% of MMHg uptake by fish, and food appears to be important for the other toxic metals and metalloids as well. For marine herbivores, ingestion is the dominant route for Cd, Cu, Ag, and Zn.[30] As a result, diet, trophic structure, and food chain length determine the total metal concentration; for example, total Hg in fish is positively correlated with trophic level as measured by $\delta\ ^{15}N$,[65] and longer food chains lead to higher Hg tissue concentrations in the top predators.[66]

For As, Se, Pb, and Cd, bioaccumulation does not occur at all levels of the food chain (see Table 6.1). There is evidence for accumulation from water into phytoplankton and microorganisms, but not at higher trophic levels. For As, phytoplankton have mechanisms for detoxifying the arsenate taken up through the phosphate accumulation channels.[50] The As(V) is reduced, methylated, and/or incorporated into organic molecules forming compounds such as arsenobentonate.[51] The highest concentrations are found in algae, with a reduction in concentration of about an order of magnitude between algae and primary consumers. Levels in fish are generally comparable with those of invertebrates.[51,55] Similarly, Se appears to be regulated and incorporated into organic compounds, predominantly proteins, via substitution for sulfur.[67] Again, the highest concentrations are found in algae, although there is evidence that, due to regulation, concentrations in algae are not strongly related to the concentrations in water. The available evidence suggests that trophic transfer from algae to primary consumers is relatively efficient for Se,[29] but that this is not the case between zooplankton and fish as a large fraction of the Se is bound in the hard parts of zooplankton.[53] It is not known whether As, Se, and other metals are taken up directly from water but the higher concentrations in the outer tissues (shell, carapace) suggest that this may be occurring. In freshwater insects, a negative correlation between size and metal concentration suggests uptake via direct absorption.[55]

Cadmium and Pb are not accumulated to any significant extent through the food chain. A number of studies have investigated the accumulation of Cd via fish gills,[49] but evidence of the dominant pathway for Cd accumulation into marine fish is limited. Additionally, most of the nonmethylated metals and metalloids have a relatively short half-life, and therefore if not continually exposed, the organisms will depurate the accumulated burden fairly rapidly. This is evidenced, for example, by the seasonal variation in tissue levels for As, Se, and Cd in both invertebrates and fish compared with the near constant concentration of MMHg.[55] Additionally, only MMHg shows a strong increase in body burden with age. For most of the metals, except MMHg, the highest concentrations are found in the detoxifying tissues (hepatopancreas, liver, kidney)[30,55] and the external tissues (shell, carapace). For MMHg, muscle tissue is the dominant reservoir, and this likely accounts for the higher trophic transfer of MMHg given its association with soft tissue and, in particular, with proteins.

6.3 BIOACCUMULATION IN BENTHIC ORGANISMS

In shallow aquatic systems, benthic invertebrates represent an important link for the transport of sediment-bound contaminants to the water column. For benthic organisms, and especially for those living in shallow, dynamic regions of the coastal zone, it is difficult to distinguish the routes of accumulation because sediment resuspension and the mixing regime result in a strong correlation among dissolved, suspended, and surface sediment concentrations. The medium (sediment, overlying water, pore water, or suspended matter) controlling bioaccumulation depends on both the metal accumulated and the composition of the sediment because it is the controlling binding phase that often determines bioavailability. In estuarine systems, the dominant metal-binding phases are either AVS or organic matter (POC). In the oxic surface layers, binding to oxide phases may be important at low organic matter content.

Lee at al.[68] have recently shown that for estuarine benthic invertebrates, even in high-AVS environments, bioaccumulation was related to the extractable sediment metal (Cd, Zn or Ni) concentration and not the pore water concentration. These results contrast with previous toxicity experiments[69,70] that used high metal concentrations and short-term exposures to demonstrate that under these conditions, pore water concentrations, controlled by metal binding to AVS, determined toxicity. However, in the environment, as shown by Lee et al.,[68] benthic invertebrates accumulate most of their metal burden from food, and metal bound to AVS is bioavailable as the metals can be solubilized during digestion of sediment.[9,69]

Mercury studies show a similar result.[6,11,12] Experiments using benthic organisms have shown preferential accumulation of MMHg (compared with inorganic Hg) in insect nymphs,[71] crustaceans,[55] polychaetes,[72] and amphipods.[6] For Hg and MMHg, it appears that in both laboratory studies and in field samples the BAF correlates best with sediment POC,[5,6] as shown in Figure 6.1 for MMHg. Studies in Lavaca Bay, Texas[7] and in the Chesapeake Bay[5,73] have shown a strong positive correlation between K_d and sediment POC for both Hg and MMHg, and a concomitant decrease in BAF for benthic organisms with increasing POC.[6] These results indicate that the decreased bioavailability of inorganic Hg and MMHg from sediment with increasing POC results from the metal being more strongly bound to the sediment with increasing POC.

While a correlation with POC is noted, it should be emphasized that POC and AVS are often correlated in surface sediments,[6] so it is difficult, except under experimental conditions, to determine the controlling phase. Further, while field and laboratory studies indicate that complexation to the solid phase is controlling bioavailability, it should be cautioned that other factors, such as the influence of POC on feeding rate, may also affect bioaccumulation.

Laboratory sediment exposure experiments[6] have demonstrated that sediment POC is an important factor in determining the sorption properties of sediment for inorganic Hg and MMHg, and thus their bioavailability. These experiments examined accumulation into amphipods under three scenarios: water only exposure; sediment

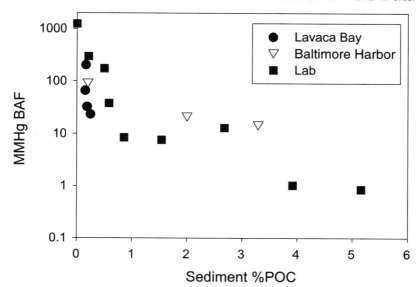

FIGURE 6.1 BAFs for amphipods for MMHg as a function of sediment organic content. Sites represented are Lavaca Bay, Texas and Baltimore Harbor, Maryland. (Data taken from Mason and Lawrence[5] and Lawrence and Mason.[6])

plus water; and sediment, water, and "food." The feeding used both contaminated and uncontaminated algae, for amphipods in contaminated and uncontaminated sediment. For the other studies, sediment POC was varied while other parameters such as grain size and exposure conditions were kept constant.[6] All sediment exposures were low-AVS environments. The results allowed determination of bioaccumulation equations, in terms of media organic content, for overlying water, pore water, sediment ingestion, and consumption of algae. The results confirmed that sediment POC controlled sediment bioavailability in these experiments, and increasing DOC decreased bioaccumulation from water.

However, because of the much higher bioavailability of Hg and MMHg from fresh algal matter, the feeding had a marked effect on the overall bioaccumulation.[6] The laboratory-derived model was applied to the field data from Lavaca Bay. The measured concentrations in field amphipods could only be mimicked by assuming that the amphipods were obtaining much of their food from fresh algal input, in agreement with the laboratory results. The notion that amphipods may preferentially feed on new organic inputs rather than in-place sediments has implications for the outcome of standard amphipod toxicity tests using these organisms. There is the potential that toxicity may be alleviated by addition of food low in contaminant concentration. The impact of food type on the outcome of amphipod toxicity tests has been noted by others,[74] and alleviation of the overall toxicity of the sedimentary environment to amphipods by feeding during toxicity tests must be considered as it could confound the outcome.

Further evidence to support the role of organic carbon in controlling Hg and especially MMHg bioaccumulation comes from solubilization studies with the intestinal fluid of benthic invertebrates.[9–12] The accepted paradigm of digestion is that only soluble compounds are absorbed across the gut lining and it has been determined for a number of contaminants that this technique provides a representative measure of bioaccumulation[10] and recent studies suggest this is also true for Hg and MMHg.[12] In these experiments, the sediment is incubated *in vitro* with the intestinal fluid of an invertebrate and the fraction of the Hg or MMHg released to solution measured[11,12] (Figure 6.2). The solubilization studies show, for both laboratory-spiked and field-collected sediments, a strong inverse correlation between the amount of MMHg released from sediment and the organic content of the sediment.[11] For inorganic Hg, the strong complexation of inorganic Hg by organic matter results in low bioavailability at all the organic contents used in the solubilization studies. The bioaccumulation studies with amphipods do, however, show high bioaccumulation of inorganic Hg in very low organic content sediments.[6] Overall, these results suggest that organic matter is binding Hg and MMHg within the sediment and renders it unavailable for solubilization within the intestinal tract, and thus for bioaccumulation.

The relative magnitude of the bioaccumulation of Hg and MMHg is in agreement with the current understanding of the relative strength of the complexes formed between Hg and MMHg and natural organic matter.[25,28,75] Comparison of the relative magnitude of the equilibrium constants for Hg and MMHg binding to organic matter with those for binding to hydroxide[25] (a surrogate measure of affinity for oxides phases in sediments) suggests that Hg will be more strongly bound to POC than MMHg. Thus, at low POC, when Hg and MMHg are largely

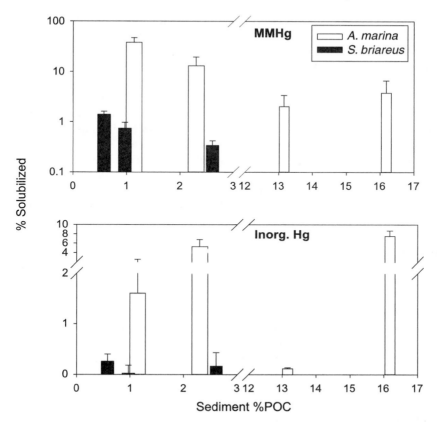

FIGURE 6.2 The extent of solubilization of MMHg (top figure) and inorganic Hg (HgI) (bottom figure) during *in vitro* extraction of sediments of differing organic content with the intestinal fluid of two benthic organisms, *Arenicola marina* and *Sclerodactyla briareus*. (Data taken from Lawrence et al.[11] and from McAloon.[12])

bound to inorganic complexes and phases, both are highly bioavailable. In contrast, at high organic content, bioaccumulation factors are small, indicating that under these conditions both forms of Hg are tightly bound and relatively unavailable for assimilation. At intermediate POC, MMHg is more bioavailable than Hg. The solubilization studies[11,12] support this contention and suggest that it is competitive ligand interactions (the strength of binding to the sediment) that control the degree of solubilization within the intestinal fluid. Chen and Mayer[9] have found similar results for Cu.

Overall, these studies provide a reasonable explanation for both the laboratory results and field studies and suggest that, over the range of POC found in the environment, bioaccumulation of Hg should be hindered to a higher degree than MMHg. The laboratory experiments and modeled results[6] indicate that fresh input of organic matter from the water column is potentially an important contamination route for surface-dwelling organisms such as amphipods. The estimated accumulation factors from algae are much higher than the comparable BAFs from sediment, except at very low sediment POC. Thus, changes in the supply of settling particulate matter/algae in the field could account for the greater than predicted bioaccumulation of both Hg and MMHg. Given that MMHg is typically <1% of total Hg in estuarine sediments[5,73] and that BAFs for MMHg are about ten times those of inorganic Hg over environmentally relevant POC concentrations,[6] it should be expected that MMHg would account for >10% of the total Hg in amphipods in the field, especially if these organisms are filter feeding as well as deposit feeding. Field data from Lavaca Bay and Baltimore Harbor show that the %MMHg in amphipods is much higher than predicted for a benthic organism.[5,6]

Bioaccumulation by amphipods and other organisms inhabiting the sediment/water interface is a complex interaction where the importance of fresh organic matter input cannot be ignored as a source of contaminants to benthic organisms. Further determinations of Hg bioaccumulation under other environmental conditions, especially those that would greatly affect Hg speciation, are therefore necessary. For example, investigation at high levels of sedimentary AVS would enhance our understanding of Hg and MMHg bioaccumulation from sediment to benthic organisms. In shallow, productive estuarine systems, the incorporation of seasonal cycles, including algal blooms, is also necessary to assess accurately the impact of most particle-associated contaminants. Finally, it appears that food and/or sediment rather than water is the most important route of inorganic Hg and MMHg uptake by amphipods. The degree to which this is true for other benthic organisms needs further investigation. Overall, MMHg is more available for bioaccumulation than inorganic Hg. Additionally, with regard to management decisions, these findings illustrate that decisions based solely on the concentration of total Hg in sediment will be inaccurate.

Of all metals, Hg, as MMHg, tends to have the highest bioaccumulation into benthic organisms, while many metals show similar trends in terms of decreasing bioaccumulation with increasing organic content of sediments. This is illustrated, for example, by the relationships between BAF and POC for clams collected from sediments around Hart-Miller Island, a dredge disposal facility outside of Baltimore, Maryland[5] (Figure 6.3). Of the elements measured, Hg, MMHg, Ag, and Cu show a reasonable relationship with POC; however, the BAF values are low (<1) for Hg and Cu. Lead also shows little bioaccumulation, whereas Cd, although it has a relatively high BAF compared with the other metals, shows little trend with POC. These sediments have relatively low AVS (1997 data only) that are typical for somewhat impacted estuarine sediments, typically <10 μmol/g.[5] For most of the metals, the relationship between AVS and BAF is not better than that for POC, and in these sediments there was not a strong relationship between AVS and POC.[5]

The AVS concentrations were greater than the total sediment Cd concentrations, on a molar basis, for all except a few sites where AVS was undetectable (<0.1 μmol/g) and thus accordingly, based on the criteria that relate bioavailability/toxicity to the relative concentration of simultaneously extractable metal (SEM), the metal should be relatively unavailable.[69,70] The SEM/AVS paradigm suggests that, if AVS is the dominant binding phase, then the metals are relatively unavailable if the ratio is less than 1. While SEM measurements were not made,

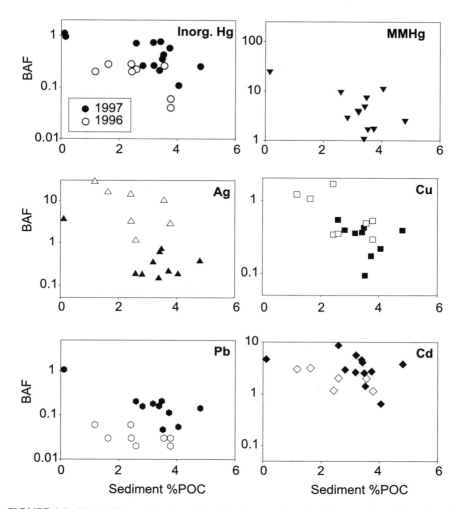

FIGURE 6.3 The BAFs as a function of sediment organic content for a variety of metals and metalloids for the clam, *Rangia cuneata*, living in sediments in the vicinity of Hart-Miller Island in the Chesapeake Bay. Data are shown for September 1996 (open symbols) and September 1997 (filled symbols).

the majority of the sites had AVS concentrations greater, on a molar basis, than those of total Cd, Pb, Ag, Hg, and MMHg, whereas for the metals Zn, Cu, Cr, and Ni the total molar concentrations were similar or greater than those of AVS. Furthermore, given the relatively low total Ag (<10 nmol/g), Hg (<2 nmol/g), and MMHg (<0.01 nmol/g) concentrations, the SEM/AVS ratio for these metals must be very low, and if sediment AVS were controlling bioavailability, the BAF values should be correspondingly small, which they are not. These bioaccumulation results concur with the data of Lee et al.[68] that while SEM/AVS may be a reasonable predictor of short-term toxicity, it does not accurately predict bioaccumulation patterns in the field because most benthic organisms obtain a large fraction of their contaminants via ingestion and not via direct accumulation from overlying water or pore water. Overall, our results suggest that the SEM/AVS model does not accurately predict bioavailability for Hg and MMHg, and probably Ag as well. Lee et al.[68] concluded that the SEM/AVS approach did not predict the accumulation of Ag, Cd, Ni, and Zn into benthic invertebrates.

Laboratory studies with sea cucumbers[12] show a similar trend of decreasing BAF with increasing POC for Hg, MMHg, Cd, Cu, and Pb. However, the AVS concentrations were low (<0.1 μmol/g). In these experiments, as in others, the BAF for MMHg was about ten times the BAF for Hg over the POC range examined (<3% POC), and the values for Hg and Cd were similar. Cu and Pb had much lower bioaccumulation. Clearly, in each exposure situation in estuarine environments, there will be changes in bioavailability dependent on the distribution of the two most important binding phases, POC and AVS, as suggested by others.[4,76] For Hg and MMHg, binding to POC likely involves complexation to thiol groups and thus the relative distribution of Hg and MMHg between AVS and POC is dependent on the relative distribution of reduced sulfur between the organic and inorganic phases. For other metals, complexation to other functional groups (ligands) in the organic phase are likely more important.[47] The greater relative binding strength to AVS compared with POC for Cu and Cd, for example, will likely result in these metals being associated with AVS except for sediments with a high POC/AVS ratio. A number of experimental studies, including some excellent reviews of the literature, have reached similar conclusions in terms of the bioavailability of metals in sediments.[4,76] The complexity of the system is such that there is unlikely to be any one "binding phase" that is controlling in all situations. Proper consideration of the sediment chemistry, including its redox state and the concentrations of the important solid phases — oxides, AVS and pyrite, and organic carbon — is needed to predict accurately the bioavailability of the metal of interest in the sediment. While much is being done to further understand these interactions, the ability to model and predict the bioavailability is limited by the lack of fundamental data, such as binding constants and adsorption constants for the metals and the various phases. Additionally, the role of organic matter in modifying adsorption to oxide and AVS or other sulfide phases needs further study. For example, there is the potential for the organic matter either to increase adsorption through formation of tertiary complexes (oxide-POC-metal),[47] or to decrease adsorption because of complexation of the metal to DOC. The resultant pore water metal concentration is obviously determined to a large degree by these interactions. This would dramatically alter uptake for those metals whose route of accumulation is mostly from water or, conversely, alter the uptake for those metals where dissolution during intestinal solubilization is a strong function of sediment POC.

6.4 MEMBRANE TRANSPORT PROCESSES

The preceding discussion of the limitation of bioavailability during intestinal solubilization assumes that this process occurs at a limited number of uptake sites and that membrane transport is not the fundamental determinant of bioaccumulation. For Hg and MMHg, it has been assumed that the opposite is true for accumulation into mammals, i.e., that MMHg is readily transferred across the gut membrane but inorganic Hg is not.[77,78] However, a variety of studies addressing the bioaccumulation of MMHg by fish in the presence of different organic ligands, and studies looking at the uptake of MMHg from food, suggest that the membrane transport across the gut is not the rate-limiting step in accumulation.[79,80] These studies rather suggest that the dissolution of MMHg from sediment or food during digestion is the main control over assimilation, and that factors that influence the extent to which MMHg is solubilized will determine the overall accumulation.

To examine further the notion of solubilization rather than membrane transport as the controlling mechanism, perfusion studies[81,82] have been performed with the gill and gut of the blue crab, *Callinectes sapidus*. In these experiments, the gill was suspended for 2 h in an external solution of known composition (10 ppt seawater) while a surrogate hemolymph solution (pH 7.5 buffered bovine serum albumin solution) was pumped through the tissue.[81] For the gut experiments, the tissue was suspended in the hemolymph solution while the external solution was pumped through the gut cavity. By collection of tissue and fluids after the experiment it was possible to determine the rate of accumulation into and across the membrane tissue.

The studies showed that the accumulation rates, under the same exposure conditions, were similar for the gill and gut, suggesting that Hg and MMHg accumulation occurs through channels or mechanisms that are general in nature and not specific to the individual tissues.[83] Further, Hg and MMHg were accumulated at similar rates in both tissues. For example, at an exposure concentration of 50 µg/l, the accumulation rate for the gut was 1220 ng/g tissue/h for inorganic Hg, 1750 ng/g tissue/h for MMHg. At the end of these short-term exposures, most of the accumulated Hg and MMHg was in the tissue, and only a few percent had crossed the membrane to the internal medium/hemolymph. The rapid accumulation into the gut tissue followed by a slow redistribution to the other tissues is similar to what was observed in exposure studies with fish.[79] It was noted in these experiments that after one feeding, the MMHg concentration in the intestinal tissue remained elevated for approximately 60 to 70 h. As the intestine MMHg concentration decreased over time, the concentration in the liver and other tissues increased. Similar results were obtained after exposure of fish to dissolved MMHg.[84] Initial uptake into the primary tissue was rapid, followed by a slower relocation to the other tissues.

While membrane accumulation rates for Hg and MMHg appear to be of the same order of magnitude, the MMHg rate is somewhat higher.[83] Furthermore, the accumulation rate does not appear to be dramatically affected by the complexation of Hg and MMHg with inorganic or organics ligands in the gut cavity solution with accumulation rates varying by about a factor of two to three (Figure 6.4). In these experiments, ligand concentrations were sufficient to complex all the Hg or MMHg. It appears that the rate is little affected by the ligand itself, with similar accumulation rates (within a factor of two) in the presence of thiourea, cysteine, and humic acids. For MMHg, accumulation was highest in the absence of any ligand, suggesting that the chloride complex is taken up somewhat more readily.

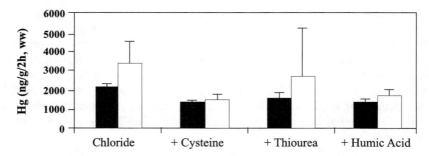

FIGURE 6.4 The accumulation of Hg and MMHg into the gut membrane of the blue crab during a perfusion experiment in the presence of different complexing ligands. See the text for details. (Solid bars = inorganic Hg; open bars = MMHg.)

Given the high molecular weight of humic acids, it is surprising that the Hg and MMHg is taken up in the presence of humic acid (Aldrich). This suggests that the complexes themselves are not being assimilated; rather, there is an exchange reaction occurring at the membrane surface and the Hg and MMHg are then internalized. It is also possible that Hg and MMHg are being taken up by a number of uptake routes, as has been suggested for mammalian cells.[85] Ongoing studies are investigating this hypothesis.

The perfusion experiments point to an active accumulation pathway for Hg and MMHg. Membrane transport has been studied in mammalian systems and the conclusions of a number of studies is that uptake involves a number of active pathways.[85,86] To examine this further, experiments were performed at different temperatures and in the presence of ouabain, a general metabolic inhibitor.[87] The results provide evidence that the uptake across the gut lining involves at least one active process.[83] The decrease in accumulation with decreasing temperature is comparable with that expected by the inhibitory effects of temperature on metabolic activity (i.e., the Q_{10} rule), with about a factor of two to three change in accumulation when the temperature was decreased from 20 to 4°C. Similarly, addition of ouabain to the medium, on either side of the gut membrane, had an inhibitory effect on accumulation.[83] Further studies are determining if this is true for invertebrates and fish. Thus, trophic transfer at higher trophic levels, which is dominated by the food pathway rather than direct uptake from water across the gills, is different from the accumulation into microorganisms. Transport across microorganism membranes involves passive diffusion of neutral complexes for Hg and MMHg.

Methylmercury accumulation in perfused crab tissue did not appear to be influenced by the presence of other metals in the gut fluid (Figure 6.5), whereas their presence influenced the accumulation of inorganic Hg to some degree. A series of experiments were conducted with either the metal alone in the gut cavity solution or in the presence of a "cocktail" containing Hg, MMHg, Cd, Zn, Pb, As, and Se. The concentration of each metal was the same, 50 μg/l. The influence of the other metals on Hg accumulation was small, compared with the effect of the mixture on the accumulation of the cations, Cd, Zn, and Pb. These results suggest, furthermore, that Cd and Pb were being taken up by a Zn transport system since

Bioaccumulation of Mercury, Methylmercury, and Other Toxic Elements

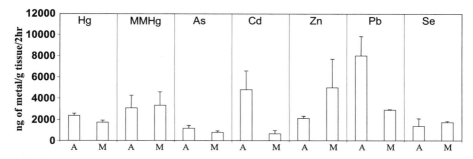

FIGURE 6.5 The accumulation of a variety of metals into the gut tissue of the blue crab after exposure to either the metal alone (A) or the metal in a mixture of the other metals (M). The results of a number of perfusion experiments are combined. Experimental details are given in the text.

uptake was less in the presence of Zn. Alternatively, all were taken up by another channel, e.g., Ca channel, but Zn is preferentially taken up. The metalloids, As and Se, do not show any strong interaction in the mixture, suggesting that their accumulation pathways are likely different from those of the metals. Although the differences between Hg and MMHg uptake are somewhat inconclusive, they indicate that Hg and MMHg have accumulation pathways that are similar.[83] If this is indeed the case, then the differences between the bioaccumulation of Hg and MMHg from food cannot be adequately explained by differences in membrane transfer. Thus, solubilization from food or sediment is the more important controlling mechanism.

Finally, the perfusion studies show that accumulation rates across the gut lining for Hg and MMHg were unsaturated at low exposure concentrations in the intestinal cavity, less than 50 ng/ml for inorganic Hg and 500 ng/ml for MMHg.[83] Concentrations of this magnitude are likely not realized during digestion. For example, for a 1:1 ratio of sediment to intestinal fluid, and 1000 ng/g Hg (dry weight) in the sediment and 1% solubilization, the gut fluid concentration would be 10 ng/ml for Hg. Similarly, in the same sediment, with 1% MMHg, but ten times more solubilization, the MMHg concentration would be 1 ng/ml. For food, which has a higher percent MMHg and expected higher solubilization (around 80%), the gut fluid concentration would approach 100 ng/ml. Thus, saturation of membrane transport sites should not be an issue under natural conditions except for food of exceedingly high MMHg content. However, it could be an issue for laboratory exposure studies using highly spiked food.

6.5 SUMMARY

This chapter has examined the mechanisms controlling Hg and MMHg bioavailability and bioaccumulation into both benthic and pelagic organisms and has contrasted these processes to those of other toxic metals and metalloids, such as Cd, Pb, As, and Se. There are many differences in the chemistry of Hg and MMHg compared with the other metals, and these account for their higher biomagnification in food webs. This is especially true for MMHg, which has a strong tendency to accumulate in the soft tissues of organisms, complexed to proteins, which likely accounts for the more efficient trophic transfer of this compound compared with the other metals. In contrast, Hg behaves similarly to Cd and Pb, which, being mostly associated with the "hard parts" of higher-trophic-level organisms, are not effectively transferred up the food chain. However, the inorganic metals may be relatively efficiently taken up into primary producers from water. For Hg and MMHg, passive uptake of neutral inorganic complexes is a potential mechanism, except in waters of high DOC. The accumulation of small neutral organic complexes of metals is another potential uptake pathway besides the route of active uptake across the cell membrane. The accumulation pathways for As and Se into phytoplankton are different as these elements exist in solution as oxyanions. These elements are often incorporated into protein and it appears that higher-trophic-level organisms have mechanisms for regulating their concentration.

For the benthic food chain, the sediment characteristics play an important role in determining uptake into invertebrates. The importance of the various potential binding phases — POC, AVS and metal-oxide phases — in controlling metal bioavailability is related to the particular chemistry of the metal. For example, Cd binds strongly to sulfide but not to POC, and thus AVS is the most likely controlling phase in sediments. For Hg and MMHg, both AVS and POC are important, and bioavailability is controlled by the amount of POC. Studies investigating the mechanisms controlling bioavailability to benthic invertebrates have demonstrated that the degree of metal solubilization during digestion is an important control over bioaccumulation and is likely the rate-determining step. For Hg and MMHg, studies of uptake across membranes using perfusion techniques, and studies of MMHg accumulation into fish, have provided complementary results. These investigations suggest that uptake into the gut membrane is rapid, followed by a slower redistribution of the MMHg from the gut to other tissues of the fish. Overall, these results challenge the previously accepted notion that the reason for the higher accumulation of MMHg over Hg was because Hg was not efficiently transported across the gut membrane. Further studies should examine the membrane transport mechanisms in more detail.

In conclusion, the accumulating information on the factors controlling the bioavailability, bioaccumulation, and trophic transfer of Hg, MMHg, Cd, Pb, As, Se, and other toxic metals should allow a more accurate assessment and modeling of bioaccumulation potential, an important factor in risk assessment. Knowledge of the mechanisms, and the chemical basis for the observed interactions, provides the framework for understanding toxicity in the environment even though more studies complicate the models as they continue to unravel the complexity of the overriding mechanisms. Additionally, increased knowledge allows the potential to estimate the impact of a contaminant in a particular system without the need for extensive monitoring. While more insight complicates the models of bioavailability and bioaccumulation, more information undoubtedly provides a more robust mechanism for impact assessment, and should allow for more informed decision making by management, which is the ultimate objective if the environment is to be protected from the insult of these contaminants.

ACKNOWLEDGMENTS

I have endeavored to cite the relevant contributions of my scientific colleagues but a chapter of this nature cannot adequately review all the research in the field. The work cited is sufficient to support the basic tenets proposed in this chapter. I acknowledge the contributions of those scientists who have not been cited and state that lack of citation does not indicate lack of importance of their contribution to this field. I would also like to thank the contributions of the students, technicians, and postdoctoral investigators in my laboratory whose work I have directly cited — Sandrine Andres, Jean-Michel Laporte, Angie Lawrence, Joy Leaner, and Kelly McAloon — and those who have indirectly contributed to the work discussed here. I thank Brenda Yates for help with the manuscript preparation. This is contribution number 3384 of the Center for Environmental Science, University of Maryland.

REFERENCES

1. U.S. EPA, Mercury Study Report to Congress. U.S. EPA-452/R-97-004, U.S. EPA Office of Air, Washington, D.C., 1997.
2. Craig, P.J., Organometallic compounds in the environment, in *Organometallic Compounds in the Environment*, Craig, P.J., Ed., Longman, Essex, 1986, chap. 1, 65.
3. Campbell, P.G.C., Interactions between trace metals and aquatic organisms: A critique of the free-ion activity model, in *Metal Speciation and Bioavailability in Aquatic Systems*, Tessier, A. and Turner, D.R., Eds., John Wiley & Sons, New York, 1995, chap. 2, 45.
4. Campbell, P.G.C. and Tessier, A., Ecotoxicology of metals in the aquatic environment: geochemical aspects, in *Ecotoxiology — A Hierarchical Treatment*, Newman, M.C. and Jagoe, C.H., Eds., Lewis Publishers, Boca Raton, FL, 1996, chap. 2, 11.
5. Mason, R.P. and Lawrence, A.L., The concentration, distribution and bioavailability of mercury and methylmercury in sediments of Baltimore Harbor and the Chesapeake Bay, Maryland USA, *Environ. Toxicol. Chem.*, 18, 2438, 1999.
6. Lawrence, A.L. and Mason, R.P., Factors controlling the bioaccumulation of mercury and methylmercury by the estuarine copepod, *Leptocheirus plumulosus*, *Environ. Pollut.*, 111, 217, 2001.
7. Bloom, N.S. et al., An investigation regarding the speciation and cycling of mercury in Lavaca Bay sediments, *Environ. Sci. Technol.*, 33, 7, 1999.
8. Mason, R.P. and Sullivan, K.A., Mercury and methylmercury transport through an urban watershed, *Water Res.*, 32, 321, 1998.
9. Chen, Z. and Mayer, L.M., Mechanisms of Cu solubilization during deposit feeding, *Environ. Sci. Technol.*, 32, 770, 1998.
10. Weston, D.P. and Mayer, L.M., *In vitro* digestive fluid extraction as a measure of the bioavailability of sediment-associated polycyclic aromatic hydrocarbons: sources of variation and implications for partitioning models, *Environ. Toxicol. Chem.*, 17, 820, 1998.
11. Lawrence, A.L. et al., Intestinal solubilization of particle-associated organic and inorganic mercury as a measure of bioavailability to benthic invertebrates, *Environ. Sci. Technol.*, 33, 1871, 1999.
12. McAloon, K., The Bioavailability of Mercury and Other Trace Metals to Benthic Invertebrates, M.S. thesis, University of Maryland, College Park, 2001, 81 pp.

13. Lindqvist, O. et al., Mercury in the Swedish environment — recent research on causes, consequences and corrective methods, *Water Air Soil Pollut.*, Special Issue, 55, 1991.
14. U.S. EPA, Listing of Fish and Wildlife Advisories, CD-ROM, U.S. EPA-823-C-97-005, 1996.
15. U.S. EPA, National Forum on Mercury in Fish, U.S. EPA 823-R-95-002, U.S. EPA Office of Water, Washington, D.C., 1995.
16. U.S. DOE, A comprehensive assessment of toxic emissions from coal-fired power plants: Phase I results from the U.S. Department of Energy Study, Final Report, September 1996.
17. Ondov, J.M., Quinn, T.L., and Han, M., Size-distribution, growth and deposition modeling of trace element bearing aerosol in the Chesapeake Bay airshed, Maryland Department of Natural Resources, Chesapeake Bay and Watershed Programs, CBWP-MANTA-AD-96-1, 1996, 111 pp.
18. Nriagu, J.O., A global assessment of the natural sources of atmospheric trace metals, *Nature*, 338, 47, 1989.
19. Pacyna, J.M., Emission inventories of atmospheric mercury from anthropogenic sources, in *Global and Regional Mercury Cycles: Sources, Fluxes and Mass Balances*, Baeyens, W., Ebinghaus, R., and Vasiliev, O., Eds., Kluwer Academic Publishers, Dordrecht, 1996, 161.
20. Mason, R.P., Fitzgerald, W.F., and Morel, F.M.M., The biogeochemical cycling of elemental mercury: anthropogenic influences, *Geochim. Cosmochim. Acta*, 58, 3191, 1994.
21. Lawson, N.M., Mason, R.P., and Laporte, J.-M., The fate and transport of mercury, methylmercury, and other trace metals in Chesapeake Bay tributaries, *Water Res.*, 35, 501, 2001.
22. Church, T.M. et al., Transmission of atmospherically deposited trace elements through an undeveloped, forested Maryland watershed, Final Report CBWP-MANTA-AD-98-2, Maryland Department of Natural Resources, Annapolis, MD, 1998, 86 pp.
23. Barnthouse, L. et al., Eds., New developments in ecological risk assessment, in *Theoretical Issues and Case Studies, Environ. Toxicol. Chem.*, Special Issue, 13, 12, 1994.
24. Sunda, W.G., Trace metal interactions with marine phytoplankton, *Biol. Oceanogr.*, 6, 411, 1989.
25. Mason, R.P., Reinfelder, J.R., and Morel, F.M.M., Uptake, toxicity, and trophic transfer of mercury in a coastal diatom, *Environ. Sci. Technol.*, 30, 1835, 1996.
26. Watras, C.J. and Bloom, N.S., Mercury and methylmercury in individual zooplankton: implications for bioaccumulation, *Limnol. Oceanogr.*, 37, 1313, 1992.
27. Back, R.C. and Watras, C.J., Mercury in zooplankton of northern Wisconsin lakes: taxonomic and site-specific trends, *Water Air Soil Pollut.*, 80, 1257, 1995.
28. Hudson, R.J.M. et al., Modeling the biogeochemical cycling of mercury in lakes, in *Mercury as a Global Pollutant: Towards Integration and Synthesis*, Watras, C.J. and Huckabee, J.W., Eds., Lewis Publishers, Boca Raton, FL, 1994, 473.
29. Reinfelder, J.R. and Fisher, N.S., The assimilation of elements ingested by marine copepods, *Science*, 251, 794, 1991.
30. Fisher, N.S. and Reinfelder, J.R., The trophic transfer of metals in marine systems, in *Metal Speciation and Bioavailability in Aquatic Systems*, Tessier, A. and Turner, D.R., Eds., John Wiley & Sons, New York, 1995, chap. 8, 363.
31. Anderson, D.M. and Morel, F.M.M., Copper sensitivity of *Gonyaulax tamarensis*, *Limnol. Oceanogr.*, 23, 283, 1978.

32. Mason, R.P., Reinfelder, J.R., and Morel, F.M.M., Bioaccumulation of mercury and methylmercury, *Water Air Soil Pollut.*, 80, 915, 1995.
33. Lawson, N.M. and Mason, R.P., Accumulation of mercury of estuarine food chains, *Biogeochemistry*, 40, 235, 1998.
34. Benoit, J.M., Mason, R.P., and Gilmour, C.C., The effect of sulfide on the octanol-water partitioning and bioavailability of Hg, *Environ. Toxicol. Chem.*, 18, 2138, 1999.
35. Reinfelder, J.R. and Chang, S.I.L., Speciation and microalgal bioavailability of inorganic silver, *Environ. Sci. Technol.*, 33, 1860, 1999.
36. Gutknecht, J., Cadmium and thallous ion permeabilities through lipid bilayer membranes, *Biochim. Biophys. Acta*, 735, 185, 1983.
37. Gutknecht, J.J., Inorganic mercury (Hg^{2+}) transport through lipid bilayer membranes, *J. Membr. Biol.*, 61, 61, 1981.
38. Phinney, J.T. and Bruland, K.W., Uptake of lipophilic organic Cu, Cd, and Pb complexes in the coastal diatom *Thalassiosira weissflogii*, *Environ. Sci. Technol.*, 28, 1781, 1994.
39. Croot, P.L. et al., Uptake of ^{64}Cu-oxine by marine phytoplankton, *Environ. Sci. Technol.*, 33, 3615, 1999.
40. Stein, W.D. and Lieb, W.R., Transport and diffusion across cell membranes, Harcourt Brace Jovanovich/Academic Press, London, 1986.
41. Watras, C.J., Morrison, K.A., and Host, J.S., Concentration of mercury species in relationship to other site-specific factors in the surface waters of northern Wisconsin lakes, *Limnol. Oceanogr.*, 40, 556, 1995.
42. Boudou, A. and Ribeyre, F., Mercury in the food web: accumulation and transfer mechanisms, in *Metal Ions in Biological Systems*, Vol. 34, Sigel, A. and Sigel, H., Eds., Marcel Dekker, New York, 1997, 289.
43. Driscoll, C.T. et al., The role of dissolved organic carbon in the chemistry and bioavailability of mercury in remote Adirondack lakes, *Water Air Soil Pollut.*, 80, 499, 1995.
44. Wren, C.D. et al., Relation between mercury concentrations in walleye (*Stizostedion vitreum vitreum*) and northern pike *(Esox lucius)* in Ontario lakes and influence of environmental factors, *Can. J. Fish. Aquat. Sci.*, 48, 132, 1991.
45. Fjeld, E. and Rognerud, S., Use of path analysis to investigate mercury accumulation in brown trout (*Salmo trutta*) in Norway and the influence of environmental factors, *Can J. Fish. Aquat. Sci.*, 50, 1158, 1993.
46. Benoit, J.M. et al., Stability constants for mercury binding by dissolved organic matter isolates from the Florida Everglades. *Geochim. Cosmochim. Acta*, in press.
47. Morel, F.M.M. and Hering, J.G., Principals and applications of aquatic chemistry, John Wiley & Sons, New York, 1993.
48. Roditi, H.A., Fisher, N.S., and Sañudo-Wilhelmy, S.A., Uptake of dissolved organic carbon and trace elements by zebra mussels, *Nature*, 47, 78, 2000.
49. Plsylr, R.C., Dixon, D.G., and Burnison, K., Copper and cadmium-binding to fish gills — modification by dissolved organic-carbon and synthetic ligands, *Can. J. Fish. Aquat. Sci.*, 50, 2667, 1993.
50. Benson, A.A., Phytoplankton solved the arsenate-phosphate problem, in *Marine Phytoplankton and Productivity*, Vol. 8, Holm-Hansen, O., Bolis, L., and Gilles, R., Eds., Springer-Verlag, New York, 1984, 55.
51. Phillips, D.J.H., Arsenic in aquatic organisms: a review, emphasizing chemical speciation, *Aquat. Toxicol.*, 16, 151, 1990.
52. Hu, M. et al., Preferential uptake of Se(IV) over Se (VI) and the production of dissolved organic Se by marine phytoplankton, *Mar. Environ. Res.*, 44(2), 225, 1996.

53. Reinfelder, J.R. and Fisher, N.S., Retention of elements absorbed by juvenile fish (*Meridia memidia, Meridia berylina*) from zooplankton prey, *Limnol. Oceanogr.*, 39, 1783, 1994.
54. Luoma, S.N. et al., Determination of selenium bioavailability to a benthic bivalve from particulate and solute pathways, *Environ. Sci. Technol.*, 26, 485, 1992.
55. Mason, R.P., Laporte, J.-M., and Andres, S., Factors controlling the bioaccumulation of mercury, methylmercury, arsenic, selenium, and cadmium by freshwater invertebrates and fish, *Arch. Environ. Contam. Toxicol.*, 38, 283, 2000.
56. Boudou, A. and Ribeyre, F., Comparative study of the trophic transfer of two mercury compounds — $HgCl_2$ and Ch_3HgCl — between *Chlorella vulgaris* and *Daphnia magna*, influence of temperature, *Bull. Environ. Contam. Toxicol.*, 27, 624, 1981.
57. Boudou, A. and Ribeyre, F., Experimental of trophic contamination of *Salmo gairdneri* by two mercury compounds: analysis at the organism and organ levels, *Water Air Soil Pollut.*, 26, 137, 1985.
58. Spry, D.J., and Wiener, J.G., Metal bioavailability and toxicity to fish in low-alkalinity lakes: a critical review, *Environ. Pollut.*, 71, 243, 1991.
59. Watras, C.J. et al., Bioaccumulation of mercury in pelagic freshwater food webs, *Sci. Total Environ.*, 219, 183, 1998.
60. Lange, T.R.M., Royals, H.E., and Connor, L.L., Influence of water chemistry on mercury concentration in largemouth bass from Florida lakes, *Trans. Am. Fish. Soc.*, 122, 74, 1993.
61. Mason, R.P. and Sullivan, K.A., Mercury in Lake Michigan, *Environ. Sci. Technol.*, 31, 942, 1997.
62. Riisgård, H.U. and Famme, P., Accumulation of inorganic and organic mercury in shrimp, *Crangon crangon, Mar. Pollut. Bull.*, 17, 255, 1986.
63. Rodgers, D.W., You are what you eat and a little bit more: bioenergetics-based models of methylmercury accumulation in fish revisited, in *Mercury Pollution: Integration and Synthesis,* Watras, C.J. and Huckabee, J.W., Eds., Lewis Publishers, Boca Raton, FL, 1994, 427.
64. Hall, B.D. et al., Food as the dominant pathway of methylmercury uptake by fish, *Water Air Soil Pollut.*, 100, 13, 1997.
65. Kidd, K.A. et al., The influence of trophic level as measured by $\delta\ ^{15}N$ on mercury concentrations in freshwater organisms, *Water Air Soil Pollut.,* 80, 1011, 1995.
66. Cabana, G. et al., Pelagic food chain structure in Ontario lakes: a determinant of mercury levels in lake trout (*Salvelinus namaycush*), *Can. J. Fish. Aquat. Sci.,* 51, 381, 1994.
67. Besser, J.M., Canfield, T.J., and La Point, T.W., Bioaccumulation of organic and inorganic selenium in a laboratory food chain, *Environ. Toxicol.*, 12, 57, 1993.
68. Lee, B.-G. et al., Influence of dietary uptake and reactive sulfides on metal bioavailability from aquatic sediments, *Science*, 287, 282, 2000.
69. Berry, W.J. et al., Predicting the toxicity of metal-spiked laboratory sediments using acid-volatile sulfide and interstitial water normalizations, *Environ. Toxicol. Chem.*, 15, 2067, 1996.
70. Ankley, G.T. et al., Eds., Metal bioavailability in sediments, Special Issue, *Environ. Toxicol. Chem.,* 15, 12, 1996.
71. Saouter, E. et al., Mercury accumulation in the burrowing mayfly *Hexagenia rigida* (Ephemeroptera) exposed to CH_3HgCl or $HgCl_2$ in water and sediment, *Water Res.*, 6, 1041, 1993.
72. Wang, W.-X. et al., Bioavailability of inorganic and methylmercury to a marine deposit-feeding polychaete, *Environ. Sci. Technol.*, 32, 2564, 1998.

73. Benoit, J.M. et al., Sources and cycling of mercury in the Patuxent estuary, *Biogeochemistry*, 40, 249, 1998.
74. Fisher, D., personal communication, 1999.
75. Hintelmann, H., Welbourn, P.M., and Evans, R.D., Measurement of complexation of methylmercury(II) compounds by freshwater humic substances using equilibrium dialysis, *Environ. Sci. Technol.*, 31, 489, 1997.
76. Mahony, J.D. et al., Partitioning of metals to sediment organic carbon, *Environ. Toxicol. Chem.*, 15, 2187, 1996.
77. Clarkson, T.W., Human health risks from methylmercury in fish, *Environ. Toxicol. Chem.*, 9, 821, 1990.
78. Magos, L., Physiology and toxicology of mercury, in *Mercury and Its Effects on Environment and Biology*, Siegel, A. and Siegel, H., Eds., *Metal Ions in Biological Systems*, Vol. 34, Marcel Dekker, New York, 1997, chap. 11.
79. Leaner, J.J. and Mason, R.P., The distribution kinetics of methylmercury in sheepshead minnows, *Cyprinodon variegatus*: an assessment of the mechanisms controlling methylmercury accumulation and redistribution in fish tissues, presented at the 1999 SETAC Conference, Philadelphia, November, 1999.
80. Rouleau, C. et al., Accumulation of waterborne mercury(II) in specific areas of fish brain, *Environ. Sci. Technol.*, 33, 3384, 1998.
81. Laporte, J.-M., Andres, A., and Mason, R.P., Factors controlling the uptake and transfer of inorganic and methylmercury across the perfused gill and gut of the blue crab, *Callinectes sapidus*, presented at the 1999 SETAC Conference, Philadelphia, November, 1999.
82. Pierrot, C., Pequeus, A., and Thuet, P., Perfusion of gills isolated from the hyper-hyporegulating crab *Pachygrapsus marmoratus* (Crustacea: Decapoda): adaptation of a method, *Arch. Physiol. Biochem.*, 103, 401, 1990.
83. Andres, S., Laporte, J.-M., and Mason, R.P., Mercury accumulation and flux across the gills and the intestine of the blue crab *Callinectes sapidus*, *Aquat. Toxicol.*, in press.
84. Leaner, J.J. and Mason, R.P., Effect of thiolate organic compounds on methylmercury accumulation and redistribution in sheepshead minnows, *Cyprindon variegatus*, *Environ. Toxicol. Chem.*, in press.
85. Wu, G., Effect of inhibitors and substrates on methylmercury uptake by rate erythrocytes, *Arch. Toxicol.*, 69, 533, 1995.
86. Kajwara, Y. et al., Methylmercury transport across the placenta via neutral amino acid carriers, *Arch. Toxicol.*, 70, 310, 1996.
87. Wood, C.M., Playle, R.C., and Hogstrand, C., Physiology and modeling of mechanisms of silver uptake and toxicity in fish, *Environ. Toxicol. Chem.*, 18, 71, 1999.

7 Dietary Metals Exposure and Toxicity to Aquatic Organisms: Implications for Ecological Risk Assessment

Christian E. Schlekat, Byeong-Gweon Lee, and Samuel N. Luoma

CONTENTS

7.1 Introduction 152
7.2 Current Status of Regulatory Approaches for Metals
    in Aquatic Systems...154
    7.2.1 The Importance of Phase and Speciation in Metal
          Risk Assessment..154
    7.2.2 Incorporation of Metal Speciation into Risk Assessment...............155
    7.2.3 The Biotic Ligand Model 156
          7.2.3.1 Mechanisms of Metal Toxicity at the Gill.......................156
          7.2.3.2 Model Assumptions and Components.............................157
    7.2.4 Limitations of Current and Projected Risk Assessment Practices 157
7.3 Processes Affecting Dietary Metal Exposure 159
    7.3.1 Metal Partitioning 159
    7.3.2 Biological Mechanisms..161
          7.3.2.1 Food Selection 161
          7.3.2.2 Feeding Rates..162
          7.3.2.3 Mechanisms of Dietary Metal Absorption 163
                  7.3.2.3.1 pH 163
                  7.3.2.3.2 Amino Acid–Rich Digestive Fluids 163
                  7.3.2.3.3 Surfactants 164
                  7.3.2.3.4 Intracellular Digestion 165
    7.3.3 Experimental Designs for Laboratory Exposures via Diet 165

7.4 The Relative Importance of Dietary vs. Dissolved Metal Uptake
     for Bioaccumulation and Toxicity 166
     7.4.1 Mass Balance Approach 166
            7.4.1.1 Deposit and Suspension Feeders 167
            7.4.1.2 Predators 168
     7.4.2 The Use of Mathematical Models in Metals Risk Assessment 168
            7.4.2.1 Background 168
            7.4.2.2 Equilibrium Models 169
            7.4.2.3 Dynamic Multipathway Bioaccumulation Model 170
                     7.4.2.3.1 DYMBAM Structure 170
            7.4.2.4 Application of Models 172
                     7.4.2.4.1 DYMBAM Case Study: Selenium
                                 in San Francisco Bay 172
     7.4.3 Comparisons among Metals and Organisms 173
7.5 Toxicological Significance of Dietary Metals Exposure 175
     7.5.1 Examples of Dietary Metals Toxicity 179
     7.5.2 Why is Dietary Toxicity Difficult to Measure? 180
     7.5.3 How Are These Subtle Effects To Be Handled in a Risk
            Assessment Framework? 180
7.6 Conclusions/Recommendations 181
References 182

7.1 INTRODUCTION

Effects of trace element contamination on coastal and estuarine ecosystems have received considerable attention over the past 50 to 60 years.[1] Risk assessment frameworks offer a means to quantify these effects, and to develop management alternatives for dealing with historical and ongoing trace element contamination. Quantifying the risk of metals to aquatic systems is now an established practice, but important uncertainties remain about specific components of the metals risk assessment process.

In both the United States and Europe, ecological risk assessments that address metal contamination in aquatic systems are conducted in accordance with the National Research Council Risk Assessment (NRC) paradigm.[2] After contaminants of concern and relevant ecological communities have been identified, the risk assessment paradigm calls for parallel characterizations of contaminant exposure and effect (see Chapter 1 for more detail). A key element of exposure characterization is estimating the dose of contaminant to which the organisms of interest is exposed *in situ*. The effects characterization, or toxicity assessment, includes a dose–response assessment, which is the dose necessary to elicit adverse effects to exposed organisms. Both dose estimation and dose–response assessment typically assume that adverse effects are caused by exposure to dissolved metals only.

The assumption that dissolved metals are responsible for toxicity has simplified the risk assessment approach. Determinations of exposure require only consideration of dissolved metal concentrations at the site, and knowing dose–response relationships for dissolved metals. Assessing risks of individual contaminants typically involves the risk characterization ratio (RCR), which is the ratio of exposure concentration to a dose–response toxicity criterion:

$$RCR = DMC/DEC \qquad (7.1)$$

where DMC is the dissolved metal concentration ($\mu g/l$) and DEC is an effects concentration ($\mu g/l$) derived from the response of aquatic organisms to dissolved metal concentrations (e.g., ambient water quality criteria). When RCR < 1, adverse effects are not expected.

Recently, several independent lines of research have challenged the underlying assumptions supporting the "dissolved only" approach by highlighting the importance of dietary metals exposure. A growing body of work demonstrates that, in conditions similar to nature, dietary exposure to metals associated with food items is at least as important as exposure to dissolved metals.[3-5] This generalization holds for most metals and metalloids, and for organisms living within different trophic levels. The findings that dietary exposures are important have implications for risk assessment. The most important is that the dissolved only assumption may lead to underestimates of metal exposure under natural conditions if animals are exposed to both dietary and dissolved sources. If dietary exposure causes adverse biological effects, the RCR needs modification to reflect the additional dietary dose (i.e., the numerator in Equation 7.1) and its toxicological concentration threshold (i.e., the denominator in Equation 7.1). The recognition of the importance of dietary metals exposure emphasizes the need to conduct effects assessments in a way that more closely approximates exposure conditions in nature. Specifically, metal concentrations in food items that are representative of the system in question need to be measured and included in estimates of dose. Similarly, the relationship between organismal response and dietary metal dose must be better understood.

This chapter discusses the current state of knowledge concerning exposure and some aspects of effects of metals and metalloids in estuarine and coastal systems. The review will be organized to address the specific questions:

1. What is the current status of regulatory approaches for metals? Are there significant limitations to these approaches?
2. What geochemical and physiological factors determine the importance of dietary metals exposure?
3. What is the relative importance of dietary metals exposure compared with dissolved metals exposure?
4. If dietary metals exposure is important at the organismal level, does this exposure result in toxicity?
5. What are the implications for risk assessment when dietary exposure is at least as important as dissolved exposure in eliciting dose effects?

We will provide geochemical and organismal evidence to demonstrate the quantitative importance of dietary metal exposure to aquatic organisms, and we will show that it is likely that such exposures can have toxicological consequences. We will also highlight the biological and geochemical uncertainties that must be addressed to establish guidelines for dietary metals exposure in risk assessment. We conclude by presenting a conceptual model that will provide interim guidance for how to incorporate dietary metals exposure into the risk assessment framework.

7.2 CURRENT STATUS OF REGULATORY APPROACHES FOR METALS IN AQUATIC SYSTEMS

By "risk assessment," we mean regulatory programs that evaluate the potential for metals to elicit negative effects to aquatic organisms under natural conditions. These include both environmental quality guidelines (e.g., U.S. EPA water quality criteria and sediment quality guidelines) and risk assessment frameworks (e.g., NRC framework, Organization for Economic Cooperation and Development, or OECD, and European Union, or EU, programs for assessing risk of existing substances). All these approaches attempt to quantify the risk of metals similarly, by comparing metal concentrations within a specific environmental phase with concentration-specific effects data achieved from laboratory toxicity tests. The goal of this section is to examine some of the findings that have contributed to the current status of risk assessment approaches and to discuss some possible future directions.

Measuring total trace element concentrations in environmental samples can be challenging, but analytical technologies and geochemical practices exist to provide accurate measurements of metal concentrations in most matrices, e.g., dissolved, particulate, sediment, and tissue. So, great uncertainties do not impede measurement of *in situ* metal concentrations from an area of interest. Most of the uncertainty in the metal risk assessment framework is manifested in the comparison of field-measured environmental concentrations to effects concentrations and in the derivation of effects concentrations. In nature, exposure to metals is complicated by a range of geochemical or biogeochemical factors that may redistribute metals among different physical phases and biotic factors that affect how an organism is exposed to the different phases in time and space. The contrast between how exposure occurs in nature and how organisms are exposed in the laboratory will serve as a continuing theme of this chapter. We will first address the observations and theories that have contributed to the way effects concentrations are currently measured. This history aids understanding of factors that are influencing the development of the next generation of tools for measuring effects concentrations and what is needed if those tools are to address natural exposures.

7.2.1 THE IMPORTANCE OF PHASE AND SPECIATION IN METAL RISK ASSESSMENT

One of the most influential findings in terms of metal ecotoxicology has been the observation that the total concentration of metals (e.g., in dissolved or sediment phases) are poor predictors of metal bioavailability, whether determined by toxicity or bioaccumulation. This awareness began in the 1970s, when it was shown that negative effects associated with metals (Cu, Cd, Zn) in the dissolved phase could be explained by the activity of the free ionic species. Although exceptions to the rule exist,[6] a body of evidence supports the notion that free ions are more bioavailable,[7–9] and toxic[9,10] than other metal species (e.g., those complexed with organic or inorganic ligands). Independent observations describing the importance of free ions were consolidated by Morel[11] into a unifying theory called the free-ion activity model (FIAM). In short, the model holds that "biological response elicited by a dissolved metal is usually a function of the free ion concentration, $M^{z+}(H_2O)_n$."[6] A general pattern was observed in studies where biological response (e.g., cell growth or toxicity) was measured in solutions that contained metals and metal-binding ligands in different concentration combinations. When biological response was normalized to the free metal ion concentration, $[Me^{2+}]$, the response curves for solutions containing different concentrations of metal-binding ligands coalesced, indicating that biological response was a function of $[Me^{2+}]$, and not $[Me]_{tot}$.

Further study showed that biological response to metals generally decreased as concentrations of complexing ligands increased, or as the conditional stabilities of metal-binding ligands increased.[6] The major implication of these results in terms of risk assessment is that metal toxicity may be influenced by site-specific and temporal differences in geochemical conditions, alone. Conditional effects concentrations are currently derived in tests conducted under laboratory conditions using rigidly controlled water quality parameters. In most natural habitats, the geochemical parameters that affect metal speciation will be complex and may vary by site and with time.

7.2.2 Incorporation of Metal Speciation into Risk Assessment

Incorporating consideration of metal speciation into risk assessment has been a slow and incomplete process. One change was to switch the way in which water quality criteria (WQC) are expressed, from "total recoverable metals" (metals recoverable from an unfiltered water sample, after weak acid digestion) to dissolved metals (those present in solution after passing through a 0.4- to 0.45-μm filter).[12] This new approach reduces the concentration of metal determined in a natural water by excluding particle-associated metals. Geochemically, separating these phases is completely logical.

The change to dissolved metal criteria does not address complexation of metals within the dissolved phase, however. The toxicity tests used to produce WQC are routinely conducted in filtered water that has relatively low concentrations of ligands. If metal speciation drives effects of dissolved metal toxicity, and if effluents are discharged into areas with high levels of dissolved ligands, then WQCs may be overprotective (i.e., if the ligands in the natural water reduce free ion activity and thereby ameliorate toxic effects). In such conditions, if all else were equal, dischargers would be asked to achieve a concentration lower than the metal concentration that causes acute toxicity. Both empirical and mechanistic approaches have been developed to account for such site-specific changes in metal speciation, bioavailability, and toxicity.

One current approach, the water effects ratio (WER), compares results of water-only toxicity tests using both a reference water source and water from the site in question.[13] Differences in bioavailability are expressed as

$$WER = LC_{50\text{site-specific}} / LC_{50\text{reference}} \tag{7.2}$$

If the site contains dissolved ligands that bind metals, metal bioavailability decreases, and the site-water LC_{50} will be higher than that of the reference water LC_{50}. A site-specific WQC is then obtained by multiplying the nominal WQC by the WER.

The WER is an operational solution to the speciation problem. It addresses site-specific toxicity but does not explicitly address site-specific geochemical conditions. A representative WER would depend on conditions at the site remaining constant, or that side-by-side bioassays must be performed whenever there is a question or concern that geochemical conditions might be dynamic. Geochemical conditions are commonly variable in nature, but rarely are the WER bioassays conducted repeatedly to account for such variability. A more mechanistic approach would offer the ability to explain the toxicological consequences that result across a range of geochemical conditions and thus predict implications of changes in chemistry in a more generic fashion. Recent progress on identifying mechanisms of metal toxicity in freshwater fish offers such a tool.

7.2.3 THE BIOTIC LIGAND MODEL

Using gills as both the site that determines metal bioavailability and a site of potential toxicity has led to a modification of the FIAM, called the biotic ligand model (*sensu* Reference 14). Both the FIAM and the biotic ligand model (BLM) use chemical equilibrium properties to estimate the proportion of dissolved metals that are in the free ionic state. Thus, both evaluate the modifying effects on toxicity of physico-chemical parameters, e.g., pH, water hardness, and dissolved organic matter.[15] But the BLM also incorporates the affinity of toxicologically relevant biological surfaces (the "biotic" ligand) for the free ion and thereby quantitatively incorporates a critical biological process into estimates of bioavailability and local toxic effects.[16,17] The model uses affinities of the gill membrane for metals to predict the molar quantity of metal that is complexed by the membrane. Above certain dissolved metal concentrations, the quantity of complexed metal impairs certain physiological processes that occur within the gill membrane.

The BLM has generated interest as a regulatory tool because it is mechanistic with regard to both geochemical and biological processes.[18] To date, model development has mostly centered on the gill of freshwater fish (models have been developed for rainbow trout and fathead minnows).[15,19-21]

7.2.3.1 Mechanisms of Metal Toxicity at the Gill

Like the FIAM, the BLM is ultimately based on the affinity of ligand molecules for specific metals. The difference is that the BLM uses ligands in a tissue of direct toxicological significance, i.e., the gill membrane. In freshwater fish, gills serve dual functions of gas exchange (influx of O_2 and efflux of CO_2 and NH_3) and ion transport (influx of Na^+, Cl^-, and Ca^{2+}).[15,17,22] Gas exchange is essential to maintain respiratory function, whereas ion transport is critical for maintaining plasma osmolality (in fish this is ~300 mosmol). These functions are carried out by specific proteins within the apical membrane of the fish gill.[22] Metal ions can interfere with these processes by complexation with functional proteins. For example, both silver and copper affect Na$^+$ and Cl$^-$ balance in fish by disrupting the function of Na$^+$/K$^+$-ATPase, which can reduce plasma sodium concentrations to critically low levels.[20,22] Mechanisms of other metal ions are summarized in Table 4-1 in Wood et al.[17]

7.2.3.2 Model Assumptions and Components

The function of the model is to predict uptake of metals into the fish gill in the presence of relevant ligands. The model requires knowledge of such parameters as: BS, log $K_{Cu\text{-}gill}$, log $K_{Ca\text{-}gill}$, log $K_{H\text{-}gill}$, log $K_{Cu\text{-}DOM}$, pH, [DOM], [Ca^{2+}], [Cu^{2+}], and water temperature, where log $K_{A\text{-}B}$ = the log of the conditional stability constant for complexes between A (ions) and B (ligands), and BS = the number of binding sites on gills. The molar quantity of Cu bound to the fish gill membrane is estimated using the speciation model approach (such as MINEQL$^+$). Of course, the uptake estimates are only as accurate as the model itself. An important limiting factor in such models is quantitative knowledge of the more complex associations like those involving organic ligands, which are best incorporated in more advanced models like WHAM.[23]

Model calculations can be performed to fit operationally defined scenarios, or to assess the effects of watershed-specific geochemical characteristics. For example, Playle[15] addressed the effects of dissolved organic matter, pH, and water hardness on the binding of Cu to the gills of rainbow trout. The toxicological consequences of the modeled gill metal concentrations are assessed by comparing model outcomes to results of water-only toxicity tests. For example, Playle et al.[20] exposed fathead minnows (*Pimephales promelas*) to Cd and Cu in six sources of fresh water that differed in pH and water hardness. Gill concentrations of Cd and Cu (both measured and modeled) were significantly related to LC$_{50}$ values for each element.[20] In another study, Meyer et al.[14] showed that gill concentrations of Ni explained toxicity to *P. promelas* across water hardness, whereas the free-ion activity of Ni did not. This is because the FIAM does not take into consideration competition between nontoxicant cations (such as Ca^{2+}) and Ni^{2+} ions for binding sites on fish gills. This competitive binding effectively ameliorated toxicity because it decreased [Ni]$_{gill}$.[14] The applicability of the BLM to Ag[19] and Co[24] was also shown.

7.2.4 LIMITATIONS OF CURRENT AND PROJECTED RISK ASSESSMENT PRACTICES

A chief goal of metals regulatory science is to develop a tool that can predict metal speciation based on site-specific geochemical conditions and relate that speciation to a toxicologically meaningful dose. The biotic ligand model appears to meet this goal for metals within one geochemical phase (the dissolved phase), which explains the interest it has generated from the regulatory community.[18] Although it is an important step forward, there are organismal and environmental considerations that limit how broadly the BLM can be used in regulation and in risk assessments. Some of these limitations may simply be data gaps that can be overcome by further study, but others are more fundamental.

At the simplest level, the range of application of the BLM is limited because it has been validated for only a limited number of metals and organisms. Especially with regard to metals, this limitation can be solved by further research. Similarly, because the physiological and ionoregulatory mechanisms addressed by the BLM are common to freshwater invertebrates as well as freshwater fish, the same mechanistic approach is theoretically applicable.[17] Application of the BLM to estuarine and marine organisms is more uncertain because the physiological constraints placed on organisms in these environments are different from those experienced by freshwater organisms. Whereas freshwater organisms use ion influx to maintain hyperosmotic conditions with respect to their ion-poor environment, marine organisms do the opposite. Saltwater fish, for example, use energy to excrete ions through the gill.[22] In general, mechanisms of dissolved metal uptake and toxicity by marine fish are poorly understood.[22] It does appear that metal uptake occurs to some degree in the intestine, and that toxicity occurs at the gill by complexation with proteins involved with ion excretion.

A more fundamental factor that could limit the robustness of the BLM is that the bioavailabilities of some dissolved metal complexes are not predicted by the thermodynamic principles that drive the FIAM concept. One example is neutral metal complexes. Silver forms a stable $AgCl°$ complex in estuarine and marine systems, and this complex is thought to diffuse across the lipid barrier in biological membranes.[25] Metals can also form bioavailable complexes with lipophilic organic ligands, such as those found in synthetic pesticides like carbamates[26] and xanthates,[27] which apparently can dissolve across the membrane. Naturally occurring and anthropogenically synthesized methylated metalloids, e.g., Hg and Sn, are also highly bioavailable, and their bioavailability is not predicted from BLM and other equilibrium-based concepts. Finally, the BLM considers only cationic metals. Bioavailability of metals and metalloids exhibiting anionic behavior (e.g., Se, As, Cr, V) is controlled by other mechanisms.[28]

Geochemically, the BLM does not yet address processes (e.g., complexation, sorption, and other reactions) that are exhibited at particle surfaces, which act to concentrate metals in the particulate phase. Nor does it consider transport into the organism from the particles or other foods ingested by aquatic organisms (e.g., bacteria cells, unicellular algae, and nonliving suspended particles and colloids). If metals in an organism's diet are assimilated, organisms will receive an additional, "hidden" exposure at the BLM-predicted, toxicologically relevant concentration in nature. In such a case the BLM toxicity assessments would underestimate the metal dose experienced by heterotrophic aquatic organisms (i.e., herbivores, detritivores, and predators). The BLM also fails to assign significance to systemic toxicity other than what occurs at the gill. Systemic adverse effects are assumed not to be significant if they originate from dietary exposure or metal transport from the gill to other locations. Thus, at its present state of development, the BLM model is most suitable for acute toxicity estimates that manifest themselves at the gill. In circumstances where diet or chronic adverse effects on other systemic processes are important, BLM predictions could be underprotective. Therefore, it is important to better understand the extent of dietary metal exposure and its implications.

7.3 PROCESSES AFFECTING DIETARY METAL EXPOSURE

Conceptually, there are geochemical and organismal reasons why dietary pathways should be important routes of metal exposure for aquatic organisms. A principal geochemical reason is that metals tend to partition preferentially to particles in aquatic systems. Thus, metal concentrations in particles and other food items tend to be enriched orders of magnitude over concentrations of dissolved metals. Many of the digestive mechanisms exhibited by aquatic organisms to acquire carbon and other nutrients from food could result in assimilation of metals from these highly concentrated sources. Yet, the importance of these sources of exposure remains controversial. It is valuable to evaluate why this is the case.

7.3.1 METAL PARTITIONING

One reason dietary metals uptake has received inadequate attention is the difficulty associated with reproducing at least some natural exposure conditions in the laboratory. One important example is metals partitioning to particles. Widely referenced studies using laboratory exposures[29] have demonstrated that pore water concentrations of metals can explain acute[30-32] and chronic[33] toxic effects to infaunal organisms. These conclusions are undoubtedly correct for the experimental conditions, but several key aspects of the experimental approaches differ both chemically and mechanistically from what occurs in nature. Experimental conditions can have a critical effect on which routes dominate bioavailability.

One of the most important experimental factors affecting the relative importance of dietary vs. dissolved metal exposure is the distribution of metals between pore water and particulate phases. Distribution coefficients, or K_D, are ratios of metal concentrations between particulate and dissolved phases.[34] When K_D values are greater than 1, metals are preferentially associated with the particulate phase for a given mass or volume. Distribution coefficients are conditional and can vary widely depending on many factors, including metal speciation in both the dissolved and particulate phases and the geochemical nature of the particulate phase.[34,35] Table 7.1 lists some K_D values that have been published for suspended particles and coastal sediments. For associations with suspended particles in marine systems, metals typically exhibit K_D values that range between 1×10^3 and 8×10^4 for Cd to 1×10^7 for Pb.[34-36] In sediments, metal K_D values range from 1×10^3 for Ag to 2×10^5 for Pb.[35]

Table 7.2 shows K_D values for several experimental studies that compared the route of metal exposure in sediment toxicity tests. The observed K_D values exhibited a broad range within certain experiments, and were consistently low in others. Most notably, K_D values were low where sediments were spiked to achieve high metal concentrations, in order to observe acute toxic effects.[3,28] The organisms in these tests were subject to a habitat that exhibited disproportionately greater distributions of metals in pore water (and correspondingly smaller distributions of particle-associated metals) than what is observed in nature.

TABLE 7.1
Distribution Coefficients from the Literature for Sediment and Suspended Particles

	K_D[a]			
Metal	Oceanic	Coastal Sediment	Seston[b]	Seston[c]
Ag	10000	1000	160000	
Cd	5000	2000		5000
Cr	50000	50000		
Ni	1000000	100000		
Pb	10000000	200000		
Zn	100000	20000		19000

[a] Reference 35.
[b] Reference 72.
[c] Reference 73.

TABLE 7.2
Distribution Coefficients (K_D) for Metals in Spiked Sediment Bioassays

Ref.	Element	Sediment Concentration ($\mu g/g$)	Pore Water Concentration ($\mu g/l$)	K_D (l/kg)
32	Cd	16	2000	8
32	Cd	72	1620	44.4
31	Cd	62.4	2500	25
31	Cd	65.5	800	81.9
30	Cd	17–19895	299–481971	41–6288
30	Cu	3.2–11194	2–40297	4–41667
30	Ni	10–33578	47–6985120	0.8–12250
30	Pb	4–16195	60–130028	32–6506
30	Zn	0.7–4859	5–2870014	0.6–191509

To demonstrate the consequences of differences in K_D on metal exposure routes, we applied the pore water and sediment metal concentrations from several published laboratory exposure studies to a dynamic multipathway bioaccumulation model for the bivalve *Macoma balthica* (the theory and elements of this model will be discussed later). For comparative purposes, K_D values were also calculated using particulate and dissolved metal concentrations that were measured from a range of naturally contaminated sediments. K_D values for natural sediments were higher than those achieved through laboratory spiking (Table 7.3). When the experimental, laboratory-spiked metal distribution data were applied to the bioaccumulation model, the majority (>92%) of Cd uptake by *M. balthica* occurred from pore water (Table 7.3). However, under conditions that more closely approximate

TABLE 7.3
Predicted Contribution of Diet toward Tissue Cd Concentrations in the Bivalve *Macoma balthica* for Cd in Spiked Sediment Bioassays and in Moderately Contaminated Estuarine Sediment

Source	Sediment Concentration (μg/g)	Pore Water Concentration (μg/l)	% Cd from Diet
32	16	2000	0.8
32	72	1620	4.3
31	62.4	2500	2.4
31	65.5	800	7.2
Natural, moderately contaminated sediment	10	1	90.9

the natural condition, the bioaccumulation model predicted that dietary exposure was more important than dissolved exposure. Thus, the partitioning conditions of exposure determined the relative importance the pathways. Experiments that do not mimic distribution conditions typical of nature will not yield results that can be widely extrapolated to nature.

7.3.2 BIOLOGICAL MECHANISMS

7.3.2.1 Food Selection

Both deposit- and suspension-feeding invertebrates ingest suspended particles or surficial sediments or both. Because the nutritious quality of these particles is generally quite low, most aquatic invertebrates exhibit selective feeding to some degree.[37] Selective feeding determines the biogeochemical features of the particles ingested; accordingly, the biogeochemical features affect metal sorption affinities and metal bioavailability from the particle. Organic carbon coatings (humic acids, microbial biofilms) and mineralogical features (iron oxyhydroxides) can tightly bind metals by complexation or other mechanisms.[38,39] The ability of pelagic diatoms[40,41] and bacterial cells[42] to adsorb metals has also been well documented. It is well established that metal concentration is often negatively correlated with particle size, which is a function of surface area. Some features of particles that increase metal binding (e.g., organic materials) can be the same features that particle-ingesting organisms select for.[37,43] For example, many benthic invertebrates selectively feed on small (e.g., <10 μm) particles[44]; many organisms employ strategies that favor ingestion of the living component of seston or surface sediment. By selecting particles that are the richest potential food source, animals may also be selecting the particles with potentially the highest bioavailable concentrations of metals.

Advances in radioisotopic techniques allow for the measurement of metal assimilation efficiencies from geochemically distinct particle types. Some generalizations are now emerging from a body of work using the radioisotope tools.

Metals associated with labile sediment coatings, for example, bacterial exopolymers, are generally assimilated with higher efficiencies by particle-ingesting invertebrates than from more recalcitrant coatings, e.g., mineralogical features and humic acids. This has been shown for bivalves[45,46] and amphipods.[47] Metals associated with phytoplankton cells are of higher bioavailability than other types of particulate materials. Lee and Luoma[48] showed that the bivalves *M. balthica* and *Potamocorbula amurensis* assimilated Cd and Zn from seston more efficiently as the proportion of phytoplankton within the seston increased. Many studies have demonstrated relationships between the proportion of trace element in algal cell cytoplasm and trace element assimilation by a diversity of herbivorous invertebrates.[36,48–50] This is particularly important for elements such as Se, which appears to be rapidly incorporated into cytoplasmic proteins of many phytoplankters.[50] The importance of the living component of the sediments means that laboratory exposures should include a realistic, metal-exposed food component to approximate the magnitude of dietary metals exposure that occurs in nature.

Other feeding behaviors suggest that complicated relationships between particle selection and metal exposure are possible. In general, particle-bound Cr is thought to be of low bioavailability to invertebrates. In fact, it can be used as an inert tracer in studies of assimilation efficiency. But Decho and Luoma[51] showed that the bivalves *P. amurensis* and *M. balthica* assimilated Cr from bacteria cells with high (>90%) efficiencies. Therefore, some types of biotransformation appear to result in significant dietary exposure of animals to Cr. Adding an additional complication, *P. amurensis* will selectively avoid digesting bacterial cells with high Cr concentrations.[52] Similarly, Schlekat et al.[53] showed that when the amphipod *Leptocheirus plumulosus* ingests particles with increasing Cd concentrations, assimilation efficiency was highest at median concentrations.

7.3.2.2 Feeding Rates

The nutritional quality of the surficial sediments and suspended particulate matter on which many aquatic invertebrates subsist is either low or inconsistent. For example, the organic component of surficial sediments is typically less than 5%.[37] Similarly, suspended matter is also a poor source of nutrition. For example, the maximum contribution of phytoplankton to the mass of suspended particulate matter in San Francisco Bay is 20%.[54] To compensate for these nutritional constraints, many aquatic invertebrates ingest large quantities of particulate food. Making accurate measurements of invertebrate feeding rates in the field is difficult, and laboratory studies are also subject to artifacts that make extrapolations to nature difficult. However, some generalities on feeding rates of aquatic organisms can be made that serve to highlight the potential importance of dietary metal exposure.

Deposit feeders can ingest a minimum of one body weight of sediment per day.[37] In general, suspension feeders ingest substantially less than deposit feeders,[5] and suspension feeding rates can vary according to several factors, including the quantity and quality of total suspended solids (TSS), and the size distribution of suspended particles.[55] Many questions concerning the feeding processes of aquatic invertebrates remain. For example, do suspension-feeding animals feed continually with respiratory ventilation? How is feeding selectivity affected by flow strength and turbulence? Resolving such unknowns is critical to modeling the quantity and type of particle ingested. Nevertheless, it is clear that a high flux rate of particle-associated metals to particle-ingesting organisms occurs.

7.3.2.3 Mechanisms of Dietary Metal Absorption

If a high flux rate of particles containing high concentrations of metals occurs in a benthic organism, then it is important to explore the mechanisms with which such metals might be absorbed in the digestive system. Digestive mechanisms have been adapted to extract carbon, nitrogen, and other nutrients from particulate material and other food items. Many of these mechanisms also act to first solubilize metals from particulate material in the gut, and then assimilate the soluble metals across the gut wall. The evolutionary forces behind the development of these mechanisms was probably not absorption of toxic metals. However, the exhibition of distinctly different mechanisms that function to assimilate metals, a limited number of which will be reviewed here, suggests that the ability to absorb metals from food is widespread among aquatic organisms.

7.3.2.3.1   pH

pH offers an obvious mechanism for solubilization of metals from particulate matter. It is well established from chemical principles that the solubility of cations increases as pH decreases, i.e., as the concentration of H^+ increases. The presence of acidic conditions within the digestive tracts of benthic invertebrates is a controversial subject, largely because it has been difficult to obtain accurate *in vivo* measurements.

Plante and Jumars[56] used microelectrodes to measure pH in the guts of several deposit-feeding polychaetes and holothurians. Gut pH of these organisms was similar to that of their neutral to slightly basic sedimentary habitats. Ahrens and Lopez (unpublished data) used epifluorescent microscopy and particles labeled with pH-sensitive fluoroscein to measure *in vivo* gut pH of polychaetes, harpacticoid copepods, and grass shrimp. All taxa exhibited slightly acidic guts, with pH values ranging from 5 to 7. The guts of bivalves are also reported to exhibit slightly acidic (pH 5 to 6) conditions.[57] Gangnon and Fisher[58] and Griscom et al.[59] showed that assimilation of cationic metals by bivalves from a range of organic and inorganic particle coatings correlated with increased metal desorption as pH dropped from 8 (pH of seawater) to 5 (pH of bivalve digestive system).

7.3.2.3.2   Amino Acid–Rich Digestive Fluids

Various attempts have been made to estimate the bioavailable fraction of particle-associated metals by using chemical extractions as surrogates of the digestive processes that extract metals from particles.[60,61] Recently, Mayer and colleagues[62] have estimated the bioavailability of sediment-associated metals and organic contaminants by extracting sediments *in vitro* with digestive fluids collected from the guts of various benthic invertebrates. Digestive fluids used as extractants include those from adult deposit- and suspension-feeding annelids and from deposit-feeding holothurians.

Metal concentrations measured in digestive fluids before extraction are high, indicating that metals are naturally solubilized from sediments in guts of these organisms. Solubilization does not demonstrate that assimilation of these metals occurs across the gut wall, but solubilization alone could be of geochemical significance because excreted soluble metals are subject to physical transport and may be available for uptake through dissolved pathways. Gut fluids from the deposit feeding worm *Arenicola marina* solubilized approximately 10% of copper from contaminated sediments.[62] Other metals, including lead and cadmium, were less susceptible to solubilization. The digestive fluids typically contained high concentrations of amino acids, and differences in amino acid concentration among different annelids and holothurians affected the degree of copper solubilization. Chen and Mayer[63] concluded that between 75 and 90% of the observed copper solubilization was due to complexation with the imadazole subunit of histidine residues, rather than a result of active enzymatic processes.

Interestingly, the mechanism utilized by polychaetes and holothurians appears to be different from that of other organisms in the degree to which metals are solubilized from different geochemical forms of metals. For example, Chen and Mayer[64] showed that the digestive fluids from three deposit-feeding species were ineffective at solubilizing copper from reduced amorphous iron sulfides relative to amorphous iron oxyhydroxides. However, these results contrast with *in vivo* results showing that the bivalve *Mytilus edulis* assimilated Cd, Co, Cr, and Zn more efficiently from anoxic, sulfidic sediments than from oxic sediments.[3,59] There are both geochemical and organismal explanations for this contrast. Copper-sulfide complexes show lower solubility coefficients than Cd or Zn complexes.[65] Particles in the gut may be sorted and undergo different digestive processes, such as intensive glandular digestive processes.[51] Finally, long residence times (48 to 72 h) combined with oxidizing conditions (for example) in the gut may result in a change in metal form during digestion *in vivo*.

7.3.2.3.3 Surfactants

Another potential mechanism by which contaminants can be solubilized from ingested food is through the action of biologically produced surfactants. Surfactants are molecules that exhibit both hydrophilic and hydrophobic characteristics, and they function by increasing the apparent solubility of compounds that would normally exhibit hydrophobic/low-solubility behavior in the absence of the surfactant. For example, lipids and other fatty acids exhibit low solubility in aqueous solutions, but when a surfactant is added, the hydrophobic and hydrophilic ends of the surfactant interact with lipid and water molecules, respectively, forming a water-soluble "micelle" in which the lipid is encapsulated.

Surfactancy has long been known to be an attribute of marine invertebrate gut fluids,[66] but the prevalence of surfactant production across taxonomic phyla and functional feeding groups remains unclear. Mayer et al.[66] measured the surfactant activity from the extracellular gut fluids of 19 species of benthic polychaetes and holothurians that included deposit feeders, suspension feeders, carnivores, and omnivores. The highest surfactant activity was in sediment-ingesting deposit feeders;[66] the lowest surfactancy was found in animals that ingested little sediment, such as predators and suspension feeders. Additionally, surfactant production has been qualitatively described for the bivalves, *Macoma balthica* and *Mytilus edulis*.[59]

Many functions of surfactants have been proposed, and some of these could solubilize metals from food particles. The surfaces of sediment particles are often coated with polymeric compounds (i.e., peptides, bacterial exopolymers, humic substances), and these compounds often exhibit high metal affinities.[38,46,67] Surfactants can desorb these polymers,[66] thus providing a linkage to the gut epithelium. Surfactants can also act to disaggregate lipid matrices, providing access to metals associated with membrane-bound proteins. The lugworm *Arenicola marina* exhibits strong surfactant activity, and the gut fluids of this organism have been shown to solubilize Cu *in vivo*.[68] Lawrence et al.[69] found a relationship between solubilization of methylmercury by gut fluids from the polychaete *A. marina* and bioaccumulation factors for the amphipod *Leptocheirus plumulosus*. However, the relationship between the presence of surfactants and solubilization of trace elements is difficult to establish because of the co-occurrence of other potential mechanisms, e.g., the action of histidine-bearing amino acids.[63]

7.3.2.3.4 Intracellular Digestion

In vitro extractions of particle-associated metals operate on the assumption that solubilization of metals from ingested particles occurs through extracellular digestion and that solubilized metals are bioavailable. However, Decho and Luoma[51] and references therein show that digestion in some bivalves is complicated, and involves both extracellular and intracellular processes. Intracellular digestion occurs within the digestive gland. Decho and Luoma[51] showed that the proportion of ingested food that is sent through this pathway differed between the bivalves *Macoma balthica* and *Potamocorbula amurensis,* and that this was consequential to metal uptake. Higher bioavailability occurred where a greater fraction of ingested material was passed through the glandular phase of digestion (and retained longer). Greater than 90% of bacterial-bound Cr was assimilated by *P. amurensis*, at least partly because nearly all ingested bacteria are subjected to intracellular digestion in this bivalve.

7.3.3 Experimental Designs for Laboratory Exposures via Diet

Experimental design can be influential in determining the outcomes of studies of dietary metal exposure. The effects of partitioning were described above. The length of time that sediments are incubated with metals before the biological exposure begins greatly affects partitioning and conclusions about exposure routes. To be environmentally relevant, duration of exposure, habitat, and food source must also be reflective of what occurs in nature. Short duration of exposure (4 to 10 days) may not be sufficient to allow for the manifestation of toxicity through dietary routes. In some situations, experimental animals may not ingest the metal-contaminated particles used as an exposure matrix. For example, test designs that offer organisms uncontaminated food are likely to underestimate exposure in an equilibrated environmental setting where food would be contaminated.[33] Lee et al.[70] showed that, when deposit-feeding invertebrates select uncontaminated food over contaminated sediment particles, metal uptake is less than when both food and pore waters are contaminated. Test organisms also may avoid or slow their ingestion of particles because they are metal contaminated.[52] Similarly, if sediments are not the food of the test organism, then sediment bioassays are unlikely to include a dietary exposure. For example, the amphipod *Rhepoxynius abronius*, which has been used widely as a sediment toxicity test organism, is described as a meiofaunal predator.[71] Other invertebrates that are carnivorous or omnivorous and are used in sediment assessments include estuarine mysids or juvenile fish that prey primarily upon small invertebrates. It is rare that equilibrated prey species are included in sediment bioassays with these animals, but in nature exposures to sediment-associated metals could be dominated by ingestion of contaminated prey. Many standard toxicity test organisms are herbivorous, e.g., freshwater cladocera can subsist on single-celled algae. It would be appropriate to investigate exposure from algae-associated metals to these organisms. Developing protocols that include dietary exposures will be more complicated than the sediment bioassays or dissolved-only exposures that are at present standard. But it is the only way to adequately address questions about exposures in nature.

7.4 THE RELATIVE IMPORTANCE OF DIETARY VS. DISSOLVED METAL UPTAKE FOR BIOACCUMULATION AND TOXICITY

As shown above, dissolved metals can be accumulated through permeable membranes,[74] and particle-associated metals can be assimilated after dietary ingestion.[5,75,76] Until recently, the relative importance of these pathways was difficult to resolve quantitatively. Most risk assessments for terrestrial mammals and birds assume that exposure occurs predominantly through the dietary route.[77] Exposure is therefore a function of metal concentration in food and the ingestion rate of the test organism. Dietary exposure of aquatic organisms to metals also has been considered experimentally for some time. For example, accumulation of Zn and Fe by herbivorous snails from macroalgae was measured by Young[78] more than 25 years ago. But separating co-occurring dietary and dissolved uptake has been challenging because contaminants can desorb from contaminated food during feeding, and can then be accumulated through dissolved pathways. Similarly, dietary uptake is possible if animals are fed during dissolved exposure. Traditionally, studies addressing this issue employed extended exposures and a mass balance approach in which exposure routes were physically separated.

7.4.1 MASS BALANCE APPROACH

Mass balance studies are conducted in the laboratory or, indirectly, *in situ*.[79] Dietary uptake was calculated by determining the difference between metal accumulation that arose from dual exposure through both routes, and metal accumulation that arose from dissolved uptake only. This approach was applied to deposit- and suspension-feeding invertebrates, and to predatory invertebrates and fish.

7.4.1.1 Deposit and Suspension Feeders

Although Boese et al.[80] identified ten potential contaminant-uptake mechanisms for the facultative deposit/suspension feeding bivalve, *Macoma nasuta*, most experimental mass balance efforts have focused on only separating dissolved and dietary uptake. Results from this literature can be found to support any point of view about exposure routes. But it has long been clear that diet cannot be ignored as a source of exposure, under many of the circumstances typical of nature. For example, Luoma and Jenne[81] separated uptake of pore water metals by placing bivalves (*M. balthica*) in dialysis bags that were buried in sediments. In complementary treatments, clams were allowed to ingest sediment particles, so both exposure routes were presumably operating. Results of this experiment showed that dietary uptake contributed between 75 and 89, 35 and 76, and 17 and 57% for Ag, Zn, and Co, respectively, depending on the particle type, assuming the contribution of dietary and dissolved exposures to overall body burden was fully additive. Selck et al.[82] subjected the deposit-feeding polychaete *Capitella* sp. I to two cadmium exposure regimes: a water column exposure, and a combination of sediment and pore water exposures. Pore water Cd concentrations in the combination treatment were similar to Cd concentrations in the water column. After 5-day exposures, worms in the combination treatment accumulated 470 μg Cd/g dry weight, compared with 26 μg Cd/g dry weight for water-only worms. Assuming the forms of dissolved Cd were the same in both treatments, dietary ingestion was responsible for 95% of Cd tissue concentration in the combination treatment.

Lee et al.[3] investigated the importance of dietary metals uptake for several invertebrates, including *Neanthes arenaceodentata* (deposit-feeding polychaete), *Heteromastis filiformis* (head-down deposit feeding polychaete), and *M. balthica* (surface-deposit feeding bivalve). The invertebrates were exposed to sediment spiked with Cd, Ni, and Zn for 18 days. By manipulating both spiked-metal and acid volatile sulfides (AVS) concentrations, pore water metal concentrations were controlled at environmentally realistic levels. Following incubation, increases in *M. balthica* and *P. amurensis* tissue metal concentrations were statistically related to concentrations of sediment-phase metals that were extractable with weak acid, but no relationship was shown with either pore water metal concentrations or to AVS-normalized extractable metal concentrations. Similar results were shown by *N. arenaceodentata* for Ag, Cd, and Zn.[83] The most reasonable explanation for the relationship between tissue metal concentration and extractable metals is that metal exposure for these organisms occurs from dietary ingestion, and subsequent assimilation of the extractable proportion of metals.

Harvey and Luoma[84] compared routes of uptake in suspension feeding (as compared to deposit feeding) *M. balthica*. Two groups of clams were placed in suspensions of metal-enriched bacteria, which served as food. The first group fed on the suspended bacteria; the second group was enclosed in filter chambers, which separated the clams from the bacterial suspension via 0.4-μm filters. The proportion of metal concentration in feeding clams attributed to dietary uptake, calculated by assuming additivity, was shown to be 95, 75, and 67% for Co, Zn, and Cd, respectively. One commonality among the selected studies cited above was that the authors manipulated partitioning to achieve distribution coefficients that were similar to those found in nature (i.e., final conclusions were dependent upon phase-specific Cd concentrations, but efforts were made to assure those concentrations were similar to natural settings).

7.4.1.2 Predators

A body of mass balance studies dating from the late 1970s evaluate the relative importance of dietary metal uptake to predators. Jennings and Rainbow[85] compared accumulation of Cd between groups of crabs (*Carcinus maenas*) that were exposed to either dissolved Cd alone or to a combination of dissolved Cd and Cd-enriched prey (*Artemia salinas*). Cadmium accumulation between the groups was similar, but the true nature of the dietary pathway was probably underestimated, as the fed crabs received only two mysids per day. Recent work has used more realistic prey consumption rates. Woodward et al.[86] compared rainbow trout fed benthic macroinvertebrates from a metal-contaminated stretch of the Clark Fork River, Montana, compared to trout fed insects from an uncontaminated river. Metal (As, Cd, and Cu) concentrations were up to 27 times higher in contaminated vs. reference invertebrates. Each feeding group was exposed to a series of water concentrations, ranging from clean, uncontaminated river water to river water that was amended with increasing concentrations of metals, which reflected concentrations in the Clark Fork. Fish accumulated Cu and As predominantly from food even in the presence of excess dissolved metals. Trout bioconcentrated dissolved Cd, but uptake was not as great as when trout were also offered contaminated food.

Several studies suggest that predatory aquatic insect larvae can also gain the majority of their metal body burden through dietary exposure. The phantom midge, *Chaoborus punctipennis*, accumulated more than 90% of its Cd from dietary ingestion of Cd-enriched cladocerans.[87] Roy and Hare[88] showed similar results in a food chain study designed to determine the relative importance of dietary metals to alderfly (*Sialas valeta*) larvae. In this study, prey items (larvae of the midge, *Cryptochironomus* sp.) were contaminated with Cd by exposure to either dissolved Cd alone, or to dissolved and dietary Cd (in the form of meiobenthos). Midges exposed through both routes showed higher tissue Cd concentrations. As a consequence, *S. valeta* accumulated Cd more rapidly and to a higher concentration from Cd-exposed *Cryptochironomus* sp. than from dissolved uptake alone. Additionally, Cd distributions in *S. valeta* exposed to both dietary and dissolved Cd more closely resembled those of field-collected insects. Results of these diverse studies highlight the need to expose predatory animals via metal-contaminated prey if an element of realism is to be brought to laboratory-based exposures.

7.4.2 THE USE OF MATHEMATICAL MODELS IN METALS RISK ASSESSMENT

7.4.2.1 Background

Mathematical models can be used to evaluate relationships between bioaccumulation and environmental toxicant concentrations, or to understand the processes that affect this relationship. Landrum et al.[89] reviewed the progressive development of bioaccumulation models and Luoma and Fisher[28] expanded on that review. In the simplest expressions, steady-state tissue concentrations are described relative to environmental concentrations by ratios. Bioconcentration factors (BCFs) describe tissue concentrations relative to water concentrations (either overlying or pore water). Bioaccumulation factors (BAFs) are ratios of tissue concentration to concentrations in ingested food or sediment. The ratios can be derived from field data or experimental data. Many studies have demonstrated that BCFs and BAFs, even when normalized to account for covarying factors (e.g., contaminant lipophilicity, organism lipid concentrations, and sediment organic carbon concentrations), are highly variable. Modeling approaches that include more sophisticated consideration of geochemistry and biology can narrow that variability.

7.4.2.2 Equilibrium Models

Equilibrium models of various types are widely employed in risk assessment. These approaches were described elsewhere.[28,65,89] In the case of metals, a widely described approach (e.g., the AVS model[65]) uses ratios to account for equilibrium partitioning to pore waters, and relates the metal activity so determined to toxicity test results. We will not further describe that approach here because it does not address the question of bioaccumulation routes. A body of work, beginning with Tessier et al.,[90] illustrates both the strengths and weaknesses of the equilibrium modeling approach with regard to multipathway exposures. These authors used multiligand equilibrium models (i.e., FIAM) to compare metal form to concentrations in the freshwater bivalve *Anadonta grandis* in lakes from Quebec.[90] Free-ion activity of Cd in overlying water was estimated using measured total dissolved Cd and concentrations of inorganic ligands. Equilibration constants for iron oxyhydroxide (FeOOH) and organic matter (OM) binding sites in sediments were used to estimate Cd concentrations specific to these phases. When clam tissue Cd concentrations, $[Cd]_{clam}$, were compared with overlying water and sediment ligand Cd concentrations, only overlying $[Cd^{2+}]$ showed a statistically significant relationship. Other studies have used correlation analysis to find similar relationships between $[Cd^{2+}]$ and $[Cd]_{org}$. For example, Hare and Tessier[91] showed that Cd tissue concentrations of the larval phantom midge, *Chaoborus punctipennis*, in 23 Canadian lakes could be explained by $[Cd^{2+}]$, pH, and concentrations of dissolved organic carbon.

Although these relationships suggest that these organisms directly bioaccumulated dissolved $[Cd^{2+}]$, they do not eliminate uptake from zooplankton, phytoplankton, or suspended organic matter equilibrated with (and thus covarying with) $[Cd^{2+}]$. Later studies indeed showed that *C. punctipennis* bioaccumulated Cd through ingestion of food. Munger and Hare[87] exposed *C. punctipennis* to dietary and dissolved Cd. Dietary exposure was accomplished by feeding *C. punctipennis* Cd-contaminated cladoceran (*Ceriodaphnia dubia*), which acquired their body burdens by feeding on Cd-contaminated algae (*Selenastrum capricornutum*). Cadmium uptake by animals exposed to both dietary and dissolved Cd was the same as that shown by animals exposed to food alone, indicating that the dissolved route was unimportant at the concentrations used. Thus, the earlier relationship between $[Cd^{2+}]$ and $[Cd]_{org}$ for *Chaloborus punctipennis*[91] is indirect under natural conditions. Uptake by this predator is dependent upon Cd concentrations in food, which are in turn dependent upon [Cd^{2+}].

7.4.2.3 Dynamic Multipathway Bioaccumulation Model

Landrum et al.[89] described three forms of mechanistically based, dynamic bioaccumulation models that can characterize dietary and dissolved exposures to metals: (1) compartmental models, (2) physiological-based pharmacokinetic models, and (3) bioenergetic models. Of these, bioenergetic models most easily allow for multiple uptake pathways. Bioenergetic models describe contaminant accumulation and loss as functions of organismal energy requirements. In one approach, a deposit-feeding organism's contact with contaminants in the different physical phases of its habitat is directly related to the flux of water across its gills to obtain oxygen (dissolved metals in overlying water) and the flux of ingested material through its gut to obtain nutrients (sediment-bound metals). A crucial assumption in this approach is that integrated organismal exposure is an additive function of dissolved and dietary uptake pathways. If this assumption is accepted, then the kinetics of each pathway can be calculated independently.

A form of bioenergetic model that specifically addresses metal accumulation is called the dynamic multipathway bioaccumulation model, or DYMBAM. Application of this model to aquatic organisms has been recently reviewed,[5,75] so we will limit our discussion to the utility of DYMBAM for risk assessment issues.

7.4.2.3.1 DYMBAM Structure

In the DYMBAM model, steady-state metal concentrations (C_{ss}) are the sum of metal concentrations from dissolved ($C_{ss,w}$) and dietary pathways ($C_{ss,F}$):

$$C_{ss} = C_{ss,w} + C_{ss,F} \tag{7.3}$$

$$C_{ss} = ((k_u \times C_W)/(k_{ew} + g)) + (AE \times IR \times C_F)/(k_{ef} + g) \tag{7.4}$$

where k_u = dissolved metal uptake rate constant (l/g/day), C_W = dissolved metal concentration (μg/l), AE = assimilation efficiency (%), IR = ingestion rate (mg/g/day), C_F = metal concentration in food (e.g., phytoplankton, suspended particulate matter, sediment) (μg/g), $k_{ew,f}$ = efflux rate from water and food, respectively (1/day), and g = growth rate constant (1/day). The critical environmental factors are C_W and C_F. The C_F can be determined directly, or can be estimated as a function of C_W and the distribution coefficient, K_D. The K_D, and subsequently C_F, is a conditional factor, and can be influenced by many parameters, including pH and ionic strength.[34]

The important organismal factors are dissolved uptake rate, assimilation efficiency, ingestion rate, and efflux rate. IR can vary considerably among organisms and can be difficult to measure, as has been discussed earlier. Uptake from dissolved phase, i.e., k_u, is most effectively measured in the laboratory using radiotracers. This approach uses short-term exposures that measure gross metal influx rates. Longer-term exposures may underestimate k_u because they measure net accumulation, the balance between uptake and efflux. Uptake rates from the dissolved phase are variable among metals and animal species. For example, in a comparison of Cd, Cr, and Zn influx rates by *Macoma balthica* and *Potamocorbula amurensis*, Lee et al.[92] demonstrated that influx rates of Zn in both clams were three to four times those for Cd and 15 times those for Cr. Further, the influx rates of all three metals were four to five times greater in *P. amurensis* than in *M. balthica*, which is probably a reflection of the latter clam's greater clearance rate.

Efflux rate is a measurement of the physiological turnover rate of assimilated metals. The critical parameter is the rate constant of loss, which is specific to metals and animal species. Multicompartment models assume that metals in different storage pools exhibit different transport and release kinetics. Generally, efflux rates have not been affected by uptake pathway for mussels,[93] which simplifies the efflux term in models using these organisms. For example, Fisher and Wang[94] found no differences among rate constants of loss in mussels (*Mytilus galloprovincialis*) when exposure routes differed. However, Wang et al.[95] showed that efflux rates of copepods were higher after dietary metal exposure than from uptake of dissolved metals.

Assimilation efficiency represents the proportion of metal within a particular food item that an organism accumulates. Metal assimilation efficiencies are now known for a wide range of aquatic invertebrates including bivalves, polychaetes, and crustaceans. They are also known from a range of food types, including phytoplankton, geochemically defined organic and inorganic particles, natural seston, natural sediment, and prey organisms.[96] Several methods have been used to determine assimilation efficiencies, including pulse-chase[5,75] and mass-balance approaches.[88,97–100] Like other parameters, variability occurs among different food types and metals. But assimilation efficiency is also constrained by species-specific physiological mechanisms, and so it varies considerably among different organisms. Wang and Fisher[5] compared the assimilation efficiencies for five metals (Ag, Cd, Co, Se, and Zn) by three taxonomically different organisms (the bivalve *M. edulis*, the polychaete *Neries succinea*, and the copepod *Temora longicornus*) from a wide range of food types and discovered few generalizations. Narrowing the comparisons to specific food types does not improve consistency. For example, the Cd assimilation efficiencies by *M. edulis* from amorphous iron oxide coatings ranged from 6%[58] to 40.5%.[59] Values for *Macoma balthica* ranged from 23%[59] to 35%.[46] Assimilation efficiency for Cd by amphipods was 5%.[47] The assimilation efficiencies of some metals by bivalves and copepods from phytoplankton can be a function of the proportion of metal in the algal cell cytoplasm.[36,50] However, additional metal forms in the phytoplankton are available in some cases and digestive features can limit availability in other cases. For example, bioavailability from phytoplankton to amphipods is limited by incomplete digestion of the plant cells.[47] Cytoplasmic metal in the phytoplankton is not completely bioavailable because not all cells are broken open. These species-specific differences illustrate the importance of measuring or estimating the assimilation efficiencies for unknown consumer/food type combinations when risk assessments are conducted on a site-specific basis.

7.4.2.4 Application of Models

The DYMBAM model has been especially effective in determining the relative contribution of dissolved and dietary pathways[4,5] for bivalves, copepods, and polychaetes.

Luoma et al.[4] demonstrated that laboratory measurements of model parameters could accurately predict Se bioaccumulation by the bivalve *Macoma balthica* in San Francisco Bay, California. Laboratory measurements showed relatively slow uptake of dissolved Se (as selenite), whereas clams assimilated between 26 and 88% of Se in two different diets. Even when clams consumed reduced sediment-associated Se, which has low bioavailability, dietary uptake contributed 98% of predicted steady-state Se concentrations.

Wang et al.[93] determined metal uptake and depuration kinetics for *Mytilus edulis* in laboratory experiments. Assimilation efficiencies for a range of metals (Ag, Cd, Se, and Zn) were measured from natural seston. Predicted steady-state metal concentrations were calculated for San Francisco Bay (SFB) and Long Island Sound, New York (LIS) using metal concentrations representative of those systems. For SFB, particulate concentrations were estimated from measured dissolved water concentrations and mean K_D values; for LIS, dissolved metal concentrations were estimated from measured phytoplankton concentrations and mean phytoplankton concentration factors. Predicted C_{ss} were close to measured tissue metal concentrations for *M. edulis* that were collected in SFB and LIS in close proximity to water and seston collection sites. The importance of dietary metals uptake varied among metals. Selenium concentrations were dominated by dietary uptake regardless of food type. Dietary uptake was of variable importance for Ag (43 to 69%), Cd (24 to 49%), and Zn (48 to 67%), depending on how efficiently mussels assimilated metals from food and on partitioning coefficients. Roughly similar patterns were observed for the copepod *Temora longicornus*.[95]

Wang et al.[101] applied the biokinetic model to a deposit-feeding polychaete, *Neries succinea*. Dietary uptake was measured based on the assimilation efficiency of metals associated with oxic and anoxic sediments that were encapsulated within gelatin. Weight-normalized uptake rates for dissolved Cd, Co, Se, and Zn by *N. succinea* were an order of magnitude less than rates exhibited by mussels and copepods.[101] This slow uptake of dissolved metals, combined with low predicted concentrations of metals in pore water, explained why dietary uptake contributed more than 98% of predicted steady-state concentrations of Cd, Co, Se, and Zn. Compared with these metals, steady-state Ag concentrations showed less influence from dietary uptake, but this pathway still explained the majority of Ag uptake (65 to 95%). In contrast, Lee et al.[3] determined that the polychaete *N. arenaceodentata* accumulated Cd predominantly from pore water during laboratory exposures.

7.4.2.4.1 DYMBAM Case Study: Selenium in San Francisco Bay

The DYMBAM model can be used to assess the risk from metals or metalloids by comparing predicted tissue metal concentrations with established residue-based threshold values for adverse effects. The risk from Se to predators through two possible food webs in SFB is an example. Predators are particularly susceptible to

Se because it accumulates progressively through trophic orders.[102] Concentrations of Se in food that are above approximately 8 μg/g are known to cause reproductive and developmental toxicity to wildlife.[102,103] So, 8 μg/g in prey can be used as an effects concentration.

Using the model, risk can be compared between generic predators in pelagic-based and benthic-based food webs. Data and model parameters are available for a pelagic-based food web that consists of phytoplankton (diatoms) to herbivorous zooplankton (copepods) to carnivorous zooplankton (mysid shrimp). In SFB, striped bass and other fish feed on mysids at various times during their life cycles. Data also exist for a benthic-based food web of phytoplankton to bivalves. The bivalve *Potamocorbula amurensis* is an invasive species that has reached high densities in SFB and is frequently eaten by bottom-feeding fish (e.g., sturgeon and glassy flounder) and diving ducks (e.g., surf scoter and scaup). The DYMBAM model was used to estimate steady-state tissue concentrations (C_{ss}) in the mysids and in *P. amurensis* (Table 7.4). Experiments were used to derive bioconcentration of Se by diatoms and bioaccumulation from the diatoms by zooplankton of two size classes. Assimilation from ingested zooplankton and loss rates were used to calculate C_{ss} for the mysid *Neomysis mercedis*. Both a low and a high Se assimilation efficiency were used to determine bioaccumulation by *P. amurensis* to account for bioavailability differences among food types (Reference 104, C. Schlekat, unpublished data). The range of Se concentrations in seston in SFB is 1.0 to 3.0 μg/g (G. Cutter, personal communication). Dissolved Se concentration averages approximately 0.25 μg/l in the bay.[105] Results show that bioaccumulation by the bivalve *P. amurensis* was two to four times higher than the highest C_{ss} predicted for *N. mercedis* (Table 7.5), primarily because rate constants of loss from the bivalve are ten times slower than loss from zooplankton and mysids. Risk characterizations were consistently below 1.0 for the mysid food web, and ranged from 1.7 to 3.0 for the benthic food web (Table 7.5). Thus, predators that feed exclusively on mysids are unlikely to experience toxicity, whereas those feeding on bivalves may approach thresholds for toxicity. The model not only allowed direct evaluation of risk but provided mechanistically based insights about which organisms were most at risk.

7.4.3 COMPARISONS AMONG METALS AND ORGANISMS

The dietary vs. dissolved exposures for a particular metal are linked to a variety of factors: assimilation efficiency, feeding rate, geochemical partitioning, and chemical species in dissolved and particulate phases. In addition, metal specific biogeochemical factors operate at the semipermeable membrane of the organism to affect dissolved metal uptake. These are expressed empirically in the rate constant for metal uptake (k_u), as derived from empirical determinations of gross influx rates.[93] Table 7.6 compares k_u with assimilation efficiencies for Ag, Cd, Cr, Se, and Zn to illustrate some basic differences among the metals. For example, dissolved uptake of silver could often be important, because of high rates of dissolved uptake and low efficiency (with exceptions) of silver assimilation from diet (Table 7.6). The consensus from a variety of studies is that both dissolved and dietary exposure of Cd are potentially important. Dissolved Cd is important at low salinities where free Cd ion is an

TABLE 7.4
Parameters Used in the Dynamic Multipathway Bioaccumulation Model for Estimating Steady-State Se Tissue Concentrations (μg/g) for the Crustacean *Neomysis mercedis* and the Bivalve *Potamocorbula amurensis*

Food Chain	Species	Dissolved Uptake		Dietary Uptake				Efflux, k_e (1/day)
		k_u (l/g/day)	C_w (μg/l)	Food Type	C_{food} (μg/g)	IR (g/g/day)	AE (%)	
Bivalve	*Potamocorbula amurensis*	0.003[a]	0.24[b]	Diatoms	1–3[b]	0.25[a]	0.45–0.8[a]	0.025[a]
	Small copepods	0.024[c]	0.24[b]	Diatoms	1–3[b]	0.42[c]	0.6[a]	0.155[c]
	Large copepods	0.024[c]	0.24[b]	Diatoms	1–3[b]	0.42[c]	0.71[a]	0.155[c]
Mysid	*Neomysis mercedis*	0.027[a]	0.24[b]	Small copepods	Model dependent	0.45[d]	0.73[a]	0.25[a]
		0.027[a]	0.24[b]	Large copepods	Model dependent	0.45[d]	0.61[a]	0.25[a]

[a] Unpublished data (C. Schlekat, USGS).
[b] Unpublished data (G. Cutter, Old Dominion University).
[c] Reference 106.
[d] Reference 107.

TABLE 7.5
Selenium Risk Characterizations for Organisms That Consume Either the Crustacean *Neomysis mercedis* or the Bivalve *Potamocorbula amurensis*

Species	Food Concentration ($\mu g/g$)	$C_{ss\text{-Water}}$ ($\mu g/g$)	$C_{ss\text{-Food}}$ ($\mu g/g$)	$C_{ss\text{-Total}}$ ($\mu g/g$)	Risk Characterization
Potamocorbula	1	0.03	4.5	4.53	0.6
amurensis (low AE)	2	0.03	9.0	9.03	1.1
	3	0.03	13.5	13.53	1.7
P. amurensis	1	0.03	8.0	8.03	1.0
(high AE)	2	0.03	16.0	16.03	2.0
	3	0.03	24	24.03	3.0
Neomysis mercedis	1.88	0.03	2.5	2.53	0.3
(small copepods)	3.72	0.03	4.8	4.83	0.6
	5.56	0.03	7.3	7.33	0.9
N. mercedis	1.96	0.03	2.2	2.23	0.3
(large copepods)	3.88	0.03	4.3	4.33	0.5
	5.81	0.03	6.4	6.43	0.8

important geochemcial species. Although influx rates of dissolved Cd are relatively low, the K_D of Cd is also relatively low, which contributes to a more pronounced dissolved phase uptake for Cd compared with other elements (Table 7.6). Dissolved Cr is accumulated slowly, regardless of whether it occurs as Cr(III) or Cr(VI) (Table 7.6). Although dietary assimilation is also low (with exceptions), diet dominates, or bioaccumulation is very low (Table 7.6). Dietary ingestion of Se and Zn are important. Influx rates of dissolved Se (as selenite) are low, but particle-ingesting organisms uniformly assimilate greater than 50% of Se from phytoplankton sources. Zinc is assimilated with high efficiencies by most herbivorous invertebrates and its K_D is high.

The DYMBAM model not only allows predictions of the relative importance of uptake routes, but sensitivity analyses allow mechanistic insights regarding the cause of those differences. The model requires both laboratory and field data acquisition, but neither is especially onerous using modern approaches. The model can be readily adapted to site-specific questions and new organisms.

7.5 TOXICOLOGICAL SIGNIFICANCE OF DIETARY METALS EXPOSURE

To be important in a risk assessment framework, dietary metals exposure should yield consistent, quantifiable, and meaningful ecological end points. Recent recognition of the importance of dietary metal exposure has stimulated efforts to quantify the toxicological significance of this pathway, and to distinguish between effects caused by dietary exposure from those caused by dissolved exposure. Traditional approaches have yielded some examples in which dietary metals exposure caused

TABLE 7.6
Relative Importance of Dietary vs. Dissolved Uptake for Metals by Invertebrates

Metal	Species	k_u, Dissolved Uptake Rate (l/g/day)	Food	Assimilation Efficiency (%)	Ref.
Ag	*Mytilus edulis*	1.79	Phytoplankton	5–18	93
			Sediment	3	58
	Marine copepods	10.42	Phytoplankton	8–19	50, 106
	Leptocheirus plumulosus (estuarine amphipod)	N.M.	Phytoplankton	4–8	47
			Mineral coatings	12	
			Organic coatings	10–27	
			Sediment	11–21	
	Chionectes opolio (snow crab)	N.M.	Fish	90	98
	Hippoglossoides platessoides (American plaice)	N.M.	Fish	9	98
Cd	*Mytilus edulis*	0.365	Phytoplankton	8–20	58, 93
			Sediment	15	
	Potamocorbula amurensis	0.125	Natural planktonic assemblage	8.1–48	48, 92
	Macoma balthica	0.032	Natural planktonic assemblage	13–21	48, 92
	Marine copepods	0.666	Phytoplankton	33–53	50, 106
	Leptocheirus plumulosus (estuarine amphipod)	N.M.	Phytoplankton	6–12	47
			Mineral coatings	6	
			Organic coatings	8–27	
			Sediment	10–16	
	Orchestia sp. (semi-terrestrial amphipods)	N.M.	Macroalgae	67–95	108
	Mystacides sp. (caddisfly larvae)	N.M	Chironomid larvae	4	100
	Limnesia maculata (water mite)	N.M.		61	
	Chaoborus sp. (insect larvae)	N.M	Cladocera	4	87

Metal	Species	Food source	Concentration	Range	References
Cr	*Sialus velata* (alderfly larvae)	Chironomid larvae	N.M.	50	88
	Mytilus edulis	Phytoplankton: Cr (III)	Cr(III): 0.034	1	109
		Phytoplankton: Cr (IV)	Cr (VI): 0.100	1–10	
		Sediment: Cr (III)		1	
		Sediment: Cr (IV)		1	
	Potamocorbula amurensis	Natural planktonic assemblage	0.028	2–5	48, 92
	Macoma balthica	Natural planktonic assemblage	0.006	3–4	48, 92
	P. amurensis	Bacteria	N.M.	65–95	51, 52
	M. balthica	Bacteria	N.M.	27–86	51, 52
Se	*Mytilus edulis*	Phytoplankton	0.035	55–71	49, 93
		Natural seston		28–34	
	Potamocorbula amurensis	Phytoplankton	N.M.	55–82	104
		Se(0)		4	Schlekat, unpublished data
	Macoma balthica	Phytoplankton	N.M.	53–90	Schlekat, unpublished data; 4
		Se(0)		26	
	Marine copepods	Phytoplankton	0.024	58–97	50, 106
		Natural seston		49	
	Leptocheirus plumulosus (estuarine amphipod)	Phytoplankton	N.M.	32–70	Schlekat, unpublished data
Zn	*Mytilus edulis*	Phytoplankton	1.044	44–48	49, 93
		Natural seston		32–41	
	Marine copepods	Phytoplankton	2.7–3.9	58–64	50, 106
		Natural seston		52	

(*continued*)

TABLE 7.6 (CONTINUED)
Relative Importance of Dietary vs. Dissolved Uptake for Metals by Invertebrates

Metal	Species	k_u, Dissolved Uptake Rate (l/g/day)	Food	Assimilation Efficiency (%)	Ref.
	Potamocorbula amurensis	0.425	Natural planktonic assemblage	13–50	48
	Macoma balthica	0.091	Natural planktonic assemblage	39–82	48
	Leptocheirus plumulosus (estuarine amphipod)	N.M.	Phytoplankton	8–11	47
			Mineral coatings	15	
			Organic coatings	5–15	
			Sediment	11–16	
	Mystacides sp. (caddisfly larvae)	N.M	Chironomid larvae	7	100
	Limnesia maculata (water mite)	N.M.		16	

Note: Data are from studies that measured dissolved uptake constant (k_u) over short periods of time (because depuration can be excluded), and dietary assimilation using pulse-chase methods. N.M. = not measured.

toxicity (e.g., death and inhibition of reproduction). The most recent studies suggest that new pathways of toxicity will become evident once a body of understanding about dietary exposure develops.

7.5.1 Examples of Dietary Metals Toxicity

Compared with the adverse effects related to waterborne exposure, which include both acute and chronic effects, dietary exposure most likely elicits chronic effects. One of the reasons for this is that dietary exposure is a prolonged and multitiered process (e.g., ingestion, digestion, and assimilation). Behavioral avoidance may reduce extreme exposures and the metal concentrations within prey organisms are constrained by the sensitivity of the latter to metals. Many aquatic organisms have detoxification mechanisms that can first complex excess metals (e.g., metallothionein and metallothionein-like proteins) and then sequester assimilated metals in metal-rich granules.[110] Detoxification by sequestration may limit toxic effects, but could also slow the expression of toxicity, or constrain expression to periods when other stresses result in release of the sequestered forms. Sequestration also limits the bioavailability of metals to predators. For the oligochaete *Limnodrilus hoffmeisteri*, only metals associated with cell cytoplasm and intracellular organelles were available to predatory shrimp.[111] The effectiveness and mechanisms of detoxification probably vary greatly among organisms.

Metal tissue concentration and toxicity are related, but that relationship is not always simple. Studies that support the biotic ligand model approach demonstrate that metal concentration of fish gills correlates well with metal toxicity.[14,15] However, using whole fish metal concentration as a metric of toxicity would not be appropriate because the method emphasizes concentration at the site of toxicity (i.e., the gill). In the amphipod *Hyalella azteca*, Borgmann et al.[112] showed that 4-week LC_{25} values (nmol/l) for the nonessential metals Cd, Hg, and Tl varied by a factor of >10 (range: 3.5 nmol/l for Cd, 48 nmol/l for Tl). Toxicity calculated on a body-concentration LB_{25} (the concentration, in nmol/g, at which 25% of the population die) were roughly similar for the three metals, around 300 nmol/g. Body concentrations of Tl proved to be a better predictor of toxicity than dissolved concentrations, even when normalized for free-ion activity.[112] This suggests that toxicity occurred after Tl was internalized. These experiments were conducted using only dissolved metal exposures. But, if toxic burden proves to be an additive consequence of dissolved and dietary exposure, then it also has implications for dietary uptake.

Examples also exist in which toxicity was caused exclusively by diet as a route of metal exposure. For example, Fisher and Hook[113] observed no lethal effects to freshwater cladocerans and marine copepods that were exposed to dissolved Ag concentrations ranging from 0.014 to 0.5 mg/l. However, reproductive effects were observed when zooplankton were fed algal cells that had been incubated with Ag at concentrations as low as 0.05 mg/l. Exposure of freshwater cladocerans to algae contaminated with Ag produced similar reproductive toxicity.[114] Trout exposed to metals by consuming metal-enriched invertebrates[86] experienced lethal toxicity as well as decreased growth, digestive impairment, and histopathological abnormalities. Dietary exposure to Se concentrations as low as 8 mg/g has caused teratogenic effects in fish, reproductive effects in wildfowl, and effects on growth and behavior in the laboratory.

Some effects, while correlated with tissue metal concentrations, may be difficult to place into traditional risk assessment frameworks. For example, Wallace et al.[115] demonstrated that dietary metal affected the behavior of the estuarine shrimp, *Paleomonetes pugio,* that were fed Cd-enriched oligochaetes. Zooplankton prey capture efficiency decreased significantly in shrimp that were exposed to increasingly higher dietary Cd concentrations, which contained increased Cd associated with nonmetallothionein.

7.5.2 WHY IS DIETARY TOXICITY DIFFICULT TO MEASURE?

Our understanding of how chronic toxicity is manifested in nature is rudimentary as a result of both intellectual gaps and technological limitations. Determination of some sublethal effects, like behavioral changes, are not immediately intuitive and require nontraditional methodologies. For example, Wallace et al.[115] suggested that the neurological impairment experienced by *P. pugio* through dietary Cd exposure could result in impaired predator avoidance. Quantitative assessment of such a hypothesis requires combining natural exposure conditions and ecological interactions in a controlled experiment. Maintaining animals, other than a few surrogates, under laboratory conditions for long periods of time, which may be necessary to elicit effects, will require new protocols and greater understanding of requirements in nature for growth and reproduction. Also, maintaining a consistent dietary exposure regime while controlling other factors (e.g., geochemical conditions that affect metal speciation and partitioning) is necessary to gain mechanistic insight into toxic effects, but reliable protocols have not been developed.

Challenges also constrain the linkages of complex biological responses to metal exposure in field studies. Addressing sublethal effects in the field can be complicated by the presence of other factors (e.g., disease) that might affect organism growth, reproduction, and behavior. Timing between the initiation of metal exposure and end point measurement can also limit the ability to detect a link between exposure and effect. Full effects of exposure might be missed or underestimated if end point measurement begins only after exposure begins, and it is difficult to anticipate why this would not be the case. Coincidental change in a variety of factors can affect the explanatory power of statistical associations. Finally, metal exposure is more likely to fluctuate as natural conditions change over different temporal scales.[36,75] Our ability to detect associations between pulses of exposure and adverse effects is poorly developed.

7.5.3 HOW ARE THESE SUBTLE EFFECTS TO BE HANDLED IN A RISK ASSESSMENT FRAMEWORK?

The relationships between environmental concentration and effects concentration are not always simple or linear; similarly, tissue metal concentration may not always be related in a simple, statistical expression to the manifestation of effects. As Landrum et al.[89] state, "Standard regulatory paradigms such as water quality criteria use the environmental concentration as a surrogate for the concentration at the receptor site. These paradigms are based on the premise that the toxicant concentration at the receptor is proportional to the organism concentration, which is proportional to the environmental concentration." The importance of dietary metal exposure complicates these paradigms. It is unlikely that a simple, rigidly consistent risk assessment framework will work well for metals until we understand exposure routes and their implications. However, it is now clear that dissolved exposure alone and simple toxicity tests are not adequate to understand likely sources of toxicity in nature, unless exposures are extreme. If we are to protect ecosystems with defensible risk assessment procedures, there is no choice but to attack the complexity of multiple route exposures, whatever the challenges.

7.6 CONCLUSIONS/RECOMMENDATIONS

Dietary uptake is an important metal exposure pathway for aquatic organisms. Consequently, risk assessment procedures that compare field-measured environmental exposure concentrations to traditional laboratory-derived effects concentrations may not be accurate if animals in nature experience metals in their diet, whereas animals in laboratory settings do not. When dissolved metal concentrations in nature reach laboratory-derived effects concentrations, metal concentrations in particles and food will be orders of magnitude higher. The consequences of this are just beginning to be explored, but examples clearly show that deleterious effects are possible for some organisms. An important need is the development of meaningful frameworks and protocols for evaluations of dietary metals toxicity.

In addition to research needs, and in lieu of immediate solutions, scientifically based correction factors might be used to account for the magnitude of dietary metal exposure. Wang et al.[93] expressed the total concentration of trace elements in the water column, i.e., dissolved plus particulate trace element concentration as:

$$C_t = C_W + (C_W \times K_D \times \text{TSS}) \tag{7.5}$$

where C_t is total concentration (μg/l), C_W is dissolved concentration (μg/l), K_D is the distribution coefficient (l/kg), and TSS is total suspended solids (mg/l). For example, if a hazard assessment were addressing effects of Cd to the eastern oyster, *Crassostrea virginica*, in an estuary, the environmental concentration would be adjusted using dissolved Cd concentration, a K_D for Cd, and the measured value for TSS. This would function to increase the exposure to account for both routes. Alone, this expression assumes that 100% of ingested particulate Cd is assimilated by oysters; but literature-derived values for Cd assimilation efficiencies could be used to modify the dietary portion of the C_t. For example, Reinfelder et al.[116] showed that *C. virginica* assimilated 68% of Cd from the diatom *Thalassiosira pseudonana*. The total exposure concentration would then be as follows:

$$C_t = C_W + (C_W \times K_D \times \text{TSS} \times \text{AE}) \tag{7.6}$$

Using a dissolved Cd concentration of 1 μg/l, a TSS concentration of 50 mg/l, a K_D of 1000, and an AE of 0.65, C_t is determined to be 32.5 μg/l. This interim value could be compared with existing toxicity information until multipathway toxicity data are available. The approach addresses the geochemical inaccuracy of measuring total metal concentrations (the original water quality criteria), the biological inaccuracies of deriving toxicity only from dissolved concentrations (as recommended by recent U.S. EPA changes), and is an advance over methods that address potential dietary exposure by multiplying exposure concentrations by an arbitrarily chosen correction factor (e.g., European Union approach to risk assessment). There are obvious limitations to any interim approach, but the point is that interim procedures must begin to incorporate exposure pathways in ways that are consistent with modern knowledge.

REFERENCES

1. Bryan, G.W. and Langston, W.J., Bioavailability, accumulation, and effects of heavy metals in sediments with special reference to United Kingdom estuaries: a review, *Environ. Pollut.*, 76, 89, 1992.
2. National Research Council, Risk Assessment in the Federal Government: Managing the Process, National Academy Press, Washington, D.C., 1983.
3. Lee, B.-G. et al., Influence of dietary uptake and acid-volatile sulfide on bioavailability of metals to sediment-dwelling organisms, *Science*, 287, 282, 2000.
4. Luoma, S.N. et al., Determination of selenium bioavailability to a benthic bivalve from particulate and solute pathways, *Environ. Sci. Technol.*, 26, 485, 1992.
5. Wang, W.-X. and Fisher, N.S., Delineating metal accumulation pathways for marine invertebrates, *Sci. Total Environ.*, 237/238, 459, 1999.
6. Campbell, P.G.C., Interactions between trace metals and aquatic organisms: a critique of the free-ion activity model, in *Metal Speciation and Biovailability in Aquatic Systems*, Tessier, A. and Turner, D.R., Eds., John Wiley & Sons, New York, 1995, 45.
7. Anderson, M.A., Morel, F.M.M., and Guillard, R.R.L., Growth limitation of a coastal diatom by low zinc ion activity, *Nature*, 276, 70, 1978.
8. Zamuda, C.D. and Sunda, W.G., Bioavailability of dissolved Cu to the American oyster *Crassostrea virginica*. I. Importance of chemical speciation, *Mar. Biol.*, 66, 77, 1982.
9. Sunda, W.G. and Huntsman, S.A., Processes regulating cellular metal accumulation and physiological effects: phytoplankton as model systems, *Sci. Total Environ.*, 219, 165, 1998.
10. Allen, H.E., Hall, R.H., and Brisbin, T.D., Metal speciation. Effects on aquatic toxicity, *Environ. Sci. Technol.*, 14, 441, 1980.
11. Morel, F.M.M., *Principles of Aquatic Chemistry*, Wiley-Interscience, New York, 1983, 301.
12. U.S. EPA, Water quality criteria: aquatic life criteria for metals, *Fed. Regis.*, 58, 32131, 1993.
13. Allen, H.E. and Hansen, D.J., The importance of trace metal speciation to water quality criteria, *Water Environ. Res.*, 68, 42, 1996.
14. Meyer, J.S. et al., Binding of nickel and copper to fish gills predicts toxicity when water hardness varies, but free-ion activity does not, *Environ. Sci. Technol.*, 33, 913, 1999.
15. Playle, R.C., Modelling metal interactions at fish gills, *Sci. Total Environ.*, 219, 147, 1998.

16. Pagenkopf, G.K., Gill surface interaction model for trace-metal toxicity to fishes: role of complexation, pH, and water hardness, *Environ. Sci. Technol.*, 17, 342, 1983.
17. Wood, C.M. et al., Environmental toxicology of metals, in *Reassessment of Metals Criteria for Aquatic Life Protection*, Bergman, H.L. and Dorward-King, E.J., Eds., SETAC Press, Pensacola, FL, 1997, 31.
18. U.S. EPA, A SAB Report: Review of the Biotic Ligand Model of the Acute Toxicity of Metals, Science Advisory Board, Washington, D.C., 2000, 16.
19. Janes, N. and Playle, R.C., Modeling silver binding to gills of rainbow trout (*Oncorhychus mykiss*), *Environ. Toxicol. Chem.*, 14, 1847, 1995.
20. Playle, R.C., Dixon, D.G., and Burnison, K., Copper and cadmium binding to fish gills: estimates of metal-gill stability constants and modelling of metal accumulation, *Can. J. Fish. Aquat. Sci.*, 50, 2678, 1993.
21. Playle, R.C., Dixon, D.G., and Burnison, K., Copper and cadmium binding to fish gills: modification by dissolved organic carbon and synthetic ligands, *Can. J. Fish. Aquat. Sci.*, 50, 2667, 1993.
22. Wood, C.M., Playle, R.C., and Hogstrand, C., Physiology and modelling of mechanisms of silver uptake and toxicity in fish, *Environ. Toxicol. Chem.*, 18, 71, 1999.
23. Tipping, E., WHAM-A chemical equilibrium model and computer code for waters, sediments, and soils incorporating a discrete site/electrostatic model of ion-binding by humic substances, *Comput. Geosci.*, 6, 973, 1994.
24. Richards, J.G. and Playle, R.C., Cobalt binding to gills of rainbow trout (*Oncorhynchus mykiss*): an equilibration model, *Comp. Biochem. Physiol.*, 119C, 185, 1998.
25. Engle, D.W., Sunda, W.G., and Fowler, B.A., Factors affecting trace metal uptake and toxicity to marine organisms. I. Environmental parameters, in *Biological Monitoring of Marine Pollutants*, Vernberg, F.J. et al., Eds., Academic Press, New York, 1981, 127.
26. Poldoski, J.E., Cadmium bioaccumulation bioassays — their relationship to various ionic equilibria in Lake Superior water, *Environ. Sci. Technol.*, 13, 701, 1979.
27. Block, M. and Wicklund Glynn, A., Influences of xanthates on the uptake of ^{109}Cd by Eurasian dace (*Phoxinus phoxinus*) and rainbow trout (*Onchorhynchus mykiss*), *Environ. Toxicol. Chem.*, 11, 873, 1992.
28. Luoma, S.N. and Fisher, N.S., Uncertainties in assessing contaminant exposure from sediments, in *Ecological Risk Assessment of Sediments*, Ingersoll, C.G., Dillon, T., and Biddinger, G.R., Eds., SETAC Press, Pensacola, FL, 1997, 211.
29. Ankley, G.T. et al., Technical basis and proposal for deriving sediment quality guidelines for metals, *Environ. Toxicol. Chem.*, 15, 2056, 1996.
30. Berry, W.J. et al., Predicting the toxicity of metal-spiked laboratory sediments using acid-volatile sulfide and interstitial water normalizations, *Environ. Toxicol. Chem.*, 15, 2067, 1996.
31. Green, A.S., Chandler, G.T., and Blood, E.R., Aqueous-, pore-water-, and sediment-phase cadmium: toxicity relationships for a meiobenthic copepod, *Environ. Toxicol. Chem.*, 12, 1497, 1993.
32. Kemp, P.F. and Swartz, R.C., Acute toxicity of interstitial and particle-bound cadmium to a marine infaunal amphipod, *Mar. Environ. Res.*, 26, 135, 1988.
33. DeWitt, T.H et al., Bioavailability and chronic toxicity of cadmium in sediment to the estuarine amphipod *Leptocheirus plumulosus*, *Environ. Toxicol. Chem.*, 15, 2095, 1996.
34. Turner, A., Trace-metal partitioning in estuaries: importance of salinity and particle concentration, *Mar. Chem.*, 54, 27, 1996.

35. International Atomic Energy Agency, Sediment K_ds and Concentration Factors for Radionuclides in the Marine Environment, International Atomic Energy Agency, Vienna, 1985, 73.
36. Fisher, N.S. and Reinfelder, J.R., The trophic transfer of metals in marine systems, in *Metal Speciation and Bioavailability in Aquatic Systems*, Tessier, A. and Turner, D.R., Eds., John Wiley & Sons, New York, 1995, 363.
37. Lopez, G.R. and Levinton, J.S., Ecology of deposit feeding animals in marine sediments, *Q. Rev. Biol.*, 62, 235, 1987.
38. Mayer, L.M., Geochemistry of humic substances in estuarine environments, in *Humic Substances in Soil, Sediment, and Water: Geochemistry, Isolation, and Characterization*, Aiken, G.R. et al., Eds., John Wiley & Sons, New York, 1985, 211.
39. Decho, A.W., Microbial exopolymer secretions in ocean environments: their roles in food webs and marine processes, *Oceanogr. Mar. Biol. Annu. Rev.*, 28, 73, 1990.
40. Fisher, N.S., Bohè, M., and Teyssiè, J.-L., Accumulation and toxicity of Cd, Zn, Ag, and Hg in four marine phytoplankters, *Mar. Ecol. Prog. Ser.*, 18, 201, 1984.
41. Fisher, N.S., On the reactivity of metals for marine phytoplankton, *Limnol. Oceanogr.*, 31, 443, 1986.
42. Ford, T. and Ryan, D., Toxic metals in aquatic ecosystems: a microbiological perspective, *Environ. Health Perspect.*, 103 (Suppl. 1), 1995.
43. Rhoads, D., Organism-sediment relations on the muddy sea floor, *Oceanogr. Mar. Biol. Annu. Rev.*, 12, 263, 1974.
44. Miller, D., Mechanical post-capture particle selection by suspension- and deposit-feeding *Corophium*, *J. Exp. Mar. Biol. Ecol.*, 82, 59, 1984.
45. Harvey, R.W. and Luoma, S.N., Effect of adherent bacteria and bacterial extracellular polymers upon assimilation by *Macoma balthica* of sediment-bound Cd, Zn, and Ag, *Mar. Ecol. Prog. Ser.*, 22, 281, 1985.
46. Decho, A.W. and Luoma, S.N., Humic and fulvic acids: sink or source in the availability of metals to the marine bivalves *Macoma balthica* and *Potamocorbula amurensis*? *Mar. Ecol. Prog. Ser.*, 108, 133, 1994.
47. Schlekat, C.E., Decho, A.W., and Chandler, G.T., Bioavailability of particle-associated silver, cadmium, and zinc to the estuarine amphipod *Leptocheirus plumulosus* through dietary ingestion, *Limnol. Oceanogr.*, 45, 11, 2000.
48. Lee, B.-G. and Luoma, S.N., Influence of microalgal biomass on absorption efficiency of Cd, Cr, and Zn by two bivalves from San Francisco Bay, *Limnol. Oceanogr.*, 43, 1455, 1998.
49. Wang, W.-X. and Fisher, N.S., Assimilation of trace elements and carbon by the mussel *Mytilus edulis*: effects of food composition, *Limnol. Oceanogr.*, 41, 197, 1996.
50. Reinfelder, J.R. and Fisher, N.S., The assimilation of elements ingested by marine copepods, *Science*, 251, 794, 1991.
51. Decho, A.W. and Luoma, S.N., Time-courses in the retention of food material in the bivalves *Potamocorbula amurensis* and *Macoma balthica*: significance to the absorption of carbon and chromium, *Mar. Ecol. Prog. Ser.*, 78, 303, 1991.
52. Decho, A.W. and Luoma, S.N., Flexible digestion strategies and trace metal assimilation in marine bivalves, *Limnol. Oceanogr.*, 41, 568, 1996.
53. Schlekat, C.E., Decho, A.W., and Chandler, G.T., Dietary assimilation of cadmium associated with bacterial exopolymer sediment coatings by the estuarine amphipod *Leptocheirus plumulosus*: effects of Cd concentration and salinity, *Mar. Ecol. Prog. Ser.*, 183, 205, 1999.
54. Luoma, S.N. et al., Metal uptake by phytoplankton during a bloom in South San Francisco Bay, *Limnol. Oceanogr.*, 43, 1007, 1998.

55. Arifin, Z. and Bendell-Young, L.I., Influence of a selective feeding behaviour by the blue mussel *Mytilus trossulus* on the assimilation of ^{109}Cd from environmentally relevant seston matrices, *Mar. Ecol. Prog. Ser.*, 192, 181, 2000.
56. Plante, C. and Jumars, P., The microbial environment of marine deposit-feeding guts characteristics via microelectrodes, *Microb. Ecol.*, 23, 257, 1992.
57. Owen, G., Digestion, in *Physiology of Mollusca*, Wilbur, K.M. and Yonge, C.M., Eds., Academic Press, New York, 1966, 53.
58. Gagnon, C. and Fisher, N.S., The bioavailability of sediment-bound Cd, Co, and Ag to the mussel, *Mytilus edulis*, *Can. J. Fish. Aquat. Sci.*, 54, 147, 1997.
59. Griscom, S.B., Fisher, N.S., and Luoma, S.N., Geochemical influences on assimilation of sediment-bound metals in clams and mussels, *Environ. Sci. Technol.*, 34, 91, 2000.
60. Luoma, S.N. and Jenne, E.A., Estimating bioavailability of sediment-bound trace metals with chemical extractants, in *Trace Substances in Environmental Health: A Symposium*, Hemphill, D.D., Ed., University of Missouri, Columbia, 1976, 343.
61. Luoma, S.N. and Bryan, G.W., A statistical assessment of the form of trace metals in oxidized sediments employing chemical extractants, *Sci. Total Environ.*, 17, 165, 1981.
62. Mayer, L.M et al., Bioavailability of sedimentary contaminants subject to deposit-feeder digestion, *Environ. Sci. Technol.*, 30, 2641, 1996.
63. Chen, Z. and Mayer, L.M., Mechanisms of Cu solubilization during deposit feeding, *Environ. Sci. Technol.*, 32, 770, 1998.
64. Chen, Z. and Mayer, L., Assessment of sedimentary Cu availability: a comparison of biomimetic and AVS approaches, *Environ. Sci. Technol.*, 33, 650, 1999.
65. DiToro, D.M et al., Toxicity of cadmium in sediments: the role of acid volatile sulfide, *Environ. Toxicol. Chem.*, 9, 1487, 1990.
66. Mayer, L.M. et al., Digestive environments of benthic macroinvertebrate guts: enzymes, surfactants and dissolved organic matter, *J. Mar. Res.*, 55, 785, 1997.
67. Decho, A.W. and Moriarty, D.J.W., Bacterial exopolymer utilization by a harpacticoid copepod: a methodology and results, *Limnol. Oceanogr.*, 35, 1039, 1990.
68. Mayer, L.M. et al., Bioavailability of sedimentary contaminants to deposit-feeder digestion, *Environ. Sci. Technol.*, 30, 2641, 1996.
69. Lawrence, A.L. et al., Intestinal solubilization of particle-associated organic and inorganic mercury as a measure of bioavailability to benthic invertebrates, *Environ. Sci. Technol.*, 33, 1871, 1999.
70. Lee, J.S. et al., Influence of reactive sulfide (AVS) and supplementary food on Ag, Cd and Zn bioaccumulation in the marine polychaete, *Neanthes arenaceodentata*, *Mar. Ecol. Prog. Ser.*, in press.
71. Oakden, J.M., Feeding and sediment selection in five species of central Californian phoxocephalid amphipods, *J. Crustacean Biol.*, 4, 233, 1984.
72. Smith, G.J. and Flegal, A.R., Silver in San Francisco Bay estuarine waters, *Estuaries*, 16, 547, 1993.
73. Kuwabara, J.S. et al., Trace metal associations in the water column of South San Francisco Bay, California, *Estuarine Coastal Shelf Sci.*, 28, 307, 1989.
74. Rainbow, P.S., Physiology, physiochemistry and metal uptake — a crustacean perspective, *Mar. Pollut. Bull.*, 31, 55, 1995.
75. Reinfelder, J.R. et al., Trace element trophic transfer in aquatic organisms: a critique of the kinetic model approach, *Sci. Total Environ.*, 219, 117, 1998.
76. Luoma, S.N., Can we determine the biological availability of sediment-bound trace elements? *Hydrobiologia*, 176/177, 379, 1989.
77. Wolfe, M.F., Schwarzbach, S., and Sulaiman, R.A., Effects of mercury on wildlife: a comprehensive review, *Environ. Toxicol. Chem.*, 17, 146, 1998.

78. Young, M.L., The transfer of ^{65}Zn and ^{59}Fe along a *Ficus serratus* (L) – *Littorina obtusata* (L) food chain, *J. Mar. Biol. Assoc. U.K.*, 55, 583, 1975.
79. Warren, L.A., Tessier, A., and Hare, L., Modelling cadmium accumulation by benthic invertebrates *in situ*: the relative contributions of sediment and overlying water reservoirs to organism cadmium concentrations, *Limnol. Oceanogr.*, 43, 1442, 1998.
80. Boese, B. et al., Comparison of aqueous and solid-phase uptake for hexachlorobenzene in the tellinid clam *Macoma nasuta* (Conrad): a mass balance approach, *Environ. Toxicol. Chem.*, 9, 221, 1990.
81. Luoma, S.N. and Jenne, E.A., The availability of sediment-bound cobalt, silver, and zinc to a deposit-feeding clam, in *Biological Implications of Metals in the Environment: CONF-750929*, NTIS, Springfield, VA, 1977.
82. Selck, H., Forbes, V.E., and Forbes, T.L., Toxicity and toxicokinetics of cadmium in *Capitella* sp. I: Relative importance of water and sediment as routes of cadmium uptake, *Mar. Ecol. Prog. Ser.*, 164, 167, 1998.
83. Lee, J.S. et al., Influence of acid volatile sulfides and metal concentrations on metal partitioning in contaminated sediments, *Environ. Sci. Technol.*, 34, 4511, 2000.
84. Harvey, R.W. and Luoma, S.N., Separation of solute and particulate vectors of heavy metal uptake in controlled suspension-feeding experiments with *Macoma balthica*, *Hydrobiologia*, 121, 97, 1985.
85. Jennings, J.R. and Rainbow, P.S., Studies on the uptake of cadmium by the crab *Carcinus maenas* in the laboratory. I. Accumulation from seawater and a food source, *Mar. Biol.*, 50, 131, 1979.
86. Woodward, D.F. et al., Effects on rainbow trout fry of a metals-contaminated diet of benthic invertebrates from the Clark Fork River, Montana, *Trans. Am. Fish. Soc.*, 123, 51, 1994.
87. Munger, C. and Hare, L., Relative importance of water and food as cadmium sources to an aquatic insect (*Chaoborus punctipennis*): implications for predicting Cd bioaccumulation in nature, *Environ. Sci. Technol.*, 31, 891, 1997.
88. Roy, I. and Hare, L., Relative importance of water and food as cadmium sources to the predatory insect *Sialis velata* (Megaloptera), *Can. J. Fish. Aquat. Sci.*, 56, 1143, 1999.
89. Landrum, P.F., Lee, H., II, and Lydy, M.J., Toxicokinetics in aquatic systems: model comparisons and use in hazard assessment, *Environ. Toxicol. Chem.*, 12, 1709, 1992.
90. Tessier, A. et al., Modelling Cd partitioning in oxic lake sediments and Cd concentrations in the freshwater bivalve *Anadonta grandis*, *Limnol. Oceanogr.*, 38, 1, 1993.
91. Hare, L. and Tessier, A., Predicting animal cadmium concentrations in lakes, *Nature*, 380, 430, 1996.
92. Lee, B.-G., Wallace, W.G., and Luoma, S.N., Uptake and loss kinetics of Cd, Cr, and Zn in the bivalves *Potamocorbula amurensis* and *Macoma balthica*: effects of size and salinity, *Mar. Ecol. Prog. Ser.*, 175, 177, 1998.
93. Wang, W.-X., Fisher, N.S., and Luoma, S.N., Kinetic determinations of trace element bioaccumulation in the mussel *Mytilus edulis*, *Mar. Ecol. Prog. Ser.*, 140, 91, 1996.
94. Fisher, N.S. and Wang, W.-X., Trophic transfer of silver to marine herbivores: a review of recent studies, *Environ. Toxicol. Chem.*, 17, 562, 1998.
95. Wang, W.-X. et al., Assimilation and regeneration of trace elements by marine copepods, *Limnol. Oceanogr.*, 41, 70, 1996.
96. Wang, W.-X. and Fisher, N.S., Assimilation efficiencies of chemical contaminants in aquatic systems: a synthesis, *Environ. Toxicol. Chem.*, 18, 2034, 1999.

97. Munger, C. and Hare, L., Influence of ingestion rate and food types on cadmium accumulation by the aquatic insect *Chaoborus*, *Can. J. Fish. Aquat. Sci.*, 57, 327, 2000.
98. Rouleau, C., Gobiel, C., and Tjälve, H., Accumulation of silver from the diet in two marine benthic predators: the snow crab (*Chinoecetes oplio*) and American plaice (*Hippoglossoides platessoides*), *Environ. Toxicol. Chem.*, 19, 631, 2000.
99. Smokorowski, K.E., Lasenby, D.C., and Evans, R.D., Quantifying the uptake and release of cadmium and copper by the opossum shrimp *Mysis relicta* preying upon the cladoceran *Daphnia magna* using stable isotope tracers, *Can. J. Fish. Aquat. Sci.*, 55, 909, 1998.
100. Timmermans, K.R. et al., Cadmium and zinc uptake by two species of aquatic invertebrate predators from dietary and aqueous sources, *Can. J. Fish. Aquat. Sci.*, 49, 655, 1992.
101. Wang, W.-X., Stupakoff, I., and Fisher, N.S., Bioavailability of dissolved and sediment-bound metals to a marine deposit-feeding polychaete, *Mar. Ecol. Prog. Ser.*, 178, 281, 1999.
102. Skorupa, J., Selenium poisoning of fish and wildlife in nature: lessons from twelve real-world examples, in *Environmental Chemistry of Selenium*, Frankenberger, J., William, T., and Engberg, R.A., Eds., Marcel Dekker, New York, 1998, 315.
103. Lemly, A.D., Pathology of selenium poisoning in fish, in *Environmental Chemistry of Selenium*, Frankenberger, J., William, T., and Engberg, R.A., Eds., Marcel Dekker, New York, 1998, 281.
104. Schlekat, C.E. et al., Bioavailability of particle-associated Se to the bivalve *Potamocorbula amurensis*, *Environ. Sci. Technol.*, 34, 4504, 2000.
105. Cutter, G.A., The estuarine behavior of selenium in San Francisco Bay, *Estuarine Coastal Shelf Sci.*, 28, 13, 1989.
106. Wang, W.-X. and Fisher, N.S., Accumulation of trace elements in a marine copepod, *Limnol. Oceanogr.*, 43, 273, 1998.
107. Johnston, N.T. and Lasenby, D.C., Diet and feeding of *Neomysis mercedis* (Crustacea, Mysidacea) from the Fraser River Estuary, British Columbia, *Can. J. Zool.*, 60, 813, 1982.
108. Weeks, J.M. and Rainbow, P.S., Interspecific comparisons of relative assimilation efficiencies for zinc and cadmium in an ecological series of talitrid amphipods (Crustacea), *Oecologia*, 97, 228, 1994.
109. Wang, W.X., Griscom, S.B., and Fisher, N.S., Bioavailability of Cr(III) and Cr(VI) to marine mussels from solute and particulate pathways, *Environ. Sci. Technol.*, 31, 603, 1997.
110. Mason, A.Z. and Jenkins, K.D., Metal detoxification in aquatic organisms, in *Metal Speciation and Bioavailability in Aquatic Systems*, Turner, K.D. and Tessier, A., Eds., John Wiley & Sons, New York, 1995, 479.
111. Wallace, W.G. and Lopez, G.R., Relationship between subcellular cadmium distribution in prey and cadmium trophic transfer to a predator, *Estuaries*, 19, 923, 1996.
112. Borgmann, U. et al., Toxicity and bioaccumulation of thallium in *Hyalella azteca*, with comparison to other metals and prediction of environmental impact, *Environ. Pollut.*, 99, 105, 1998.
113. Fisher, N.S. and Hook, S.E., Silver accumulation and toxicity in marine and freshwater zooplankton, in *Proceedings: Fifth International Conference on the Transport, Fate, and Effects of Silver in the Environment*, Hamilton, Ontario, Canada, 1997.

114. Schmittschmitt, J.P., Shaw, J.R., and Birge, W.J., The effects of silver on green algae and prospects for trophic transfer, in *Proceedings: Fourth International Conference on the Transport, Fate, and Effects of Silver in the Environment*, Madison, WI, 1996.
115. Wallace, W.G. et al., Alterations in prey capture and induction of metallothioneins in grass shrimp fed cadmium-contaminated prey, *Environ. Toxicol. Chem.*, 19, 962, 2000.
116. Reinfelder, J.R et al., Assimilation efficiencies and turnover rates of trace elements in marine bivalves — a comparison of oysters, clams and mussels, *Mar. Biol.*, 129, 443, 1997.

8 Endocrine Disruption in Fishes and Invertebrates: Issues for Saltwater Ecological Risk Assessment

Kenneth M.Y. Leung, James R. Wheeler, David Morritt, and Mark Crane

CONTENTS

8.1 Introduction 190
8.2 Effects of Endocrine Disrupting Chemicals on Saltwater Fishes and Invertebrates 190
    8.2.1 Fishes 190
        8.2.1.1 Modes of Action 191
        8.2.1.2 Effects of EDCs on Fishes 191
        8.2.1.3 Limitations of Current Approaches 193
    8.2.2 Invertebrates 194
        8.2.2.1 Modes of Action 194
        8.2.2.2 Effects of EDCs on Aquatic Invertebrates 195
        8.2.2.3 Limitations of Current Approaches 198
8.3 Developing a Coherent and Cost-Effective Risk Assessment Strategy for Saltwater Endocrine Disrupters 199
    8.3.1 Prospective Risk Assessment 199
        8.3.1.1 Structure–Activity Relationships 199
        8.3.1.2 Molecular and Biochemical Techniques 200
        8.3.1.3 Toxicity Testing for EDCs with Saltwater Organisms 200
        8.3.1.4 Protection of Aquatic Assemblages: TBT Case Study 202
    8.3.2. Retrospective Risk Assessment 203
        8.3.2.1 Assessment of EDCs by Field Monitoring 205
            8.3.2.1.1 Morphological Indicators and Biomarkers 205
            8.3.2.1.2 *In Situ* Bioassays 205
            8.3.2.1.3 Population and Assemblage Monitoring 205
8.4 Conclusions 206
References 208

8.1 INTRODUCTION

This chapter considers some of the issues associated with risk assessment of endocrine-disrupting chemicals (EDCs) in the saltwater environment. Endocrine disrupting chemicals have been defined in the following way: "An endocrine disrupter is an exogenous substance that causes adverse health effects in an intact organism, or its progeny, secondary to changes in endocrine function."[1] In other words, an EDC is a substance that *interacts* with an animal's endocrine system, thereby altering processes under hormonal control. These substances, as a class, were first linked to potentially widespread reproductive and developmental disorders in both humans and wildlife in the early 1990s,[2] although earlier studies had also implicated other environmental pollutants as a cause of reproductive failure (e.g., see Reference 3). Over the past decade there has been considerable interest in methods to measure the biological effects of potential EDCs.[1,4-12]

Much of the early interest in EDCs focused on vertebrates, but this bias has become less acute recently, with greater consideration of potential EDC effects on invertebrates, including those found in saltwater systems.[13] Fear of widespread and possibly severe EDC effects on saltwater wildlife, after recognition of worldwide problems with tributyltin (TBT), has stimulated funding for research programs across the globe. In Europe, there has been the development of laboratory toxicity testing protocols with marine copepods and funding for surveys of coastal waters.[13] In North America, both the U.S. Environmental Protection Agency (U.S. EPA) and Environment Canada have introduced regulations and research programs to quantify EDC effects in saltwater systems.[4,13,14] In other regions and countries, such as Japan, interest in issues such as the effect of TBT on marine gastropods, remains high and attracts research funding. Research on endocrine disruption may therefore be one of the few areas of ecotoxicological research in which saltwater environments could become as well investigated as freshwater environments.

This chapter reviews current knowledge about the modes of action in, and effects of EDCs on, saltwater fishes and invertebrates. It also identifies some limitations in current approaches and argues for development of a wider array of screening tools, plus greater investment in monitoring of saltwater systems for EDC effects. We argue that the peculiar nature of EDCs and their potential biological effects require far greater emphasis on environmental monitoring than is normally the case with other chemical substances discharged into saltwater habitats.

8.2 EFFECTS OF ENDOCRINE DISRUPTING CHEMICALS ON SALTWATER FISHES AND INVERTEBRATES

8.2.1 FISHES

Scientists and environmental regulators in the United Kingdom were first alerted to the possibility that chemical contaminants were affecting normal endocrine function in fishes by the appearance of intersexuality in some common riverine species. Alarmingly, such effects were confirmed on a national scale using caged rainbow trout.[15] Consequently, considerable research effort was expended to identify and assess the impacts of these contaminants.[16–18] As in most areas of environmental toxicology, emphasis was, and currently still is, firmly focused on freshwater species, resulting in relatively little data concerning marine and estuarine species.[19] This is of concern as estuaries are likely to have high contamination levels due to the historical location of industries in these areas, with associated adverse biological effects. Indeed, this is borne out by a recent study of flounder (*Platichthys flesus*) in the United Kingdom in which fish exhibited a variety of responses associated with endocrine disruption in eight out of ten estuaries surveyed.[20] This initial study raised the profile of endocrine disruption studies in saltwater fish species, encouraging further estuarine and marine surveys and the development of test methods. One of these initiatives is a major new European research program, Endocrine Disruption in the Marine Environment (EDMAR).

The purpose of this section is to present a selection of the major biological effects of EDCs that have been observed in saltwater fish species. Major end points measured in fishes are the occurrence of intersex; effects on gonad growth, sex steroid levels, sperm motility, and metabolism; induction of egg yolk protein (vitellogenin); and gross indicators of fecundity.

8.2.1.1 Modes of Action

The fish reproductive endocrine system is complex, and mediated by several hormones interacting with several discrete tissues. Consequently, it is susceptible to disruption at one or more stages.[21] EDCs interfere with normal hormonal processes and regulation in one of two ways:

1. Agonistic or estrogenic substances, such as the alkylphenols, can bind to hepatic estrogen receptors mimicking natural endogenous estrogens. This can have the effect of feminizing male fish or altering the normal hormonal control in females. An agonist may also compete with the natural estrogen, estradiol, for pituitary-hypothalamic feedback receptors that regulate egg development.
2. Antagonistic or antiestrogenic substances, such as the phenylethylenes, may block hepatic estrogen receptor sites, preventing the normal interaction of estradiol. In addition, other interactions may occur, affecting the synthesis and metabolism of hormones[22] and alteration of hormone receptor levels.[23]

8.2.1.2 Effects of EDCs on Fishes

The agonistic (estrogenic) process outlined above has been shown to directly affect fish tissues and normal development. A recent survey in Japan has implicated salt waters known to be contaminated with nonylphenol and sewage effluent in causing intersex in the flounder *Pleuronectes yokohame*. Some 15% of males sampled off Haneda, in Tokyo Bay, contained primary egg cells in their testicular tissue.[24] Similarly, egg cell growth has been observed in the testes of the native flounder *P. flesus* in British estuaries.[20] Decreased testicular growth has also been observed in response to EDC exposure of the freshwater rainbow trout, *Oncorhynchus mykiss*.[18] A common measurement end point in these studies is the gonadal somatic index (GSI), where the weight of the gonads is expressed as a percentage of the total body weight.[25] Although this is a useful measure of effect, it must be remembered that, unlike intersex, there is not a direct causal relationship between decreased GSI and endocrine disruption per se.

Many test systems measure the concentrations of sex steroid levels and compare these to levels expressed in control animals. Typically the fish estrogen 17β-estradiol and the androgen 11-ketotestosterone are measured. Antiestrogenic effects have been demonstrated by dietary exposure of flounder to polycyclic aromatic hydrocarbons. Phenanthrene and chrysene did not cause any morphological changes, but a dose-dependent decrease in plasma 17β-estradiol levels was recorded.[26]

Fish sperm have also been used as an end point to assess the effect of EDCs on fish reproduction. The quality and quantity of sperm are dependent upon hormonal control and consequently can provide a useful measure of endocrine disruption. Kime and Nash[27] have developed methods to assess the number, duration, and velocity of sperm cells. However, these data do not provide direct evidence of fertilization rates. More recently, a technique developed to measure the metabolic activity of mammalian sperm has been adapted to measure sperm fertilization capacity in marine fishes.[28] The system uses a redox dye, resazurin, to measure dehydrogenase activity. Hamoutene et al.[28] were able to measure decreases in spermatozoan metabolism after exposure to tributyltin in the capelin (*Mallotus villosus*) as well as in two invertebrate species. This may provide an effective tool in establishing adverse effects on reproductive effects beyond sperm motility. Nonetheless, it is worthy of note that other factors such as nutritional status and disease-related deformity of reproductive systems may also influence sperm quality and quantity.

Vitellogenin (Vtg), the fish egg yolk precursor protein, has been extensively used as a biomarker response to EDCs in both monitoring and laboratory testing. Vtg is normally produced by liver cells of female fish in response to estradiol that has been secreted by the pituitary gland. It is released into the blood plasma where it circulates until reaching the ovaries, where it is then taken up by the developing oocytes. Interestingly, male fish also carry the Vtg gene, although, because circulating levels of estrogen are very low in male blood plasma, the Vtg protein is not expressed.[29] However, the capability of males to express Vtg remains, and male fish are known to produce the protein under the influence of EDCs.[30] There is great interest in the use of Vtg expression as a quantifiable end point in hazard identification programs[31,32] and in standard bioassay protocols[33] because male Vtg induction is a clear-cut measure of estrogenic stimulation. Although most Vtg studies have been performed on freshwater species, there have been some recent studies with marine species.[34,35] Again, most have used flounder.[24,36-38] For example, Lye et al.[38] demonstrated elevated levels of serum Vtg associated with high levels of testicular abnormalities in flounder from the Tyne estuary (northern England). A later study[39] suggested that the cause could be the biodegradation products of some nonionic surfactants, such as alkylphenols and alkylphenol monoethoxylates, accumulated in the tissues of mature male flounder. One of the few saltwater fishes used as a standard test species is the sheepshead minnow (*Cyprindon variegatus*), for which a Vtg induction test for males has been developed. A comparative test for the estrogenicity of three compounds showed that the assay clearly demonstrates dose dependency.[40] In summary, the Vtg biomarker response has been shown to be a sensitive tool in establishing estrogenic responses to EDCs in freshwater fish, and recently some saltwater species methods have become available, particular for flounder species.

Other gross reproductive end points have also been used in a variety of test systems, i.e., subchronic, chronic, and full life-cycle tests. End points assessed include number of eggs, embryo survival, time to hatch, and fry/juvenile survival. For example, the effect of bistributyltin oxide on the life cycle of the sheepshead minnow has been investigated.[41] Exposure effects on hatch rate, growth, and reproductive success were measured in different generations. Significant mortality and reduced growth were observed in the embryos and juveniles of the F_0 generation, while fecundity (number of viable eggs) was unaffected in all treatments. All the measured end points indicated no effect in the F_1 generation.

8.2.1.3 Limitations of Current Approaches

Although all of the effects mentioned above have serious consequences at the level of the individual organism, it is still unclear what the ecological effects of EDC exposure may be for populations or assemblages of fish species. Further work is necessary to relate these end points to significant higher-level effects.[42,43] This would help reduce uncertainty in decision making during ecological risk assessment.[44] There is a need to develop higher-tier fish tests, possibly through multigenerational experiments with ecologically relevant end points,[45] although these are technically difficult and expensive to perform with fishes. It may therefore be necessary to modify experimental designs so that species are tested at particular life-stage events, such as sexual differentiation, which may be most sensitive to EDCs.[31–32,45] There has been much interest in using monosex cultures of fishes so tests may be completed before sexual maturity.[46,47] In addition, a fuller understanding of the consequences and associated threshold levels of Vtg induction may provide an effective biomarker approach for studying EDC-mediated reproductive impairment.

Furthermore, there is the need to establish a robust rationale for extrapolating from standard test species to species in wild fish populations. This may be of particular importance because, to date, examination of the effects of EDCs on saltwater fish has focused on very few species. In comparison to the array of common freshwater test species, saltwater species are grossly underrepresented. Expansion of test methods and species may be necessary to take into account interspecies differences in hormonal mechanisms, including those that control sexual differentiation, which may be affected by EDCs.[48] This potential interspecific difference in mechanisms also reinforces the need at this stage of test development for saltwater species-specific data, rather than reliance on simple extrapolation from freshwater responses. In addition, it has been suggested that a carefully selected set of saltwater fish species, for which basic endocrinology is understood, should be incorporated into standardized test guidelines.[31]

In conclusion, it is clear that EDCs are having pronounced effects on individual fishes in the saltwater environment. There are test methods in place (but not international standards), although far fewer than are available for freshwater species. Further marine tests need to be developed, with a variety of test species, to address some of these significant gaps in our understanding. However, neither freshwater nor saltwater methods for detecting EDC effects in fishes have yet been linked to significant ecological effects. Whether demonstration of such a link is necessary for decision making within an ecological risk assessment framework is likely to be a political rather than a scientific decision.

8.2.2 INVERTEBRATES

Invertebrates constitute 95% of all species in the animal kingdom and they are key components of marine and estuarine ecosystems.[49] The potential impact of EDCs on these aquatic invertebrates must be investigated and assessed to safeguard biodiversity and ecosystem sustainability. There are over 19 different phyla of invertebrates present in estuarine and marine environments.[50] Such a phylogenetically diverse fauna has widely differing endocrine systems, which are likely to be affected differently by potential EDCs. In lower invertebrates, for example, the sponges, there are no classical endocrine glands, as these animals do not possess neurons or neurosecretory cells, whereas hydrozoans (coelenterates) have neurosecretory cells whose activity is associated with normal growth, asexual reproduction, and regeneration.[51] In contrast, there are relatively well-developed nervous, circulatory, neuroendocrine, and endocrine systems present in the higher invertebrates such as annelids, insects, mollusks, and crustaceans. The endocrine systems of insects are the most widely studied and described, but there is only sparse information on the endocrinology of the other phyla.[52] The rather fragmentary knowledge of invertebrate endocrinology often prevents an adequate understanding of the mechanisms involved in chemically mediated endocrine disruption,[53] and also makes risk assessment of EDCs difficult for aquatic invertebrates.

8.2.2.1 Modes of Action

The effects of endocrine disrupters on aquatic invertebrates can be due to several different processes.

1. Disruption in the levels of sex-associated hormones, e.g., increased testosterone and decreased estradiol tissue levels in estuarine clams, *Ruditapes decussata*, and in freshwater mussels, *Mariso cornuarietis*, after exposure to TBT.[54,55]
2. Interference with steroid metabolism, e.g., reduced metabolic clearance of testosterone in *Daphnia magna* by exposure to diethylstilbestrol and 4-nonylphenol, respectively,[56] and increased production of oxido-reacted derivatives of testosterone in *D. magna* by TBT.[57] Exposure to TBT can also result in increased testosterone in marine neogastropods such as *Nucella lapillus* and *Hinia reticulata* because TBT may inhibit the normal function of a cytochrome P-450-dependent aromatase and thus reduce the normal conversion of testosterone to estradiol.[58]

3. Interference with sex determination and development of secondary sex characteristics, as in the widely reported occurrence of imposex or intersex in marine gastropods such as *N. lapillus* and *Littorina littorea* exposed to water contaminated with TBT (e.g., see Reference 59). In crustaceans, sex ratio was altered in *Daphnia* spp. exposed to 4-nonylphenol,[60] while exposure to diethylstilbestrol or methoprene stimulated development of the abdominal process in female *D. magna* and exposure to androstenedione stimulated development of the first antennae in male *D. magna*.[61] In the amphipod *Corophium volutator*, an increase in the length of the second antennae of the male was observed when animals were exposed to 4-nonylphenol.[62]
4. Possible developmental effects in embryonic and larval stages. For example, exposure to pentachlorophenol resulted in abnormal embryonic development of sea urchin, *Paracentrotus lividus*.[63] Larval development of the estuarine shrimp, *Palaemonetes pugio*, was inhibited by the pesticide methoprene, which is thought to mimic the action of the steroid juvenile hormone.[64] Similarly, larval development to D-shape was delayed in the oyster, *Crassostrea gigas*, by exposure to waterborne 4-nonylphenol.[65]
5. Inhibition of molting hormones (ecdysteroids) and thus reduction in molting success in crustaceans, e.g., barnacles, *Balanus amphitrite*, exposed to cadmium or 4-nonylphenol[66,67] and *D. magna* exposed to PCBs, diethylphthalate, diethylstilbestrol, and endosulfan.[68,69]
6. Possible reductions in growth and reproductive success. For example, exposure to 4-nonylphenol resulted in reduced survivorship of offspring, depressed population growth, and reduced egg production in the copepod *Tisbe battagliai*.[70] Nonylphenol also caused reduced egg viability in the polychaete *Dinophilus gyrociliatus*.[71] Shell growth in bivalve mollusks is affected by TBT.[72–75]
7. Other potential effects such as interference with metabolic activity, e.g., increased levels of nitrogen oxide in the hemolymph of mussels, *Mytilus edulis*, after exposure to TBT,[76] and inhibited metabolic activity in freshwater mussels, *Elliptio complanata*, exposed to estrogen mimics.[77]

8.2.2.2 Effects of EDCs on Aquatic Invertebrates

Potential endocrine disruption has been reported in aquatic invertebrates (Copepoda, Crustacea, Echinodermata, Mollusca, Annelida, and Insecta), mainly based on laboratory studies.[43,78] Most recent literature on EDC effects on saltwater invertebrates is summarized in Table 8.1 (readers should refer to DeFur et al.[6] for a more comprehensive review of older literature). Table 8.1 shows that TBT and 4-nonylphenol are the most frequently studied chemicals, with Crustacea and Mollusca the most common phyla involved in laboratory tests for EDCs.

In addition to these laboratory tests, field investigations on naturally occurring aquatic invertebrates showed that the effects of EDCs can extend to the population level. Classic examples are the remarkable reductions in oyster *C. gigas* and dogwhelk *N. lapillus* populations caused by exposure to organotin compounds leached

TABLE 8.1
A Summary of the Effects of Environmental EDCs on Saltwater Invertebrates

Taxonomic Group	Species	Test Chemical (effective concentration)	Effects	Ref.
Echinodermata				
Seastar	*Asterias rubens*	Cadmium (25 µg/l); or fed with mussels containing 26 µg PCBs/g lipid	Reduced progesterone and testosterone levels in the pyloric caeca; increased testosterone level in the gonads and decreased cytochrome P-450 and cytochrome b5 in pyloric caeca microsomes	89
		Cadmium (100 µg/l)	Influenced the sterol composition and reduced the sterol/phospholipid ratio	90
Sea Urchin	*Paracentrotus lividus*	Tributyltin (EC_{50}: 3.4–4.7 µg/l)	Decrease in the cleavage rate; reduced production of DNA and echinochrome	91
Mollusca				
Gastropod	*Nassarius obsoletus*	Tributyltin (field study)	Females developed imposex (i.e., pseudohermaphroditic condition)	92
	Nucella lapillus	Tributyltin (2 ng/l; 1 year)	Females developed imposex and lost weight	93
		Tributyltin (>1 ng/l)	Females could be sterilized; this may result in collapse or extinction of population	80, 81
		Tributyltin (40 ng Sn/l)	Increased testosterone titers together with an increase in penis length in imposex females	94
	Lepsicilla scokina	Tributyltin (0.01 µg/l)	Females developed imposex	95
	Hinia reticulata	Tributyltin (5–100 ng Sn/l)	Increased testosterone titers together with an increase in penis length in imposex females	58

Group	Species	Chemical (dose)	Effect	Ref
Bivalve	Ruditapes decussata	Tributyltin (24 ng Sn/l)	Increased testosterone titers by 30% and decreased estradiol levels	54
	Crassostrea gigas	4-Nonylphenol (0.1 µg/l)	Delayed larval development to D-shape and reduced survival of the larvae	65
	Mytilus edulis	4-Nonylphenol (56 µg/l)	Reduced byssus strength and reduced scope for growth	96
	M. edulis	Tributyltin (2.3 ng Sn/l)	Reduced shell growth of post larvae	75
Crustacea				
Barnacle	Balanus amphitrite	4-Nonylphenol (0.1 µg/l)	Inhibited larval settlement	97
		4-Nonylphenol (1.0 µg/l)	Increased in the level of cypris major protein	66
		Cadmium (0.25 mg/l)	Reduced molting success of stage II larvae and inhibited larval settlement	67
		Cadmium (0.1 mg/l) or Phenol (10 mg/l)	Inhibited larval settlement	98
Amphipod	Corophium volutator	4-Nonylphenol (10 µg/l)	Reduced survival and growth, but increased fertility of females; males developed longer second antennae	62
Mysid	Americamysis bahia	Methoprene (2–8 µg/l)	Delayed the release of first brood and reduced number of young produced per female	99
Decapod	Palaemonetes pugio	Methoprene (1 mg/l)	Inhibited larval development	64
		Endrin (0.03 mg/l)	Delayed the onset of spawning and reduced viability of embryo	100
Copepod	Tisbe battagliai	4-Nonylphenol (20–41 µg/l)	Reduced survival of offspring, population growth, and egg production	70

from antifouling paints during the mid-1970s to early 1980s.[73,74] Marine antifouling paints, containing organotin compounds, were first introduced in the mid-1960s and became widely used because of their effectiveness.[79] These organotins, particularly TBT, are highly toxic to aquatic animals. Concentrations of TBT exceeding 2 ng/l were responsible for shell calcification anomalies in *C. gigas*, while higher TBT levels (≥20 ng/l) reduced the reproductive success of bivalve mollusks.[74] At 1 to 2 g TBT/l, female *N. lapillus* developed imposex, and they were effectively sterilized by blockage of the oviduct at concentrations above 3 ng/l, leading to population decline and even local extinction.[80,81] High levels of TBT were present in European coastal waters (50 to 1000 ng/l) before implementation of restrictions on the use of TBT, so significant declines of oyster and dogwhelk populations associated with TBT contamination were noticed in France and the United Kingdom.[73,80] Similar population declines of the clam *Scrobicularia plana* attributed to TBT were also noticed in the United Kingdom during the same period.[82,83] To reduce the impact of TBT on the environment, during the period 1982 to 1989 countries including France, the United Kingdom, the United States, Australia, Japan, and Canada subsequently banned the use of TBT-based marine antifouling paints for boats under 25 m.[73] This ban on TBT use in antifouling paints was an effective way of reducing TBT inputs in coastal environments, and resulted in the recovery of oyster spatfall of *C. gigas* in France[73] and of populations of *N. lapillus* throughout European waters.[84-86]

Another example of apparent endocrine disruption in naturally occurring aquatic invertebrates was reported by Moore and Stevenson,[87] who discovered abnormal levels of intersexuality in marine harpacticoid copepods along sewage-contaminated coasts of Scotland. Recently, Gross et al.[88] reported that there was a significantly higher incidence of abnormal oocyte development in female freshwater amphipods *Gammarus pulex* collected from sites below sewage treatment works. Water from the same site is known to elicit high estrogenic responses in vertebrates. Similar studies on saltwater amphipods such as *Corophium volutator* would be of interest.

These field observations indicate that chemically mediated endocrine disruption already occurs in aquatic invertebrates and such effects should not be ignored.[20]

8.2.2.3 Limitations of Current Approaches

There are several unanswered questions regarding endocrine disruption in aquatic invertebrates. First, inter- and intraspecific differences appear to exist in organism responses to the same EDC. Evans et al.[101] recently showed that female *N. lapillus* developed imposex after exposure to nonylphenol, although another study demonstrated *inhibition* of imposex development in the same species caused by estrogens.[58] Exposure to monophenyltin caused an increase in the penis length of imposex female *Ocenebra erinacea* collected from Torquay in the United Kingdom, but a decrease in length in those collected from the Solent, also in the United Kingdom.[102] These studies indicate that the effects of EDCs on invertebrates can be very unpredictable, and raise a question about whether toxicity test results based on a single species can represent the responses of the remaining untested species (or phyla).

Another important consideration is that natural factors may cause endocrine disruption in animals. For example, sexual development in neogastropods can be affected by the presence of parasites (e.g., trematode larvae).[101] In addition, other natural compounds such as natural estrogens (e.g., 17β-estradiol), phytoestrogens (e.g., genistein), and mycoestrogens (e.g., Zearalenol) may also elicit endocrine-disrupting effects on aquatic invertebrates.[103] Particularly important, and not always fully appreciated, is that the effects of EDCs on endocrine systems in invertebrates could be very different from those occurring in vertebrates. For example, injection of steroids, which had pronounced androgenic effects on vertebrates, had no effect on crustaceans.[43]

Certainly, endocrine disruption caused by a combination of two or more EDCs could be antagonistic or synergistic. For example, the adverse effects of TBT on sexual development in female neogastropods were enhanced by addition of testosterone in the test water, but were inhibited by adding estrogen.[58] Apart from TBT, other chemicals such as copper can also induce imposex in *Lepsiella vinosa*.[104] However, there is very little information on the combined effects of different EDCs on aquatic invertebrates and, as the presence of multiple EDCs in the environment is the norm, this may make interpretation of field-monitoring studies more difficult.

8.3 DEVELOPING A COHERENT AND COST-EFFECTIVE RISK ASSESSMENT STRATEGY FOR SALTWATER ENDOCRINE DISRUPTERS

The Endocrine Disrupter Screening and Testing Advisory Committee (EDSTAC)[8] of U.S. EPA and the European Center for Ecotoxicology and Toxicology (ECETOC)[7] both agree with the concept of a tiered evaluation program for EDCs, in which prioritization and initial assessment are performed on the basis of short-term, less-complex screening and testing protocols.[7,105] In addition, ECETOC also proposed that emphasis should be placed on establishing appropriate triggers for the conduct of higher-tier, long-term, and complex tests.[105] In this section, we outline some of the issues presented when testing for EDCs during risk assessment of new chemicals in the laboratory (prospective risk assessment) and when attempting to identify EDCs that are already impacting the environment (retrospective risk assessment).

8.3.1 PROSPECTIVE RISK ASSESSMENT

8.3.1.1 Structure–Activity Relationships

The cost of running long-term ecotoxicity tests to examine potential EDC effects on hormonally mediated development and reproduction may be very expensive. There is therefore a pressing need for more rapid approaches that can be used to screen out chemicals that are of no concern, yet reliably identify potential EDCs for further testing in a higher tier of risk assessment.

Information on chemical structure and effects can be used to establish structure–activity relationships (SARs)[106] that allow prediction of the toxic effects of new chemicals based on their structure alone. Recent development of a three-dimensional SAR can predict the estrogenicity of alkylphenolic compounds, based on whether there is gene activation of the estrogen receptor by the test chemical.[107] Further advances in characterizing different hormone receptors in fish and invertebrates, as well as their potential reaction with chemical functional groups, would enhance the usefulness of SAR as a screening tool for endocrine disrupters. However, this is likely to involve considerable toxicity testing to develop relationships between chemical structure and toxicity in a diversity of marine organisms. Because of this, the effective use of SARs to predict EDC effects on all susceptible marine organisms probably lies some distance in the future.

8.3.1.2 Molecular and Biochemical Techniques

Ingersoll et al.[78] point out that substances can only be identified positively as EDCs through knowledge of their mode of action. Molecular and biochemical techniques are indispensable for this purpose, and there is the potential to develop them into screening tools, as is increasingly the case for the Vtg assay in fishes. It has also been suggested that biomarkers of endocrine disruption should be developed for aquatic invertebrates, especially if a change in the level of the biomarker can be linked to effects at the population level.[34,43,45] Oberdorster et al.[108] have developed Vtg antibodies, based on the vitellin purified from grass shrimp *Palaemonetes pugio*, which can be used to detect levels of lipovitellin in various Crustacea. Cyprid major protein (CMP), produced by barnacles during their development from nauplii to cyprids, has also been suggested as a potential biomarker for EDCs, because CMP is associated with cyprid settlement and metamorphosis.[66,97] *In vitro* screening tests such as the yeast two-hybrid assay,[109,110] and reporter gene or protein-binding assays[111,112] are additional approaches available for testing of estrogen mimics. In the near future, similar assays are likely to be developed for other types of EDCs to allow screening of new chemicals for both estrogenic and other EDC properties. If positive results were obtained in SAR and/or *in vitro* screening tests, a higher-tier *in vivo* chronic test would be triggered.

Use of standard laboratory toxicity tests with validated biomarkers incorporated into normal procedures could be useful. For example, increased levels of Vtg in the blood or liver of male fish, such as flounder or sheepshead minnow, exposed during standard tests would indicate estrogenic effects.

Eventually, biomarkers might be applied as reliable monitoring tools in laboratory and field situations. However, all these biomarkers, particularly for aquatic invertebrates, remain under trial and require further development and validation before they can be used routinely.

8.3.1.3 Toxicity Testing for EDCs with Saltwater Organisms

Do we need to develop saltwater fish and invertebrate tests for EDCs, particularly if SAR and rapid molecular or biochemical techniques become available? At present, new chemicals that are considered to require toxicity testing are generally tested with standardized *in vivo*, acute toxicity tests using organisms from three different trophic levels (usually a freshwater alga, an invertebrate, and a fish). However, end points such as short-term mortality and growth inhibition cannot indicate the more subtle endocrine disrupting properties of a test chemical that may exert these effects at very low concentrations over prolonged periods. Most aquatic toxicity assessments use water fleas, such as *D. magna*, to represent all aquatic invertebrates. Whether the results of these tests adequately protect against the potential threats of EDCs to other species or phyla of invertebrates is a matter of considerable uncertainty and the subject of much discussion.[6] There is a patented 6-day *Daphnia* reproductive bioassay,[113] which not only tests the overall toxicity of aqueous samples but also indicates the presence of certain endocrine disrupters. The end points of this test include survivorship, number of female/male offspring (sex ratio), number of resting eggs, and number of offspring that display developmental deformities.

Additional ecotoxicity tests are required that ideally encompass a broader range of phyla than those currently employed, in order to reflect the phylogenetic diversity of marine biota. The following criteria have been proposed recently for identifying suitable aquatic species as indicators and test animals for EDCs:[45,78]

1. Common or widespread organisms should be used.
2. Test organisms should be ecologically important.
3. They should be likely to receive significant exposure.
4. Choice of organisms should include a range of lifestyles and feeding habits.
5. The biology of test organisms should be well understood.
6. Organisms should be relatively insensitive to conventional toxicants.
7. Organisms should be experimentally amenable and readily cultured in the laboratory.
8. Organisms should be sedentary or territorial or have a local home range.
9. Operation of the endocrine system of test organisms should be known, at least in part.
10. Organisms should reproduce sexually and preferably show sexual dimorphism.
11. Ideally there should be a rapid generation time (a few weeks).

Ingersoll et al.[78] identified multiple-generation, life-cycle (transgenerational) toxicity tests as the "gold standard" for EDC testing. Such tests were considered most likely to identify compounds with toxicity that *may* be due to endocrine disruption, but it was recognized that not all organisms were amenable to such testing. Transgenerational tests with at least two generations of fish can provide more information on hormonally mediated toxicity, and more relevant end points for prediction of population effects, although these are usually long and costly tests to run. Currently, there is no regulatory test specifically tailored for evaluating the risk of EDCs to saltwater invertebrates, although a three-generation, life-cycle test with the harpacticoid copepod, *Nitrocera spinipes*, is currently under development for this purpose in Europe.[13] Existing toxicity test guidelines that with only minimal modification would allow transgenerational testing of invertebrates are available for only a limited number of marine invertebrate groups, including rotifers, polychaetes, amphipods, the brine shrimp *Artemia*, copepods, mysids, and grass shrimp.[78] None of these groups meets all of the criteria suggested above.[45,78] For example, small crustaceans may have a rapid generation time, but their endocrinology is not particularly well known,[52] and they are likely to be sensitive to non-EDC contaminants.

For short-term EDC screening with higher organisms, monitoring of selected developmental stages of test animals may be more useful. For example, delayed development of oyster larvae to their typical D-shape is caused by exposure to nonylphenol.[65] Measurement of spermatozoan metabolism in aquatic invertebrates exposed to EDCs under laboratory conditions may provide another screening tool.[28] However, for this strategy to be successful, identification of sensitive stages such as these is a research priority.[78] This is likely to involve a substantial program of life-cycle testing with reference compounds exhibiting different modes of toxic action. There are a number of likely end points of interest in association with EDCs and all are known to be under some form of hormonal control. These include sex ratio (of adults or offspring), mating success, egg hatching success and offspring viability, morphological abnormalities, molt frequency and timing, metamorphic or larval developmental success, and pigmentation.[78]

8.3.1.4 Protection of Aquatic Assemblages: TBT Case Study

Once adequate laboratory toxicity test data have been generated for any chemical, including EDCs, the next stage is to assess its hazard to diverse assemblages of organisms in natural habitats. Laboratory toxicity tests are traditionally used to identify the toxic effects of chemicals and to estimate predicted no-effect concentrations (PNEC) that can be compared with predicted environmental concentrations. The results of laboratory toxicity tests on TBT, combined with some field observations on effects, were used to derive a U.K. environmental quality standard (EQS, a synonym for PNEC) for organotin of 2 ng organotin per liter (0.8 ng/l as tin).[34] Usually an EQS or PNEC is based on the lowest reliable chronic toxicity end point derived from the most sensitive test organism, with application of a safety factor. However, this approach may over- or underestimate the true no-effect concentration, and it provides little information on the likely magnitude of toxic effects experienced by assemblages of organisms that are exposed to concentrations above the PNEC.

This problem can be at least partially resolved by using a species sensitivity distribution (SSD) approach. The SSD approach not only considers all available toxicity effects on different species, but also explicitly estimates uncertainties within a toxicity data set. Here, we constructed an SSD for TBT based on chronic and subchronic data available in the literature (Figure 8.1, Table 8.2). Using the bootstrap method recommended by Newman et al.,[114] the HC_5, which is the concentration protecting 95% of species, was estimated as 2 ng of TBT/l (which is the same as the present U.K. EQS). The lower 95% confidence interval of the HC_5, which may be a better estimate of the PNEC, is 0.7 ng of TBT/l.

This PNEC derived from the SSD is more precautionary, uses all of the available test data, and considers variability in the data set. We can also use the SSD in Figure 8.1 to identify the most sensitive phyla (Copepoda and Mollusca in this case) for further study, and estimate possible effects of TBT on whole assemblages. For example, peak values of TBT in water close to a marina off the North Sea coast of Sweden decreased from 706 ng/l in 1988 to 88 ng/l in 1991.[134] We can estimate from the SSD that the percentage of affected species would decrease from 87% in 1988 to 50% in 1991, and test these predictions through

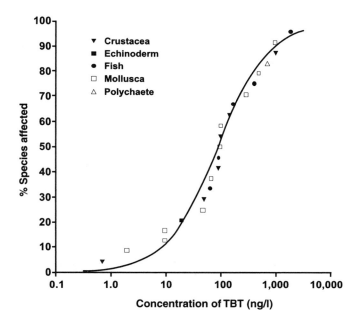

FIGURE 8.1 Species sensitivity distribution of saltwater species to tributyltin, based on chronic or subchronic toxicity data (see Table 8.2). The distribution is fitted to a log–logistic regression ($y = 1/[1 + e^{(-(x - 1.946)/0.466)}]$; $r^2 = 0.964$).

environmental monitoring. SSDs such as this may also help determine cost–benefit trade-offs in removing or reducing environmental contaminants. A similar SSD approach based on acute toxicity data has been successfully applied in a recent ecological risk assessment of TBT in surface waters of Chesapeake Bay.[135] We should, however, sound a note of caution in that reliance on laboratory-derived data to predict field scenarios does not take into consideration the potential (indeed, likely) complex mixture of EDCs and other contaminants and their interactions, nor do they consider contaminant bioavailability.

8.3.2. Retrospective Risk Assessment

It remains an unfortunate truism that, in the main, we find only what we look for. This is undoubtedly true when testing for potential EDCs. It may be possible to develop SARs, design rapid screens, run multigeneration tests with higher organisms, and construct species sensitivity distributions. These approaches will help to quantify the effects of the EDCs that they are designed to detect. However, our poor understanding of nonhuman endocrine systems, particularly those of invertebrates, and of all the modes of action that might occur between EDCs and receptors, means that we cannot be certain that these test systems will detect all EDCs. Negative results from prospective laboratory toxicity tests for EDCs should therefore be treated as null hypotheses, and tested further by examining the results from environmental monitoring programs.

TABLE 8.2
Chronic and Subchronic Toxicity Values of Tributyltin (ng/l; listed in an ascending order) for Different Marine Organisms

Group	Species	Chronic Value (ng/l)	Exposure Duration (days)	Type of Chronic Value	End Point of Toxicity	Ref.
Copepod	Acartia tonsa	0.7	8	EC_{10}	Inhibition of larval development	115
Gastropod	Nucella lapillus	2	365	LOEC	Females lost weight and developed imposex	93
Bivalve	Mytilus edulis	10	60	LOEC	Reduced shell growth (shell width)	116
Bivalve	Ostrea edulis	10	4	LOEC	Reduced digestive cell volume and potentially reduced assimilation and growth	117
Sea urchin	Paracentrotus lividus	20	2	NOEC	Larval development	118
Bivalve	Crassostrea gigas	48	Field study	LOEC	Growth inhibition	119
Amphipod	Gammarus sp.	49	24	NOEC	Growth rate	120
Fish	Morone saxatilis	67	6	LOEC	Reduced larval growth (body depth)	121
Oyster	Saccostrea commercialis	68	Field study	LOEC	Growth inhibition	119
Mysid	Acanthomysis sculpta	90	63	NOEC	Reproduction	122
Fish	Menidia beryllina	93	28	LOEC	Inhibition of larval growth	120
Bivalve	Mercenaria mercenaria	100	8	NOEC	Inhibition of growth and metamorphosis	123
Copepod	Eurytemora affinis	100	13	LOEC	Reduced survival of neonates	124
Bivalve	Scrobicularia plana	102	30	LOEC	Reduced larval shell growth and survivorship	125
Copepod	Temora longicornis	150	14	EC_{50}	Reduced biomass	126
Fish	Oncorhynchus mykiss	180	110	LOEC	Reduced growth of yolk sac fry	127
Amphipod	Gammarus oceanicus	300	56	LOEC	Reduced survivorship	128
Fish	Cyprinodon variegatus	410	145	NOEC	Survivorship of F_0 generation (life-cycle study)	41
Bivalve	Isogno mucum	500	2	LOEC	Alterations in embryogenesis	129
Polychaete	Armandia brevis	700	42	LOEC	Growth inhibition	130
Lobster	Homarus americanus	1000	23	NOEC	Survivorship of larvae	131
Oyster	Crassostrea virginica	1000	9	LOEC	Growth inhibition	132
Fish	Citharichthys stigmaens	1890	65	NOEC	Survivorship	133

8.3.2.1 Assessment of EDCs by Field Monitoring

There are several ways in which field monitoring programs can be constructed:[20]

8.3.2.1.1 Morphological Indicators and Biomarkers

Laboratory development and subsequent field validation of reliable indicators of endocrine disruption in fish and invertebrates are indispensable.[43] Specific indicators of endocrine disruption, such as imposex and intersex, can be used to monitor the presence and/or the effects of a candidate EDC. For example, widespread declines in the degree of imposex in neogastropods, associated with declines in environmental TBT levels, have been observed throughout European waters since the late 1990s because of restrictions on the use of antifouling paints.[136] Other morphological measurements might indicate estrogenic disruption, such as antennal elongation in amphipods.[62] Specific biochemical biomarkers, such as Vtg or Vtg-like products, have also proved useful for examining the exposure and effects of estrogenic mimics with fish or crustaceans. Further work is required to develop suitable biomarkers for other types of endocrine disruption (e.g., androgen mimics).

8.3.2.1.2 In Situ Bioassays

Field transplantations, or *in situ* studies, with organisms such as fish, dogwhelks, and clams are highly useful in assessing the current risks of EDCs in the aquatic environment.[15,54,137] *In situ* deployments are easy to perform and allow flexible monitoring program designs (e.g., cages of test organisms can be deployed along a gradient of pollution levels when native organisms may be excluded from some sites). End points such as survivorship, growth, reproductive success, sex ratio, biomarker expression, and embryonic and larval development can be measured, with associated quantification of the candidate EDC(s). For example, the use of *N. lapillus* transplanted in cages has proved a useful tool for monitoring TBT contamination.[137]

8.3.2.1.3 Population and Assemblage Monitoring

Field monitoring for the effects of TBT, either on particular species or on entire saltwater assemblages, is a good example of the value of environmental monitoring in assessing temporal and spatial trends caused by EDCs[138] (and see review in Matthiessen et al.[20]). The Oslo and Paris Commissions in Europe now include assessment of TBT and investigation of imposex and intersex in mollusks in their Joint Assessment and Monitoring Program, while the Japanese Environmental Agency has stored frozen specimens of different organisms, so that retrospective analyses can be performed, if necessary.

Matthiessen et al.[20] identify the general advantages of monitoring changes in the structure of natural populations and assemblages of organisms:

1. They integrate the effects of exposure over time, chemicals, media, and pathways.
2. Important biogeochemical factors, such as microbial transformation, are incorporated.

3. Relevant species are included, and uncertainty is therefore reduced when compared with reliance on results from only a few "representative" species from laboratory testing.
4. All relevant population life stages and processes are represented, which again contrasts to the situation in most laboratory tests.
5. Adverse behavioral effects that are under hormonal control are integrated in the measured response, which may be difficult to achieve in the laboratory.
6. Interindividual and interspecies interactions are represented.
7. It is far easier to attribute "ecological significance" to responses measured in the field.

Field monitoring of the structure of important populations and the assemblages that they belong to is probably the only way to catch EDCs that have slipped through the net of prospective laboratory testing. While the marine environment is complex and many factors (e.g., fishery pressures, disease, and parasites) may obscure clear cause-and-effect relationships, field monitoring should be developed further. This goal requires an extensive rather than intensive program for sampling and enumeration of fish and invertebrate assemblages. Where feasible, organisms should also be sexed so that skewed sex ratios can be detected. Populations of numerically or economically important species such as fish species and bivalve mollusks should also be examined in more detail to determine whether life history traits at the population level, or individual level traits such as morphology, physiology, or biochemistry have been affected. If potential EDC effects are measured in populations or assemblages, then more-focused approaches using *in situ* exposures and toxicity identification evaluation (TIE) techniques should be considered.[20]

8.4 CONCLUSIONS

It is clear that in the latter part of the last decade there have been two major developments in the consideration of the biological effects of potential EDCs. The first is that the problem is not confined to freshwater environments, and the second is that invertebrates, as well as vertebrates, are potential targets for EDCs. While there are now test methods in place for a very few marine organisms, these are not as yet international standards and the range of organisms used should be expanded considerably to reflect the phylogenetic diversity of marine biota. The actual mechanism of action of a given EDC may operate very differently in different saltwater phyla and here we are hampered by the fact that the endocrinology of all but a few aquatic phyla are insufficiently understood. For example, while nonylphenol may have a genuinely estrogenic effect on the sex-steroid hormonal system in fish and some mollusks, any effects observed in arthropods may be mediated by interference with the steroid-based molting system. Consequently, whether the results of a limited number of tests on a few model species adequately protect against a wide range of species or phyla remains a question of great uncertainty.

In this chapter, we have addressed recent developments and identified areas where protocols should be developed and validated to encompass a wider range of saltwater organisms, e.g., Vtg expression in male fish and the use of field transplantations. It will also be necessary to develop more novel approaches to effectively screen the large number of substances of concern. In many cases, a large-scale toxicological approach is not feasible, partly because of the great taxonomic diversity in the marine environment and the concomitant wide range of endocrine systems involved. Having said this, it is clear that, because of the unusual nature of EDCs, environmental monitoring should play an increasingly important role, especially in indicating chronic effects and effects at higher tiers of biological organization. The value of this approach is well illustrated by the well-documented effects of TBT on neogastropod populations in coastal waters, and also in monitoring the recovery of these populations once TBT input was removed. Similar field observations indicating chemically mediated endocrine disruption in aquatic organisms should be, in themselves, enough reason for concern. Indeed, the arguments for monitoring changes in field populations in relation to potential EDCs are very persuasive. While this would require an extensive program of sampling, it would provide a basis for identifying potential EDC effects (at the important population level), which could then be followed up with focused *in situ* or TIE techniques.

We endorse the suggestion that further work is required to relate conventional biomarker and reproductive end points to higher-level effects. This process is potentially very expensive, especially if considering multigenerational tests in higher organisms such as fish. Consequently, future testing may need to focus on specific developmental stages that are predicted to be particularly susceptible, e.g., oyster larval development, or sperm motility tests. Identification of sensitive developmental stages and life-cycle testing with reference compounds are important research priorities. Although a number of invertebrate groups do have potential for EDC testing, for example, the harpacticoid copepods, very few, if any, meet all the desirable criteria for saltwater test organisms for EDCs.

The concept of a tiered evaluation program is a sensible approach. There may be considerable value in using rapid screening methods, e.g., SARs, molecular, or biochemical tests, in identifying potential EDCs. Unfortunately, because the use of SARs, especially with marine organisms, is a comparatively recent development, the large amount of toxicity testing required to establish a useful relationship precludes the immediate implementation of this approach. The use of molecular and biochemical biomarkers is potentially very useful and further development and validation of techniques such as the yeast two-hybrid assay and protein binding assays should be undertaken. These could then provide a useful screen before initiating higher-tier *in vivo* chronic testing with those chemicals proving positive at the screening stage.

Whatever combination of toxicity tests and field observations are used to generate estimates of "safe" concentrations of EDCs, the resultant value provides little information on the likely effect on assemblages of organisms. One possible, although partial, solution is the use of SSDs. Not only does this approach use all available data and take into consideration variation in the data, it can also be used to identify potentially sensitive species and possible effects on assemblages. Consequently, we suggest that further development of the SSD approach may be a useful tool in predicting higher-tier effects. The approach does, however, depend on having access to a data-rich database which, as for many chemicals, may not be the case for EDCs in the marine environment. Perhaps this neatly encapsulates the primary problem: lack of information, whether it be toxicity, endocrinology, or basic life history data, for many groups of marine organisms.

In summary we recommend:

1. Expansion of currently available test protocols to include a wider range of saltwater fish and invertebrates;
2. Continued development of extensive environmental monitoring programs to determine whether there is a potential EDC problem in estuarine and marine assemblages;
3. Continued development of molecular and biochemical, and to a lesser extent, SARs as potential screening tools for identifying EDCs;
4. Continued research to identify sensitive developmental stages of potential test organisms;
5. Continued development of the SSD approach as a tool for predicting higher-tier effects of EDCs.

REFERENCES

1. Holmes, P. et al., European workshop on the impact of endocrine disrupters on human health and wildlife, 2–4 December 1996, MRC Institute for Environment and Health, Report EUR 17549, Weybridge, Surrey, U.K., 1997.
2. Colborn, T. and Clement, C., Chemically induced alterations in sexual and functional development: the wildlife/human connection, in *Advances in Modern Environmental Toxicology*, Vol. 21, Colborn, T., and Clement, C., Eds., Princeton Scientific, Princeton, NJ, 1992.
3. Gilbertson, M. and Fox, G.A., Pollutant-associated embryonic mortality of Great Lakes herring gulls, *Environ. Pollut.*, 12, 211, 1977.
4. Crisp, T.M. et al., Special Report on Environmental Endocrine Disruption: An Effects Assessment and Analysis, U.S. Environmental Protection Agency, U.S. EPA/630/R-96/012, Washington, D.C., 1997.
5. Daston, G.P. et al., Environmental estrogens and reproductive health: a discussion of the human and environmental data, *Reprod. Toxicol. Rev.*, 11, 465, 1997.
6. DeFur, P.L., Crane, M., and Tattersfield, L.J., Conclusions and recommendations, in *Endocrine Disruption in Invertebrates: Endocrinology, Testing, and Assessment*, DeFur, P.L., Crane, M., Ingersoll, C., and Tattersfield, L., Eds., SETAC Press, Pensacola, FL, 1999, chap. 5.
7. ECETOC, Environmental Oestrogens: A Compendium of Test Methods, European Centre for Ecotoxicology and Toxicology of Chemicals, Document 33, Brussels, Belgium, 1996.
8. EDSTAC, Endocrine Disrupter Screening and Testing Advisory Committee Report to U.S. EPA, United States Environmental Protection Agency, Washington, D.C., 1998.
9. Kavlock, R.J., and Ankley, G.T., A perspective on the risk assessment process for endocrine-disruptive effects on wildlife and human health, *Risk Anal.*, 16, 731, 1996.
10. Kendall, R. et al., *Principles and Processes for Evaluating Endocrine Disruption in Wildlife*, SETAC Press, Pensacola, FL, 1998.

11. Pinder, L.C.V. et al., Endocrine function in aquatic invertebrates and evidence for disruption by environmental pollutants, Report to the Environmental Agency and Endocrine Modulators Steering Group, Technical Report E67, Windermere, U.K., 1999.
12. Rolland, R.M., Gilbertson, M., and Peterson, R.E., Eds., *Chemically Induced Alterations in Functional Development and Reproduction of Fishes*. SETAC Press, Pensacola, FL, 1998.
13. Stahl, R.G., Jr. et al., Introduction to the workshop on endocrine disruption in invertebrates: endocrinology, testing, and assessment, in *Endocrine Disruption in Invertebrates: Endocrinology, Testing, and Assessment*, DeFur, P.L., Crane, M., Ingersoll, C., and Tattersfield, L., Eds., SETAC Press, Pensacola, FL, 1999, chap. 1.
14. Ankley, G.T. et al., Development of a research strategy for assessing the ecological risk of endocrine disrupters, *Rev. Toxicol.*, 1, 231, 1997.
15. Harries, J. et al., Survey of estrogenic activity in United Kingdom inland waters, *Environ. Toxicol. Chem.*, 15, 1993, 1996.
16. Jobling, S. and Sumpter, J., Detergent components in sewage effluent are weakly estrogenic to fish — an *in vitro* study using Rainbow trout (*Oncorhynchis mykiss*) hepatocytes, *Aquat. Toxicol.*, 27(3–4), 361, 1993.
17. Sumpter, J., Feminized responses in fish to environmental estrogens, *Toxicol. Lett.*, 82/83, 737, 1995.
18. Jobling, S. et al., Inhibition of testicular growth in rainbow trout (*Oncorhynchus mykiss*) exposed to estrogenic alkylphenolic chemical, *Environ. Toxicol. Chem.*, 15, 194, 1996.
19. Hutchinson, T., Scholz, N., and Guhl, W., Analysis of the ECETOC aquatic toxicity (EAT) database. IV. Comparative toxicity of chemical substances to freshwater versus saltwater organisms, *Chemosphere*, 36, 143, 1998.
20. Matthiessen, P. et al., Field assessment for endocrine disruption in invertebrates, in *Endocrine Disruption in Invertebrates: Endocrinology, Testing, and Assessment*, DeFur, P.L., Crane, M., Ingersoll, C., and Tattersfield, L., Eds., SETAC Press, Pensacola, FL, 1999, chap. 4.
21. Kime, D., Nash, J., and Scott, A., Vitellogenesis as a biomarker of reproductive disruption by xenobiotics, *Aquaculture*, 177, 345, 1999.
22. Andersen, M. and Barton, H., Biological regulation of receptor–hormone complex concentrations in relation to dose–response assessments for endocrine-active compounds, *Toxicol. Sci.*, 48, 38, 1999.
23. Soto, A., Sonnenschein, C., and Chung, K., The E-Screen as a tool to identify estrogens: an update on estrogenic environmental pollutants, *Environ. Health Perspect.*, 103 (Suppl. 7), 113, 1995.
24. Hashimoto, S. et al., Elevated serum vitellogenin levels and gonadal abnormalities in wild male flounder (*Pleuronectes yokohamae*) from Tokyo Bay, Japan, *Mar. Environ. Res.*, 49, 37, 2000.
25. Panter, G., Thompson, R., and Sumpter, J., Adverse reproductive effects in male fathead minnows (*Pimephales promelas*) exposed to environmentally relevant concentrations of the natural oestrogens, oestradiol and oestrone, *Aquat. Toxicol.*, 42, 243, 1998.
26. Monteiro, P., Reis-Henriques, M., and Coimbra, J., Plasma steroid levels in female flounder (*Platichthys flesus*) after chronic dietary exposure to single polycyclic aromatic hydrocarbons, *Mar. Environ. Res.*, 49, 453, 2000.
27. Kime, D. and Nash, J., Gamete viability as an indicator of reproductive endocrine disruption in fish, *Sci. Total Environ.*, 233, 123, 1999.

28. Hamoutene, D., Rahimtula, A., and Payne, J., Development of a new biochemical assay for assessing toxicity in invertebrates and fish sperm, *Water Res.*, 34, 4049, 2000.
29. Flouriot, G. et al., Monolayer and aggregate cultures of rainbow trout hepatocytes long term and stable liver specific expresssion in aggregates, *J. Cell Sci.*, 105, 407, 1993.
30. Sumpter, J. and Jobling, S., Vitellogenesis as a biomarker for estrogenic contamination of the aquatic environment, *Environ. Health Perspect.*, 103 (Suppl. 7), 173, 1995.
31. Tattersfield, L. et al., SETAC-Europe/OECD/EC Expert Workshop on Endocrine Modulators and Wildlife: Assessment and Testing, EMWAT, Veldhoven, the Netherlands, 1997.
32. Ankley, G. et al., Overview of a workshop on screening methods for detecting potential (anti-)estrogenic/androgenic chemicals in wildlife, *Environ. Toxicol. Chem.*, 17, 68, 1998.
33. OECD, Organisation for Economic Consultation and Development Expert Consultation on Testing in Fish, London, October 28–29, 1998.
34. IEH, *IEH Assessment on the Ecological Significance of Endocrine Disruption: Effects on Reproductive Function and Consequences for Natural Population (Assessment A4)*, MRC Institute for Environment Health, Leicester, U.K., 1999.
35. Taylor, M. and Harrison, P., Ecological effects of endocrine disruption: current evidence and research priorities, *Chemosphere*, 39, 1237, 1999.
36. Janssen, P., Lambert, J., and Goos, H., The annual ovarian cycle and the influence of pollution on vitellogenesis in the flounder, *Pleuronectes flesus*, *J. Fish Biol.*, 47, 509, 1995.
37. Allen, Y. et al., Survey of estrogenic activity in United Kingdom estuarine and coastal waters and its effects on gonadal development of the flounder *Platichthys flesus*, *Environ. Toxicol. Chem.*, 18, 1791, 1999.
38. Lye, C., Frid, C., Gill, M., and McCormick, D., Abnormalities in the reproductive health of flounder *Platichthys flesus* exposed to effluent from a sewage treatment works, *Mar. Pollut. Bull.*, 34, 34, 1997.
39. Lye, C. et al., Estrogenic alkylphenols in fish tissues, sediments, and waters from the UK Tyne and Tees estuaries, *Environ. Sci. Technol.*, 33, 1009, 1999.
40. Folmar, L. et al., Comparative estrogenicity of estradiol, ethynyl estradiol and diethylstilbestrol in an *in vivo*, male sheepshead minnow (*Cyprinodon variegatus*), vitellogenin bioassay, *Aquat. Toxicol.*, 49, 77, 2000.
41. Manning, C. S. et al., Life-cycle toxicity of bis(tributyltin) oxide to the sheepshead minnow (*Cyprinodon variegatus*), *Arch. Environ. Contam. Toxicol.*, 37, 258, 1999.
42. Kramer, V. et al., Reproductive impairment and induction of alkaline-labile phosphate, a biomarker of estrogen exposure, in fathead minnows (*Pimephales promelas*) exposed to waterborne 17 beta-estradiol, *Aquat. Toxicol.*, 40, 335, 1998.
43. Depledge, M. H. and Billinghurst, Z., Ecological significance of endocrine disruption in marine invertebrates, *Mar. Pollut. Bull.*, 39, 32, 1999.
44. Arcand-Hoy, L. and Benson, W., Fish reproduction: an ecologically relevant indicator of endocrine disruption, *Environ. Toxicol. Chem.*, 17, 49, 1998.
45. Taylor, M. et al., A research strategy for investigating the ecological significance of endocrine disruption: report of a UK workshop, *Sci. Total Environ.*, 233, 181, 1999.
46. Gimeno, S. et al., Feminisation of young males of the common carp, *Cyprinus carpio*, exposed to 4-tert-pentylphenol during sexual differentiation, *Aquat. Toxicol.*, 43, 77, 1998.

47. Gimeno, S. et al., Demasculinisation of sexually mature male common carp, *Cyprinus carpio*, exposed to 4-tert-pentylphenol during spermatogenesis, *Aquat. Toxicol.*, 43, 93, 1998.
48. Campbell, P. and Hutchinson, T., Wildlife and endocrine disruptors: requirements for hazard identification, *Environ. Toxicol. Chem.*, 17, 127, 1998.
49. Barnes, R.D., *Invertebrate Zoology*, W.B. Saunders, Philadelphia, 1980.
50. Hayward, P.J. and Ryland, J.S., *Handbook of the Marine Fauna of North-West Europe*, Oxford University Press, Oxford, 1995, 800.
51. Withers, P.C., *Comparative Animal Physiology*, Saunders HBJ College Publishers, Fort Worth, TX, 1992, 493.
52. LeBlanc, G.A. et al., The endocrinology of invertebrates, in *Endocrine Disruption in Invertebrates: Endocrinology, Testing, and Assessment*, DeFur, P.L., Crane, M., Ingersoll, C., and Tattersfield, L., Eds., SETAC Press, Pensacola, FL, 1999, chap. 2.
53. LaFont, R., The endocrinology of invertebrates, *Ecotoxicology*, 9, 41, 2000.
54. Morcillo, Y. and Porte, C., Evidence of endocrine disruption in clams — *Ruditapes decussata* — transplanted to a tributyltin-polluted environment, *Environ. Pollut.*, 107, 47, 2000.
55. Schulte-Oehlman, U. et al., *Marisa cornuarietis* (gastropoda, prosobranchia): a potential TBT bioindicator for freshwater environments, *Ecotoxicology*, 4, 372, 1995.
56. Baldwin, W.S. et al., Metabolic androgenization of female *Daphnia magna* by the xenoestrogen 4-nonylphenol, *Environ. Toxicol. Chem.*, 16, 1905, 1997.
57. LeBlanc, G.A. and McLachlan, J.B., Changes in the metabolic elimination profile of testosterone following exposure of the crustacean *Daphnia magna* to tributyltin, *Ecotoxicol. Environ. Saf.*, 45, 296, 2000.
58. Bettin, C., Oehlmann, J., and Stroben, E., TBT-induced imposex in marine neogastropods is mediated by an increasing androgen level, *Helgoländer Meeresuntersuchungen*, 50, 299, 1996.
59. Nehring, S., Long-term changes in Prosobranchia (Gastropoda) abundances on the German North Sea coast: the role of the anti-fouling biocide tributyltin, *J. Sea Res.*, 43, 151, 2000.
60. Shurin, J.B., and Dodson, S.I., Sublethal toxic effects of cyanobacteria and nonylphenol on environmental sex determination and development in *Daphnia*, *Environ. Toxicol. Chem.*, 16, 1269, 1997.
61. Olmstead, A.W., and LeBlanc, G.A., Effects of endocrine-active chemicals on the development of sex characteristics of *Daphnia magna*, *Environ. Toxicol. Chem.*, 19, 2107, 2000.
62. Brown, R.J., Conradi, M., and Depledge, M.H., Long-term exposure to 4-nonylphenol affects sexual differential and growth of the amphipod *Corophium volutator* (Pallas, 1766), *Sci. Total Environ.*, 233, 77, 1999.
63. Ozretic, B. and Krajnovic-Ozretic, M., Morphological and biochemical evidence of the toxic effect of pentachlorophenol on the developing embryos of the sea urchin, *Aquat. Toxicol.*, 7, 255, 1985.
64. McKenney, C.L. and Matthews, E., Influence of an insect growth-regulator on the larval development of an estuarine shrimp, *Environ. Pollut.*, 64, 169, 1990.
65. Nice, H.E. et al., Development of *Crassostrea gigas* larvae is affected by 4-nonylphenol, *Mar. Pollut. Bull.*, 40, 491, 2000.
66. Billinghurst, Z. et al., Induction of cypris major protein in barnacle larvae by exposure to 4-n-nonylphenol and 17β-oestradiol, *Aquat. Toxicol.*, 47, 203, 2000.

67. Lam, P.K.S., Wo, K.T., and Wu, R.S.S., Effects of cadmium on the development and swimming behavior of barnacle larvae *Balanus amphitrite* Darwin, *Environ. Toxicol.*, 15, 8, 2000.
68. Zou, E. and Fingerman, M., Effects of estrogenic xenobiotics on molting of the water flea, *Daphnia magna*, *Ecotoxicol. Environ. Saf.*, 38, 281, 1997.
69. Zou, E. and Fingerman, M., Synthetic estrogenic agents do not interfere with sex differentiation but do inhibit molting of the cladoceran *Daphnia magna*, *Bull. Environ. Contam. Toxicol.*, 58, 596, 1997.
70. Bechmann, R.K., Effect of the endocrine disrupter nonylphenol on the marine copepod *Tisbe battagliai*, *Sci. Total Environ.*, 233, 33, 1999.
71. Price, L.J. and Depledge, M.H., Effects of the xenooestrogen nonylphenol on the polychaete *Dinophilus gyrociliatus*, Abstracts of the 8th Annual Meeting of Society of Environmental Toxicology and Chemistry — Europe, Brussels, Belgium, 1998.
72. Alzieu, C., Environmental problems caused by TBT in France: assessment, regulations, prospects, *Mar. Environ. Res.*, 32, 7, 1991.
73. Alzieu, C., Tributyltin: case study of a chronic contaminant in the coastal environment, *Ocean Coastal Manage.*, 40, 23, 1998.
74. Alzieu, C., Impact of tributyltin on marine invertebrates, *Ecotoxicology*, 9, 71, 2000.
75. Stenalt, E. et al., Mesocosm study of *Mytilus edulis* larvae and postlarvae, including the settlement phase, exposed to a gradient of tributyltin, *Ecotoxicol. Environ. Saf.*, 40, 212, 1998.
76. Smith, K.L., Galloway, T.S., and Depledge, M.H., Neuro-endocrine biomarkers of pollution-induced stress in marine invertebrates, *Sci. Total Environ.*, 262, 185, 2000.
77. Cheney, M.A., Fiorillo, R., and Criddle, R.S., Herbicide and estrogen effects on the metabolic activity of *Elliptio complanata* measured by calorespirometry, *Comp. Biochem. Physiol.*, 118C, 159, 1997.
78. Ingersoll, C.G. et al., Laboratory toxicity tests for evaluating potential effects of endocrine-disrupting compounds, in *Endocrine Disruption in Invertebrates: Endocrinology, Testing, and Assessment*, DeFur, P.L., Crane, M., Ingersoll, C., and Tattersfield, L., Eds., SETAC Press, Pensacola, FL, 1999, chap. 3.
79. Gibbs, P.E. and Bryan, G.W., Biomonitoring of tributyltin (TBT) pollution using the imposex response of neogastropod molluscs, in *Biomonitoring of Coastal Waters and Estuaries*, Kremer, K.J.M., Ed., CRC Press, Boca Raton, FL, 1994, 205–226.
80. Bryan, G. W. et al., The decline of the gastropod *Nucella lapillus* around south-west England: evidence for the effect of tributyltin from antifouling paints, *J. Mar. Biol. Assoc. U.K.*, 66, 611, 1986.
81. Gibbs, P.E., Pascos, P.L., and Burt, G.R., Sex change in the female dogwhelk *Nucella lapillus*, induced by tributyltin from antifouling paints, *J. Mar. Biol. Assoc. U.K.*, 68, 715, 1988.
82. Langston, W.J., Burt, G.R., and Mingjiang, Z., Tin and organotin in water, sediments, and benthic organisms of Poole Harbour, *Mar. Pollut. Bull.*, 18, 634, 1987.
83. Langston, W.J. et al., Assessing the impact of tin and TBT in estuaries and coastal regions, *Funct. Ecol.*, 4, 433, 1990.
84. Evans, S.M., Evans, P.M., and Leksono, T., Widespread recovery of dogwhelks, *Nucella lapillus* (L.) from tributyltin contamination in the North Sea and Clyde Sea, *Mar. Pollut. Bull.*, 32, 263, 1996.
85. Folsvik, N. et al., Quantification of organotin compounds and determination of imposex in populations of dogwhelks (*Nucella lapillus*) from Norway, *Chemosphere*, 38, 681, 1999.

86. Svavarsson, J., Imposex in the dogwhelk (*Nucella lapillus*) due to TBT contamination: improvement at high latitudes, *Mar. Pollut. Bull.* 40, 893, 2000.
87. Moore, C.G., and Stevenson, J.M., The occurrence of intersexuality in harpacticoid copepods and its relationship with pollution, *Mar. Pollut. Bull.*, 22, 72, 1991.
88. Gross, M.Y. et al., Effects of sewage effluent exposure on sexual development of *Gammarus pulex* (L.), *Environ. Toxicol. Chem.*, in press.
89. Den Besten, P.J. et al., Effects of cadmium and polychlorinated biphenyls (Clophen A50) on steroid metabolism and cytochrome P-450 mono-oxygenase system in the sea star *Asterias rubens* L, *Aquat. Toxicol.*, 20, 95, 1991.
90. Voogt, P.A. et al., Effects of cadmium on sterol composition in the aboral body-wall of the sea star *Asterias rubens* L, *Comp. Biochem. Physiol.*, 104C, 415, 1993.
91. Ozretic, B., Petrovic, S., and Krajnovic-Ozretic, M., Toxicity of TBT-based paint leachates on the embryonic development of the sea urchin *Paracentrotus lividus* Lam., *Chemosphere*, 37, 1109, 1998.
92. Smith, B.S., Tributyltin compounds induce male characteristics on female mud snails *Nassarius obsoletus = Illyanassa obsoletus*, *J. Appl. Toxicol.*, 1, 141, 1981.
93. Davies, I.M. et al., Sublethal effects of tributyltin oxide on the dogwhelk *Nucella lapillus*, *Mar. Ecol. Prog. Ser.*, 158, 191, 1997.
94. Spooner, N. et al., The effect of tributyltin upon steroid titres in the female dogwhelk, *Nucella lapillus*, and the development of imposex, *Mar. Environ. Res.*, 32, 37, 1991.
95. Smith, B.S. and McVeagh, M., Widespread organotin pollution in New Zealand coastal waters as indicated by imposex in dogwhelks, *Mar. Pollut. Bull.*, 22, 409, 1991.
96. Granmo, A. et al., Lethal and sublethal toxicity of 4-nonylphenol to the common mussel, *Environ. Pollut.*, 59, 115, 1989.
97. Billinghurst, Z. et al., Inhibition of barnacle settlement by the environmental oestrogen 4-nonylphenol and the natural oestrogen 17β-oestradiol, *Mar. Pollut. Bull.*, 36, 833, 1998.
98. Wu, R.S.S., Lam, P.K.S., and Zhou, B.S., A settlement inhibition assay with cyprid larvae of the barnacle *Balanus amphitrite*, *Chemosphere*, 35, 1867, 1997.
99. McKenney, C.L. and Celestial, D.M., Modified survival, growth and reproduction in an estuarine mysid (*Mysidopsis bahia*) exposed to a juvenile hormone analogue through a complete life cycle, *Aquat. Toxicol.*, 35, 11, 1996.
100. Tyler-Schroeder, D.B., *Use of Grass Shrimp*, Paleomonetes pugio, *in a Life-cycle Toxicity Test*, American Society for Testing and Materials, Philadelphia, 1979.
101. Evans, S.M., Kerrigan, E., and Palmer, N., Causes of imposex in the dogwhelk *Nucella lapillus* (L.) and its use as a biological indicator of tributyltin contamination, *Mar. Pollut. Bull.*, 40, 212, 2000.
102. Hawkins, L.E. and Hutchinson, S., Physiological and morphologic effects of monophenyltin trichloride on *Ocenebra erinacea* (L.), *Funct. Ecol.*, 4, 449, 1990.
103. Rooney, A.A. and Guillette, L.J., Jr., Contaminant interactions with steroid receptors: evidence for receptor binding, in *Environmental Endocrine Disrupters: An Evolutionary Perspective*, Guillette, L.J. Jr. and Crain, D.A., Eds., Taylor & Francis, New York, 2000, chap. 4.
104. Nias, D.J., Mckillup, S.C., and Edyvane, K.S., Imposex in *Lepsiella vinosa* from Southern Australia, *Mar. Pollut. Bull.*, 26, 380, 1993.
105. ECETOC, Screening and testing methods for ecotoxicological effects of potential endocrine disrupters: response to the EDSTAC recommendations and a proposed alternative approach, European Centre for Ecotoxicology and Toxicology of Chemicals, Document 39, Brussels, Belgium, 1999.

106. McKinney, J.D. et al., The practice of structure activity relationship (SAR) in toxicology, *Toxicol. Sci.*, 56, 8, 2000.
107. Schmieder, P.K. et al., Estrogenicity of alkylphenolic compounds: a 3-D structure-activity evaluation of gene activation, *Environ. Toxicol. Chem.*, 19, 1027, 2000.
108. Oberdorster, E., Rice, C.D., and Irwin, L.K., Purification of vitellin from grass shrimp *Palaemonetes pugio*, generation of monoclonal antibodies, and validation for the detection of lipovitellin in Crustacea, *Comp. Biochem. Physiol.*, 127C, 199, 2000.
109. Graumann, K., Breithofer, A., and Jungbauer, A., Monitoring of estrogen mimics by a recombinant yeast assay: synergy between natural and synthetic compounds? *Sci. Total Environ.*, 225, 69, 1999.
110. Nishihara, T. et al., Estrogenic activities of 517 chemicals by yeast two-hybrid assay, *J. Health Sci.*, 46, 282, 2000.
111. Lutz, I. and Kloas, W., Amphibians as a model to study endocrine disruptors: I. Environmental pollution and estrogen receptor binding, *Sci. Total Environ.*, 225, 49, 1999.
112. Miller, S. et al., A rapid and sensitive reporter gene that uses green fluorescent protein expression to detect chemicals with estrogenic activity, *Toxicol. Sci.*, 55, 69, 2000.
113. Dodson, S.I., Shurin, J.B., and Girvin, K.M., *Daphnia* reproductive bioassay for testing toxicity of aqueous samples and presence of an endocrine disrupter, U.S. Patent: 5,932,436, 1999.
114. Newman, M.C. et al., Applying species-sensitivity distributions in ecological risk assessment: assumptions of distribution type and sufficient numbers of species, *Environ. Toxicol. Chem.*, 19, 508, 2000.
115. Kusk, K.O. and Petersen, S., Acute and chronic toxicity of tributyltin and linear alkylbenzene sulfonate to the marine copepod *Acartia tonsa*, *Environ. Toxicol. Chem.*, 16, 1629, 1997.
116. Guolan, H. and Young, W., Effects of tributyltin chloride on marine bivalve mussels, *Water Res.*, 29, 1877, 1995.
117. Axiak, V. et al., Laboratory and field investigations on the effects of organotin (tributyltin) on the oyster, *Ostrea edulis*, *Sci. Total. Environ.*, 171, 117, 1995.
118. Edouard, H. et al., A comparison between oyster (*Crassostrea gigas*) and sea urchin (*Paracentrotus lividus*) larval bioassays for toxicological studies, *Water Res.*, 33, 1706, 1999.
119. Batley, G.E. et al., Accumulation of tributyltin by the Sydney rock oyster, *Saccostrea commercialis*, *Aust. J. Mar. Freshwater Res.*, 40, 49, 1989.
120. Hall, L.W., Jr. et al. Chronic toxicity of tributyltin to Chesapeake Bay biota, *Water Air Soil Pollut.*, 39, 365, 1988.
121. Pinkney, A.E., Matteson, L.L., and Wright, D.A., Effects of tributyltin on survival, growth, morphometry, and RNA-DNA ratio of larval striped bass, *Morone saxatilis*, *Arch. Environ. Contam. Toxicol.*, 19, 235, 1990.
122. Davidson, B.M., Valkirs, A.O., and Seligman, P.F., Acute and chronic effects of tributyltin on the mysid *Acanthomysis sculpta* (Crustea, Mysidacea), in *Ocean 86, Proc. Int. Organotin Symposium*, Vol. 4, Marine Technology Society, Washington, D.C., 1986, 1219.
123. WHO, World Health Organization, *Environmental Health Criteria 116. Tributyltin Compounds*, Geneva, Switzerland, 1990.
124. Hall, L.W., Jr. et al., Acute and chronic effects of tributyltin on a Chesapeake Bay copepod, *Environ. Toxicol. Chem.*, 7, 41, 1988.

125. Ruiz, J.M., Bryan, G.W., and Gibbs, P.E., Acute and chronic toxicity of water tributyltin (TBT) to pediveliger larvae of the bivalve *Scrobicularia plana, Mar. Biol.*, 124, 119, 1995.
126. Jak, R.G. et al., Effects of tributyltin on a coastal north sea plankton community in enclosures, *Environ. Toxicol. Chem.*, 17, 1840, 1998.
127. Seinen, W. et al., Short term toxicity of tri-n-butyltin chloride in rainbow trout (*Salmo gairdneri* Richardson) yolk sac fry, *Sci. Total Environ.*, 19, 155, 1981.
128. Laughlin, R.B., Nordlund, K., and Linden, O., Long-term effects of tributyltin on the Baltic amphipod, *Gammarus oceanicus, Mar. Environ. Res.*, 12, 243, 1984.
129. Ringwood, A.H., Comparative sensitivity of gametes and early development stages of a sea urchin (*Echinometra mathaei*) and a bivalve species (*Isognomon californicum*) during metal exposures, *Arch. Environ. Contam. Toxicol.* 22, 288, 1992.
130. Meador, J.P. and Rice, C.A., Impaired growth in the polycheate *Armandia brevis* exposed to tributyltin in sediment, *Mar. Environ. Res.*, 51, 113, 2001.
131. Laughlin, R.B., Jr., and French, W.J., Comparative study of the acute toxicity of a homologous series of trialkyltins to larval shore crabs, *Hemigrapsus nudus*, and lobster, *Homarus americanus, Bull. Environ. Contam. Toxicol.*, 25, 802, 1980.
132. Roberts, R. and His, E., Action de l'acétate de tributylr-étain sur les oeufs et les larves D de deux mollusques d'intérêt commercial: *Crassostrea gigas* (Thunberg) et *Mytilus galloprovincialis* (Lemark), Int. Counc. Explor. Sea Paper CM 1981:F42 (mimeo), 1981, 16.
133. Valkivs, A., Davidson, B., and Seligman, P., Sublethal Growth Effects and Mortality to Marine Bivalves and Fish from Long-Term Exposure to Tributyltin, Report, Naval Oceans System Center, San Diego, CA, 1985, 36.
134. Blanck, H. and Dahl, B., Recovery of marine periphyton communities around a Swedish marina after the ban of TBT use in antifouling paint, *Mar. Poll. Bull.*, 36, 437, 1998.
135. Hall, L.W., Jr. et al., A probabilistic ecological risk assessment of tributyltin in surface waters of the Chesapeake Bay watershed, *Hum. Ecol. Risk Assess.*, 6, 141, 2000.
136. Evans, S.M., Birchenough, A.C., and Brancato, M.S., The TBT ban: out of the frying pan into the fire? *Mar. Pollut. Bull.*, 40, 204, 2000.
137. Quintela, M., Barreiro, R., and Ruiz, J.M., The use of *Nucella lapillus* (L.) transplanted in cages to monitor tributyltin (TBT) pollution, *Sci. Total. Environ.*, 247, 227, 2000.
138. Abel, P.D., TBT—towards a better way to regulate pollutants, *Sci. Total Environ.*, 258, 1, 2000.

9 The Use of Toxicity Reference Values (TRVs) to Assess the Risks That Persistent Organochlorines Pose to Marine Mammals

Paul D. Jones, Kurunthachalam Kannan, Alan L. Blankenship, and John P. Giesy

CONTENTS

9.1 Overview 218
9.2 Introduction 218
9.3 Problem Formulation 219
9.4 Exposure Assessment 221
    9.4.1 Exposure Assessment Methods 221
    9.4.2 Estimating Exposure through Modeling 221
    9.4.3 Measuring Internal Dose Using Tissue Residues 224
9.5 Effects Assessment 225
    9.5.1 Adverse Effects in Marine Mammals 225
    9.5.2 Immunotoxicological Studies in the Harbor Seal 226
    9.5.3 Toxicological Studies in Cetaceans 227
    9.5.4 Exposure Studies in Mustelids 227
    9.5.5 Toxicity Reference Values 228
    9.5.6 Toxicity Threshold Evaluation 229
    9.5.7 Uncertainties in TRV Determination 231
9.6 Risk Characterization 232
    9.6.1 Risk Assessment Based on New Zealand Data 232
9.7 Conclusions 237
Acknowledgments 238
References 238

9.1 OVERVIEW

Marine mammals are known to accumulate relatively high concentrations of persistent organochlorine contaminants (POCs). These stores of contaminants have the potential to act as a continuing source of elevated exposure to these organisms. Although a considerable amount is known about the concentrations of POCs in marine mammals and about the processes that lead to their accumulation, little is known about the potential these contaminants have to cause adverse effects in exposed animals. Although several anecdotal studies have measured relatively high POC concentrations in marine mammals associated with mass mortality events, in all cases, it has been difficult to demonstrate a cause–effect relationship.[1,2] Similarly, several semifield studies have been conducted by feeding naturally contaminated fish to captive animals and assessing adverse effects.[3,4] It is also difficult to attribute effects of organochlorines in these studies due to small sample sizes and the presence of co-contaminants in the food source used for feeding. To determine possible adverse effect levels in marine mammals, we previously compiled a number of the most relevant and rigorous studies to derive toxicity reference values (TRVs) for marine mammals.[5] In this chapter, we use these TRVs to evaluate the possibility of adverse effects in marine mammals at current levels of exposure. The data chosen for the assessment were collected in New Zealand. These data were chosen because they provide detailed information on a wide range of dioxin-like contaminants for a variety of species and are coupled with equivalent information for a variety of other environmental matrices. The New Zealand data represent one of the lower levels of exposure known to occur for marine mammals, providing a conservative estimate of possible risks to other marine mammal populations. Risks seem to be greatest for marine mammals feeding in inshore habitats presumably due to the higher concentrations of anthropogenic pollutants in these locations. Since there are identifiable levels of risk to marine mammals in the relatively pristine southern oceans, there appears to be little global capacity for the dissipation of additional POCs.

9.2 INTRODUCTION

The U.S. EPA has developed a framework for ecological risk assessment (ERA) that consists of four phases: (1) problem formulation, (2) exposure assessment, (3) effects assessment, and (4) risk characterization.[6] The problem formulation step is a formal process to develop and evaluate a preliminary hypothesis concerning the likelihood and causes of ecological effects that may have occurred, or may occur.[6] A key step in the problem formulation phase is the development of a conceptual model detailing exposure pathways and key receptor organisms. In the exposure assessment phase, the potential for adverse effects to ecological receptors due to chemical stressor exposure is assessed by evaluating the probability of co-occurrence of the stressors and the ecological receptors considered.[6] The effects assessment evaluates effects data to assess (1) the link between elicited effects and stressor concentrations, (2) the relationship between the elicited effects and the associated assessment end point, and (3) the validity of the exposure model (i.e., are conditions under which the effects occur consistent with the conceptual model?[6]). In the risk characterization phase, the results of the exposure and effects assessment are used to estimate risk to the assessment end points identified in problem formulation, and the risk is interpreted and conclusions are reported.[6] Specifically, information obtained during the exposure and the effects assessment is combined to evaluate the relationship between environmental concentrations of chemical stressors and observed adverse biological effects.

Although this framework was developed for the assessment of contaminated sites of a relatively limited geographical scale, the central paradigm of the framework can be applied to problems of larger geographical or temporal scales. In this chapter, we will utilize the U.S. EPA risk assessment paradigm to assess the potential for POCs to cause adverse effects in marine mammal populations.

9.3 PROBLEM FORMULATION

Estimates of the number of chemicals humans release to the environment range from tens of thousands to hundreds of thousands. Although many of these chemicals are relatively nontoxic and short lived in the environment, some chemicals show significant toxicological effects at relatively low concentrations, and also persist and are transported in various environmental media. A large proportion of marine mammal species, especially cetaceans (whales and dolphins), live in open ocean environments and so are not greatly subject to direct chemical exposures due to human activity. Marine mammals are, however, exposed to persistent anthropogenic chemicals that are transported in air and water during global redistribution processes.[7] These processes are based on the low but measurable volatility of POCs, which means these chemicals can enter the gaseous phase and be transported globally by air movements.[8] In addition, some shorter-distance transport of POCs bound to atmospheric particulates is possible. Atmospheric transportation of POCs results in deposition of these chemicals in areas remote from human activity where they may be accumulated by wildlife species. For this reason, many POCs are now regarded as ubiquitous global contaminants.

The life history parameters of many marine mammals result in their accumulating relatively great concentrations of POCs. Relatively long life-span, high trophic status, and use of extensive lipid reserves make marine mammals efficient at accumulating large quantities of POCs. Of the compounds studied in marine mammals, POCs such as polychlorinated dibenzo-p-dioxins (PCDDs), polychlorinated dibenzofurans (PCDFs), and polychlorinated biphenyls (PCBs) appear to accumulate to the greatest concentrations in the widest range of species.[9] While marine mammals accumulate significant concentrations of metals in various tissues, mercury is the only metal that shows both biomagnification at all levels of the food chain and a positive correlation with age at all stages during the cetacean's life (reviewed in Reference 10). An association between mercury contamination and liver damage has also been suggested in cetaceans.[11] Data on the effects of metal toxicity in cetacean species are sparse. Effects of toxicity may be different depending on species, age, and sex of the animal, but indications of toxic effects have been reported.[11] Species-specific sensitivities to the toxic effects of metals tend to vary less than those to POCs; therefore, it is likely that standard toxicological risk assessment procedures can be used to assess relatively accurately the risks posed to marine mammals by metals.[12]

Marine mammals are particularly vulnerable to the effects of POCs for several reasons. First, they inhabit aquatic environments that are the ultimate sinks for many of these compounds. Marine mammals have a unique lifestyle that requires thick layers of fatty blubber to provide thermal insulation and energy reserves for fasting periods in their life cycles. These fatty tissues act as a reservoir for the accumulation of POCs and also act as a continual source "resupplying" the rest of the body with these contaminants when fats are metabolized. The long life span and generally predatory feeding habits of marine mammals also lead to high levels of POCs in blubber. In addition, marine mammals seem to be limited in the biochemical processes required to metabolize and eliminate these chemicals.[13] Finally, because of the high lipid content of marine mammal milk, POCs can be passed by lactation to the developing young.

As previously mentioned, the POCs of most concern are the dioxins, furans, PCBs, and other dioxin-like chemicals. These chemicals consist of two linked aromatic rings with chlorine substituted around the rings. These chemicals are persistent and bioaccumulative. Many have become ubiquitous environmental contaminants as a result of global redistribution processes that have led to significant concentrations accumulating in marine wildlife in remote locations.[7,14] Some of these chemicals, notably PCBs, were deliberately manufactured and others are by-products of various processes using chlorine or are products of incomplete combustion.[15] The most biologically potent of the dioxin-like chemicals have a planar structure that allows binding to the cellular Ah-receptor (AhR) through which the most sensitive biological effects are expressed.[16,17] These planar compounds have been demonstrated to be potent reproductive and developmental toxins and their accumulation in marine mammals has been the focus of some concern.[18,19]

PCBs and dioxins express their most toxic biological effects through a common mechanism of action modulated by the cellular aryl-hydrocarbon receptor (AhR). The overall biological potency of a mixture of these chemicals can be expressed as "toxic equivalents" (TEQs). TEQs relate the potency of the mixture to that of 2,3,7,8-tetrachlorodibenzo-p-dioxin (TCDD), the most potent of these chemicals. Previous studies have generally reported a better correlation between adverse effects and TEQs than with PCBs concentrations. Although studies on wildlife in the Great Lakes and elsewhere have found a causal link between adverse health effects and POCs,[20-22] the observed toxicity usually correlates better to TEQs than to total PCBs.[21,23]

The relatively high concentrations of POCs accumulated by marine mammals have led to concerns that these contaminants may be having adverse effects on these animals, possibly by adversely affecting their immunocompetence.[1,24,26] Exposure of developing young to POCs is also possible by placental transfer[27] or lactation,[28,29] although the latter route has been shown to be the most significant.[28,30] The exposure of such sensitive early life stages to elevated contaminant concentrations may be of particular toxicological concern.

Previous studies have focused primarily on top predators in the marine environment, particularly marine mammals[9,10,31,32] and sea birds,[14,33-36] while fewer studies are available for species such as polar bears[37,38] and sea turtles.[39] It has been observed that marine mammals in all environments accumulate significant concentrations of POCs even if exposure concentrations are relatively low.[40]

Although the focus of this review is the risk assessment for marine mammals, it should be noted that open ocean birds and turtles are also exposed to POC compounds through the diet. However, they do not seem to accumulate concentrations as high as those of marine mammals. This is presumably due to their lack of a large lipid pool in which POCs can accumulate. Nevertheless, exposures of albatross are of concern because of their reliance on the ocean food chain and the high sensitivity of some bird species to the effects of POCs.[21] In addition, it has previously been observed that albatross in the North Pacific Ocean accumulate relatively high concentrations of POCs in their tissues.[14,41] Concentrations of dioxins and PCBs in these birds are near the threshold where adverse effects on reproduction could be expected.[14,21]

Considering the above discussion, the major question for this risk assessment is: "Do current concentrations of persistent organochlorine contaminants pose risks to the health of marine mammals?"

9.4 EXPOSURE ASSESSMENT

9.4.1 Exposure Assessment Methods

It is important to understand all possible sources and pathways of exposure for a particular exposure situation. Whenever possible, contributions from each complete exposure pathway should be evaluated. In the case of lipophilic and bioaccumulative chemicals such as POCs, exposure from the water column is usually considered to be negligible, whereas dietary exposure from contaminated prey items constitutes the primary exposure pathway.[42] For benthic invertebrates, exposure is primarily from ingestion and absorption from contaminated sediment. Thus, it may be possible to predict exposure through equilibrium partitioning.[43] For fish, exposure potentially results from both water passing across the gills and ingestion of contaminated food items. Although nearly all of the toxicity data for POCs to fish are expressed in terms of aqueous concentrations, fish body burdens probably provide an exposure metric that is more closely correlated with effects because the majority of exposure is through the diet.[44] For higher-trophic-level wildlife such as marine mammals, daily exposure is usually estimated through the use of exposure models or internal exposure (e.g., tissue residues) as measured directly in target tissues. Both of these approaches — exposure modeling and tissue residues — will be discussed in more detail below.

9.4.2 Estimating Exposure through Modeling

The characteristics of the ecosystem and receptors must be considered to reach appropriate conclusions about exposure. Three aspects should be considered when estimating exposure: intensity, space, and time. Intensity is the most familiar aspect for chemical and biological stressors, and may be expressed as the amount of chemical contacted per day. Spatial extent is another aspect of exposure and is most commonly expressed in terms of area. However, at large spatial scales, the shape or arrangement of exposure may be an important issue, and area alone may not be the appropriate descriptor of spatial extent for risk assessment.

Considerations for the temporal aspects of exposure include duration, frequency, and timing. Duration can be expressed as either the time over which exposure occurs or some threshold intensity is exceeded. If exposure occurs as repeated discrete events of about the same duration, then frequency may be the most important temporal dimension of exposure (e.g., regularity of migration through highly contaminated areas). If repeated events have significant and variable durations, both duration and frequency should be considered. Abiotic attributes may also increase or decrease the frequency and amount of a stressor contacted by receptors. For example, naturally anoxic areas above contaminated sediments in an estuary may reduce the time bottom-feeding fish spend in contact with sediments and thereby reduce their exposure to contaminants. In addition, the timing of exposure can be an important factor (e.g., exposure of receptors during a sensitive life stage). In large water bodies with PCB-contaminated sediments, duration is usually long and exposure is fairly constant. However, significant tidal action, strong currents, storm events, or other episodic events in some of these areas can affect exposure profiles.

The above considerations of area, extent, and intensity of exposure have particular significance if considering risk assessment of marine mammals in coastal and estuarine systems. While the majority of marine mammals frequent the open ocean, some species live in more coastal environments and many species are sporadic visitors to coastal areas.[45] Many marine mammals also roam large areas of ocean, making exposure estimation difficult. Although the frequency and duration of visits to coastal regions may be low, the relatively high concentrations of POCs encountered on those occasions could represent a significant portion of the organism's exposure. In addition, the accumulative nature of POCs suggests that any contaminants accumulated on these brief occasions may be sequestered into fat reserves to be released at a later time. Large home ranges and infrequent high concentration exposures make estimating POC exposure in marine mammals by modeling from environmental media or from dietary estimates problematic.

Exposure analysis for POC contaminant mixtures containing PCBs and dioxins should be based on a congener-specific analysis in which the concentration of each individual POC chemical, particularly the most toxic chemicals, is measured. In this way, individual congeners and/or a summation of PCB congeners can be modeled between exposure media and receptors. The use of congener-specific analysis in ecological risk assessment will also provide a better understanding of the toxicity of complex mixtures in the environment. Under certain conditions (e.g., sediments), the PCB mixture may become less toxic than would be indicated by simple comparison with total PCBs.[40,42] In other biota, toxic congeners may become enriched at higher trophic levels in the food chain.[40] The use of congener-specific analyses permits these differences in toxicity to be better estimated.

The final product of exposure analysis is an exposure profile that includes a summary of the paths of POCs and other stressors from the source(s) to the receptors, completing the exposure pathway. If exposure can occur through many pathways, it is useful to rank them, perhaps by their contribution to total exposure. It is

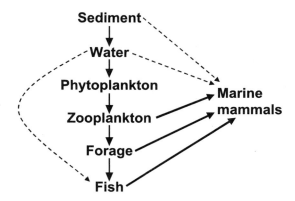

FIGURE 9.1 Conceptual model for the accumulation of POCs by marine mammals. See text for explanation. Solid lines indicate major pathways of contaminant movement; dashed lines represent minor or species-specific pathways.

recommended for top-level predators (e.g., marine mammals and piscivorous birds) that dietary exposure models be utilized in which concentrations of persistent organic contaminants be determined in as many potential prey items as practical. Exposure should be described in terms of intensity, space, and time in units that can be combined with the effects assessment.

A conceptual model can be created to facilitate interpretation of the pathways for POC uptake by marine mammals (Figure 9.1). In this model, marine sediments act as the primary source for POC contaminants to the marine ecosystem. Desorption of POCs from sediment to sediment pore water and ultimately to the water column represents the major point of entry of POCs into the food chain. From the water column, the POCs can bioaccumulate (direct transfer of dissolved phase POCs into tissue) into phytoplankton that form the base of the food chain. The POCs biomagnify (increase in concentration of POCs in a higher trophic level compared with a lower trophic level) with each subsequent trophic level transfer from phytoplankton to zooplankton to small fish to large fish. As a result, marine mammals feeding at the highest trophic levels receive the greatest exposure. Because of the lipophilic nature of POCs, the dietary exposure pathway can be expected to be the predominant route of exposure for aquatic organisms.[42] The conceptual model also includes some "minor" pathways of uptake such as direct dermal absorption from water and pathways such as direct uptake from sediment that might be more significant in certain species such as gray whales (*Eschrichtius robustus*), which ingest relatively great quantities of sediment during feeding activities.

A possible source of exposure not included in the conceptual model (Figure 9.1) is exposure from air. The major possible exposure pathways for POCs in the environment to organisms are through ingestion of water, soil, and food, or direct dermal contact with contaminated media. The relatively low volatility of POCs results in concentrations in air that are generally low. To estimate the significance of airborne POC intake, biometric data for a range of species were determined from the literature.[46] These data were used to estimate daily intakes of a 'model' POC using concentrations typically found in environmental media (Table 9.1). It was concluded

TABLE 9.1
Estimated Daily Intakes of POCs from Different Media

Species	Body Weight (g)	Water Ingestion (g/g/d)	Food Ingestion (g/g/d)	Air Inhalation (m³/d)	Total Intake (pg/g/d)	% of Intake from Air
Great Blue Heron	2,200	0.045	0.18	0.76	1,337	0.0074
Kingfisher	140	0.11	0.5	0.094	3,715	0.0052
Mink	1,000	0.11	0.13	0.55	1,300	0.021
Seal	80,000	0.0048	0.05	18	2,105	0.0031

Note: Intakes are based on a water concentration of 1 pg/ml, food 10 ng/g, air 500 pg/m³. All assimilation factors were assumed to be 1 to provide a worst case scenario.

from this assessment that the contribution of airborne POCs to daily intake was insignificant. Therefore, no further consideration was given to the effects of airborne POCs in this assessment. It is also possible that marine mammals may receive exposure to POCs by inhalation of the water surface microfilm, as some POCs are known to accumulate in this lipophilic material.[47,48] There is currently no method available to estimate the amount of surface microfilm that is inhaled or the extent to which this contributes to POC exposure.

From the conceptual model it should theoretically be possible to calculate exposure concentrations for marine mammals; however, in practice the degree of uncertainty in such estimations is considerable. Each step in the modeling process (e.g., transfer from sediment to water) introduces a degree of uncertainty and the wide spatial and temporal ranges, discussed above, make approximating environmental concentrations problematic. A more direct approach, if data are available, is to use tissue POC concentrations as measures of exposure.

9.4.3 MEASURING INTERNAL DOSE USING TISSUE RESIDUES

Tissue residues are a particularly useful means of measuring internal dose if exposure across many pathways must be integrated and if site-specific factors influence bioavailability. This method is particularly useful if the stressor–response relationship is expressed using tissue residue concentration. Tissue residue effect concentrations are becoming increasingly available for a number of chemicals and organisms. Specifically, tissue residue effect level data for PCBs are gaining increasing ecotoxicological acceptance[44,49,50] and regulatory acceptance as evidenced in the "Canadian tissue residue guidelines (TRG) for polychlorinated biphenyls for the protection of wildlife consumers of aquatic biota."[51] Similar toxicological information for marine mammals is very limited and most stressor–response relationships express the amount of stressor in terms of media concentration or potential dose rather than internal dose. In addition, few models can accurately predict uptake in a field situation.[52] Thus, tissue residues can provide valuable confirmatory evidence that exposure occurred and can provide a means of comparison to an effect level. Tissue residues in prey organisms can also be used for estimating risks to their predators. Again congener-specific analysis is the preferred method for dioxin-like chemicals as it allows the internal dose to be expressed in terms of either total PCBs (sum of congeners) or TEQs.

9.5 EFFECTS ASSESSMENT

In this section, we review available information on the effects of POCs on marine mammals and how this information was used to derive toxicity reference values (TRVs). A limited number of studies have been carried out that adequately measure biological effects and POC concentrations in the same marine mammals. Even in these few studies, the presence of contaminants other than POCs was not evaluated and, in some cases, the effects measured cannot be adequately linked to adverse effects in the organisms *in vivo*. Toxicity reference values are used as the best available estimates of contaminant concentrations that would be expected to cause adverse effects. As such, they are benchmark concentrations that can be compared to concentrations in the environment to determine whether the observed environmental concentrations pose a risk of adverse effects.

9.5.1 ADVERSE EFFECTS IN MARINE MAMMALS

For effective protection of marine mammals, it is necessary to know the potential hazard of persistent, bioaccumulative, and toxic pollutants to which they are exposed. It has been contended that, since 1968, 16 species of aquatic mammals have experienced population instability, major stranding episodes, reproductive impairment, endocrine and immune system disturbances, or serious infectious diseases.[53] The same authors also suggest that organochlorine contaminants, particularly PCBs and DDTs, have caused reproductive and immunological disorders in aquatic mammals.[53] The presence of high concentrations of PCBs in tissues have been associated with the high prevalence of diseases and reduced reproductive capability of the Baltic gray seal (*Halichoerus grypus*) and the ringed seal (*Phoca hispida*),[54] reproductive failure in the Wadden Sea harbor seal (*P. vitulina*)[3] and the St. Lawrence estuary beluga whales (*Delphinapterus leucas*),[55] viral infection and mass mortalities of the U.S. bottlenose dolphin (*Tursiops truncatus*),[56,57] the Baikal seal (*P. sibirica*)[58] and the Mediterranean striped dolphin (*Stenella coeruleoalba*).[59,60] However, because of the existence of confounding factors that limit the ability to extrapolate results from field studies, unequivocal evidence of a cause–effect linkage between disease development and mass mortalities in marine mammals is lacking. Apart from chemical contaminants, exposure to natural marine toxins has been hypothesized as a possible cause for the mortality of bottlenose dolphins along the Atlantic coast of North America[61]; however, later studies have indicated that this evidence is circumstantial.[25] Morbillivirus infection appears to have been at least a contributing factor in the Atlantic bottlenose dolphin mortality.[52] Similarly, factors such as population density, migratory movement, habitat disturbance, and climatological factors have been proposed to play a role in mass mortalities of marine mammals.[62] Another hypothesis is that synthetic chemicals, specifically AhR-active POCs, render marine mammals susceptible to opportunistic bacterial, viral, and parasitic infection.[25] Debilitating viruses such as morbillivirus may result in further immunosuppression, starvation, and death.[25] Conclusions about causality are complicated by the fact that marine mammals are exposed simultaneously to a number of synthetic halogenated hydrocarbons, many of which are not quantified or identified. Despite the high accumulation and possible adverse effects of PCBs in marine mammals, tissue concentrations of PCBs that would affect the immune system in marine mammals have not been established.

9.5.2 IMMUNOTOXICOLOGICAL STUDIES IN THE HARBOR SEAL

A semifield study was conducted, in which immune function was compared in two groups of wild-caught captive harbor seals that were fed herring originating from either the Baltic Sea ($n = 12$), an area of high contamination with a number of pollutants including POCs, or from the Atlantic Ocean ($n = 12$), a less contaminated area.[26] Seals had been caught as recently weaned pups in a relatively uncontaminated area, and were allowed an acclimation period of 1 year before the commencement of the feeding study, which lasted for 93 weeks. Seals of both groups remained healthy and exhibited normal growth patterns during the study. Blood of seals fed Baltic Sea fish contained significantly lower concentrations of vitamin A, lower natural killer (NK) cell activity, and exhibited less lymphocyte proliferation following exposure to mitogens compared with the seals fed Atlantic fish. The effect on immune function was observed within 4 to 6 months of the start of the experiment.[4,63–65] The presence of a variety of co-contaminants in the diet precludes the assumption that PCBs were the only cause for the observed immune dysfunction. Nevertheless, based on the results of other laboratory studies involving exposure of rats to AhR active compounds,[65,66] reduction in the lymphocyte proliferative response in the seals was consistent with an AhR-mediated mechanism of action. The PCBs accounted for 80 to 93% of the AhR active compounds in the diet of seals.[4,63–65] Thus, the observed effects in seals were attributed primarily to PCBs[63–65] even though other immunotoxic contaminants could have been present. Based on this, a dietary NOAEL (no observable adverse effect level) for PCBs of 5.2 µg/kg bw/day or 0.58 ng TEQ/kg bw/day was derived; the corresponding LOAEL (lowest observable adverse effect level) values were 29.2 µg/kg bw/day or 5.8 ng TEQ/kg bw/day. The NOAEL and LOAEL values for blubber TEQ concentrations were 90 and 286 ng/g on a lipid weight basis, respectively.

In another feeding study, fish collected from the Dutch Wadden Sea were fed to one group of captive harbor seals while the control group was fed less contaminated fish from the Atlantic Ocean for approximately 2 years.[67,68] Blood from seals exposed to 30 µg PCBs/kg bw/day contained significantly lower retinol and thyroid hormone.[68] Furthermore, reproductive success of the seals fed Wadden Sea fish was significantly lower than those fed the less contaminated fish.[3] Based on these results, a dietary NOAEL of 0.1 µg PCBs/g, wet weight, in fish or 5 µg PCBs/kg bw/day intake in seals or an NOAEL for maximum allowable toxicant concentration (MATC) of 5.2 µg PCBs/g, lipid weight (4.5 ng/g, wet weight), in seal blood, were derived. The corresponding LOAEL values were 0.2 µg PCBs/g in the diet, 30 µg PCBs/kg bw/day intake by seals and 25 μg PCBs/g, lipid weight (16 ng/g, wet weight), in seal blood.

9.5.3 Toxicological Studies in Cetaceans

The relationship between exposure to chemical contaminants and immune function has been examined in free-ranging bottlenose dolphins along the central west coast of Florida.[25] This study showed a negative relationship between lymphocyte proliferative responses to mitogens and concentrations of PCBs and DDTs in the blood of dolphins. The effects of *in vitro* exposure to different POCs, including PCB congeners, on immune functions of beluga whale peripheral blood leukocytes and splenocytes have been examined.[24] In addition to the relatively high concentrations that were used in this *in vitro* study, the final dose delivered to cells could not be determined, preventing derivation of tissue residue based on NOAEL or LOAEL values. Moreover, only a few PCB congeners have been tested in these *in vitro* studies that also use relatively short incubation times (several hours), which may not be sufficient to observe toxic effects. Therefore, effect concentrations of PCBs could not be derived from this study.

9.5.4 Exposure Studies in Mustelids

Mustelids such as mink and otter are sensitive to the toxic effects of PCBs and other organochlorine chemicals.[69] Over the last decades, populations of European otters (*Lutra lutra*) have declined dramatically.[70,71] PCB pollution is considered to be a major factor in this decline, although several possible causes such as habitat destruction, drowning in fishing nets, traffic accidents, eutrophication, acidification, and toxic chemicals have also been implicated for the otter population decline.[72] The toxicological plausibility of the contaminant hypothesis is supported by numerous studies in mink (*Mustela vison*) that have demonstrated the sensitivity of this species to the adverse effects of POCs.[73–75]

A semifield study that examined hepatic retinoids and corresponding total PCB concentrations in environmentally exposed feral and captive otters was used to derive threshold PCB concentrations in otters.[75,76] A TEQ concentration of 2 ng/g lipid weight in the liver or blood was considered as a LOAEL for hepatic retinoid concentrations.[76] A dietary NOAEL has been estimated based on the diet-specific biomagnification of TEQs from fish to otters.[71] Although the hepatic retinoid concentrations in European otters were negatively correlated with both TEQs and total PCBs, the relationship with total PCB concentrations was less pronounced.[75,76]

For otters, a total PCB concentration of 50 μg/g lipid weight was proposed as a critical level in the early 1980s.[70,77] This value was approximately an order of magnitude greater than the NOAEL for vitamin A deficiency of 4 μg PCBs/g lipid weight[75] in the liver but consistent with the concept that physiological effects usually occur at lower concentrations than those that cause effects at the individual or population level.

Several studies have demonstrated that the mink is among the most sensitive species to the toxic effects of AhR-active compounds.[73,77,78] For this reason, there have been a considerable number of studies of PCB effects on mink.[74,78,79] Several authors have critically reviewed the toxic effects of PCBs to mink to derive NOAEL values.[23,69,80,81]

9.5.5 TOXICITY REFERENCE VALUES

It is clear from the studies described above and studies of other species that exposure to POCs above threshold concentrations can result in adverse biological outcomes. Unfortunately, most of the evidence for POC-caused adverse effects in marine mammals remains anecdotal. The evidence for a cause–effect relationship is strong and has been extensively reviewed recently.[82] But it will not be possible to predict the probability or degree of adverse effects from measured POC concentrations until a dose–response relationship has been characterized, e.g., until TRVs are derived.

Toxicity reference values are the best available estimates of the concentrations of chemical contaminants that are likely to cause adverse effects. Conclusions can be drawn about the likelihood of the observed concentrations causing adverse effects by taking measured environmental concentrations of contaminants and comparing them to TRVs. The TRVs are ideally derived from chronic toxicity studies in which an ecologically relevant end point was assessed in the species of concern, or a closely related species. While TRVs can be based on or defined as NOAELs, the use of LOAELs is generally preferred as NOAELs by definition incorporate greater uncertainty than LOAELs. Alternatively, TRVs can be based on the geometric mean of the NOAEL and LOAEL to provide a conservative estimate of a threshold of effect.[78] There are three potential problems with the extrapolation of laboratory toxicity data to species exposed to POCs in the environment. The first is the wide range of sensitivities that even closely related species show to AhR-active chemicals.[83] For example, there is a 5000-fold difference in the toxicity of TCDD between hamsters and guinea pigs. The second difficulty is that most laboratory studies are based on exposure to complex mixtures of AhR-active compounds such as technical PCB mixtures that may be substantially different from the PCB congener mixture to which animals in the environment are exposed.[84,85] The third problem applies to the toxic equivalency factor/TCDD equivalent (TEF/TEQ) approach. When using a TEF/TEQ approach, all possible effort should be made during the literature review of TRVs based on TEQs to ensure that the TEQs are based on the most appropriate set of TEFs. For example, the TEQ-based TRVs for bald eagles derived by Elliot et al.[86] were calculated from a mammal-based set of TEFs because bird-specific TEFs were not available. In such situations, an appropriate TEQ-based TRV can be recalculated provided the congener-specific data are available. Although mammalian TEF values are available in the case of marine mammals, there are no marine mammal–specific studies available that demonstrate that marine mammal TEFs are comparable to those for the laboratory species commonly used for TEF derivation. The TEFs recently adopted by the World Health Organization were used in the studies described here.[87,88] These two TEF sets are based on a wide range of mammalian species and end points and, consequently, are those most likely to be applicable to marine mammals. Only slight differences exist between the two sets of TEF values and we indicate where the different sets were used in this assessment.

It is essential to perform a critical evaluation of the applicability of the toxicological data to the site-specific receptors of concern and exposure pathways. For the majority of wildlife receptors, TRVs derived in the same species are not available and it is therefore necessary to derive them using toxicological data for surrogate species in combination with uncertainty factors (UFs). Uncertainty concerning interpretation of the toxicity test information among different species, different laboratory end points, and differences in experimental design, age of test animals, and duration of test is addressed by applying UFs to the toxicology data to derive the final TRV. Adjustment to accommodate uncertainty is particularly difficult with PCBs and related chemicals because of the relatively great interspecies differences in sensitivity mentioned above.

A large database of toxicological studies for the effects of POCs is available; however, for each group of biota, considerations must be made regarding the appropriateness and usefulness of data for ecological risk assessments relative to marine mammals. Thus, for each group of biota, recommendations have been provided separately. In general, a weight-of-evidence approach should be utilized in which multiple measurement end point approaches (dietary TRVs, tissue residue-based TRVs, and field studies) provide separate lines of evidence.

To develop tissue residue effect levels, controlled laboratory exposure studies involving marine mammals are needed. For ethical, logistical, and practical reasons, only a few controlled "laboratory" exposure studies of marine mammals with small numbers of individuals have been conducted. Application of tissue residue guidelines derived from laboratory mammals such as the rat or guinea pig for the assessment of risks of POCs to marine mammals is inappropriate because of differences in pharmacokinetics[89–91] and potential differences in responsiveness to POCs of these classes of animals.[92] Even among marine mammal species, differences exist in cytochrome P-450 mono-oxygenase activities that metabolize POCs.[89,93,94] Tissue residue guidelines derived for the protection of other fish-eating aquatic mammals such as mink or otter, which are thought to have similar reproductive physiologies to those of certain marine mammals, provide a possible means of estimating the risks POCs present to marine mammals. For example, delayed embryo implantation seen in mink is similar to the reproductive process found in marine mammals.[95,96] It should be remembered that the dynamics of POC concentrations in a small mustelid with a relatively constant weight and food intake will be very different from the dynamics of POCs in a 50-tonne cetacean that may lose 50% of its lipid reserves annually.

9.5.6 TOXICITY THRESHOLD EVALUATION

The NOAEL and LOAEL values for toxic effects of PCBs in seals, dolphins, otter, and mink were used to derive a threshold dose for total PCBs and TEQs. Threshold values based on tissue residues (i.e., MATC values) can be applied in risk assessments because monitoring studies usually report concentrations in specific body tissues. The threshold dose for adverse effects was estimated as the geometric mean of the NOAEL and LOAEL. The rationale for selecting the threshold dose instead of the NOAEL or LOAEL is that the latter two parameters could be strongly influenced by study design and may not reflect the specific point of the dose–response relationship. Although the application of NOAEL as a reference dose could be overprotective, the LOAEL could be underprotective for the observed effects. Detailed information for TRV derivation used in this report are provided elsewhere and are only summarized here.[5]

Overall, the threshold values for the liver or blood concentrations of PCBs in seal, otter, and mink range from 6.6 to 11.0 µg/g lipid weight. The geometric mean of these values, 8.7 µg/g lipid weight, is suggested as a threshold concentration for PCBs in marine mammal liver or blood. Overall, the minimal concentrations of PCBs found in livers of diseased or dead marine mammals were in the range of 0.06 to 7 µg/g on a wet weight basis.[55,97–103] Assuming a liver lipid content of 5% in marine mammals, a reasonable threshold concentration derived from these studies would be 0.44 µg PCBs/g (wet weight basis). These results suggest that the concentrations of PCBs in diseased marine mammals were greater than the threshold values estimated in this study, supporting the estimated threshold value.

The majority of reports of PCBs in marine mammals have been from blubber samples. Therefore, threshold values that are based on blubber concentrations would expedite the risk assessment process. We extrapolated the concentrations of total PCBs in the blood to the blubber of marine mammals based on the observed relationships between blubber blood and liver PCB concentrations. The lipid-normalized concentrations of PCBs in the liver, blood, and blubber have been reported to be within a factor of two in seals.[104] By applying a factor of two to account for the differences in the lipid-normalized concentrations for PCBs in blood and blubber, a threshold concentration for PCBs in the blubber of marine mammals of 17 µg PCBs/g lipid weight was derived.

The geometric mean of the NOAEL and the LOAEL for reproductive effects observed in mink that were fed carp from Saginaw Bay, Lake Huron, in the United States, was 60 pg TEQ/g wet weight in liver.[78] This value is only two- to threefold less than the EC_{50} values for the relative litter size and kit survival, i.e., 160 and 200 pg/g wet weight, respectively.[23] Considering the EC_{50} values, which were derived based on the compilation of data from several controlled laboratory exposure studies in mink, the earlier reported value of 60 pg TEQ/g wet weight appears to be less conservative. The U.S. EPA recommends that a range of 1 to 10 be used as the LOAEL to NOAEL uncertainty factors depending upon the magnitude and severity of the effect.[105] A fivefold safety factor was applied to the EC_{50} for relative litter size to derive a threshold value of 32 pg TEQ/g wet weight of liver. A lipid-normalized threshold value of 640 pg/g for TEQs in mink liver was determined by applying the average lipid content of 5% observed in healthy mink.[106] This value is fourfold greater than the threshold value of 160 pg TEQ/g lipid weight in seal blubber (geometric mean of NOAEL and LOAEL for effects on NK cell function), and twofold less than the threshold value for hepatic vitamin A reduction in European otters of 1400 ng/g lipid weight. Thus, the threshold values for TEQs in seals, mink, and otter ranged from 160 to 1400 pg TEQ/g lipid weight. The geometric mean of the three values was 520 pg/g lipid weight. The estimated threshold concentrations for TEQs in this report are supported by the results of several field studies that showed a TEQ concentration of >300 pg/g wet weight in the blubber of diseased and stranded marine mammals.[60,84,107,108]

Evaluation of dietary threshold concentrations for POCs in marine mammals requires species information on the composition of prey items in the diet, BMFs of POCs, trophic status in the food chain, and the lipid content of the diet: therefore, the threshold concentrations derived for diet should consider BMFs in the predator of concern. Exposure studies with mink have been used to calculate a dietary NOAEL of 17 to 72 ng PCBs/g wet weight, depending on the daily ingestion rates for diet used in these calculations,[80,109] which assume either a daily ingestion rate of 0.25 kg to yield a NOAEL of 72 ng/g wet weight of PCBs in the diet[80] or an assumed ingestion rate of 1.5 kg/day to yield a dietary NOAEL of 17 ng/g wet weight.[109] These results also imply that the dietary threshold concentrations for PCBs in marine mammals cannot be represented by a single value but require a range of values instead. Overall, the concentration of total PCBs in the diet ranging from 10 to 150 ng/g wet weight has been shown to exert toxic effects in the aquatic mammals studied. The geometric mean of the values was 89 pg/g wet weight. The dietary threshold for TEQ concentrations for mink and otter were 1.9 and 1.4 pg/g wet weight, respectively. The geometric mean of these two values (1.6 pg/g wet weight) is suggested as an estimate for risk assessment purposes. The threshold values estimated in our analysis are within the range of 0.79 to 2.4 pg/g wet weight for dietary TEQs proposed for mustelids and pinnipeds by Environment Canada.[110]

9.5.7 Uncertainties in TRV Determination

A degree of uncertainty is inherent in all areas of ecological risk assessment. Uncertainties concerning interpretation of the toxicity test information among different species, different laboratory end points, and different experimental designs (e.g., age of test animals, duration of test) are typically addressed by applying UFs to literature-based toxicity data to calculate the final threshold concentrations. Uncertainty factors may need to be applied to these threshold doses derived for PCBs and TEQs depending upon the objectives of the risk assessment and site- and species-specific exposure scenarios. The U.S. EPA recommends application of UFs for intertaxon variability (for the extrapolation of data from surrogate laboratory species to the receptor of concern), exposure duration (acute or chronic exposures in the light of longevity of the receptor), and toxicological end points.[81,105]

The toxicity end points used in most mink studies were reproductive effects, whereas those in harbor seal studies were immune system effects. There could be other, more sensitive end points that were not studied. It has been shown recently that bacculum size in mink was negatively correlated with hepatic PCB concentrations above 0.02 μg/g wet wt.[111] This concentration is approximately ten times lower than the concentration affecting mink survival.

Threshold concentrations derived in this study are based primarily on exposure to contaminated diet. Even though this approach has an advantage of mimicking exposures under field conditions, the threshold values derived for PCBs could be conservative if one considers the likelihood of co-contaminants in the exposure diet. While immunotoxicity observed in seals was hypothesized to be mediated by an AhR-mediated mechanism, other immunotoxicants such as butyltin compounds that act through non-AhR mediated mechanisms of action, can also contribute to the observed effects.[112–114]

Toxic effects of POCs were compared for a range of aquatic mammals in this assessment because of limited data availability for individual classes of mammals. There is considerable variation in aquatic mammal sensitivity to PCBs. For instance, literature data on the toxicity of PCBs reveal large differences in sensitivity even among different mustelid species. Mink, otters, weasel (*M. nivalis*), and stoat (*M. erminea*) are less tolerant to the toxic effects of PCBs than ferrets (*M. furo*) and polecats (*M. putorius*).[115,116] The differences in sensitivities could be attributed to differences in diet, and selective biomagnification of toxic PCB congeners, and biotransformation capacities of individual species. Therefore, when toxicological data for individual species are available, threshold values derived specifically for the species should be applied in hazard assessment.

Mink and otter are among the aquatic mammals most sensitive to toxic effects of PCBs. A compilation of toxicity data from the literature indicates that harbor seals are comparably sensitive to toxic effects of PCBs. The tissue residue guidelines proposed by Environment Canada for mustelids (otters and mink) and pinnipeds (seals) were within a factor of two,[110] which corroborates our estimates for various sensitive physiological end points.

9.6 RISK CHARACTERIZATION

9.6.1 Risk Assessment Based on New Zealand Data

Over the past several years, data have been accumulating for concentrations of persistent organic contaminants in the New Zealand marine environment, particularly marine mammals.[29,31,117] These data are particularly interesting because they represent some of the lower levels of exposure for marine mammals to persistent organic contaminants. Industrial activity is limited in the Southern Hemisphere; therefore, the major source of POC to the southern oceans appears to be atmospheric deposition.[8] Atmospheric transport of POCs from the more contaminated Northern Hemisphere seems to be limited; thus, POC concentrations are considerably lower in the Southern Hemisphere.[118] As a result, any risks indicated for Southern Hemisphere species are likely lower than those for Northern Hemisphere species.

Concentrations of total PCBs, PCB congeners, dioxin, and furan congeners have been measured in marine organisms from a range of trophic levels from the nearshore ocean environment off the east coast of New Zealand (Table 9.2). Samples analyzed included a range of zooplankton, forage fish, commercial fish species, and blubber samples from New Zealand fur seal (*Arctocephalus forsteri*).[117] The PCBs were detected in all fur seal samples with concentrations in some samples exceeding 1 µg/g wet weight. Concentrations in other biota samples were generally in the low ng/g range. A number of sediment samples from a major city harbor and the east coast of the South Island were also analyzed. Concentrations of PCB congeners were measured in more than 70 samples of blubber from pilot whale (*Globicephala melaena*) collected from various locations around New Zealand.[29] Pilot whale samples contained from 33 to 931 ng/g wet weight of PCBs with a mean of 311 ng/g wet weight. The congener patterns in all pilot whale samples were similar, suggesting a common source, i.e., atmospheric deposition.

TABLE 9.2
Organochlorine Concentrations in the New Zealand Marine Environment

	Location	ΣPCB (ng/g)	ΣDDT (ng/g)	TEQ (pg/g)	Ref.
Inshore sediment	Banks Peninsula	0.36–7.3		0.99–4.14	40
Inshore sediment	Wellington	8.2–42	0.05–786	—	17
Planktonic crustacea	East coast	0.14–0.29	0.07–0.21	—	17
Small fish	East coast	0.37–0.73	0.23–0.98	—	17
Hoki/mackerel	Cook Strait	0.47–2.19	1.27–6.16	—	17
Assorted fish	Banks Peninsula	0.36–7.2	—	0.1–1.3	40
Fur seal blubber	Wellington	48.3–1069	92.1–8650	—	17
Hector's dolphin	Banks Peninsula	300–4500	—	18–200	40
Pilot whale blubber	Various	33–931	—	—	29
Albatross egg	Chatham Islands	18–66.7	20.6–75.82	3.03–6.32	120

Concentrations of dioxins and PCBs measured in the blubber of New Zealand marine mammals have previously been reported.[31] PCB congeners were detected in all samples; the average sum of PCBs was lowest (<50 ng/g wet weight) in the open ocean baleen or mysticete plankton-feeding whales, minke (*Balaenoptera acutorostrata*), blue (*B. musculus*), and pygmy right whale (*Caperea marginata*), and intermediate (100 to 500 ng/g wet weight) in open ocean toothed or odontocete whales and dolphins that consume fish and squid. The average sum of PCBs was highest (750 to >1000 ng/g wet weight) in the inshore Hector's dolphin (*Cephalorhynchus hectori*) (Table 9.3).[119]

The PCDD and PCDF congeners were commonly detected in the inshore feeding dolphins only while most congeners were below detection limits in the open ocean baleen whale species.[31] In the open-ocean dolphins and beaked whales, hepta- and octa-chlorinated PCDDs and PCDFs were the most commonly detected congeners. TEFs[88] were used to calculate TEQs for the dioxin and PCB mixtures found in the marine mammal samples (Table 9.4). As anticipated, TEQ concentrations were lowest in the baleen whales, higher in the open-ocean odontocetes, and highest in the inshore species that showed the greatest concentrations of dioxin congeners. Pattern recognition techniques were also used in these studies to demonstrate that open ocean and inshore species were exposed to distinct sources of contaminants.[31] Such studies can be used to assess the relative risks in different portions of the marine environment.

These studies illustrate the large extent to which POCs biomagnify in marine mammals. Specifically, biomagnification ratios (concentration in marine mammals compared with those in food) in New Zealand marine mammals appear to be greater than in more contaminated Northern Hemisphere environments (Table 9.2).[40,117] As a consequence, although environmental concentrations of POCs are several orders of magnitude lower than in similar Northern Hemisphere locations, the concentrations accumulated in New Zealand marine mammals were generally within an order of magnitude of those in similar species in the Northern Hemisphere (see Table 9.3). This is particularly the case for those species feeding at higher trophic levels such as the open-ocean odontocetes (e.g., pilot whales) than for those feeding at lower trophic levels such as the mysticetes.

TABLE 9.3
Concentrations of Total PCBs Measured in New Zealand Cetaceans Compared with Cetaceans from Other Geographical Regions

Species	Ref.	Location	PCBs (µg/g)
Bottlenose dolphin	119	South Africa	13.8
	56	East United States	81.4
Dall's porpoise	121	North Pacific	8.6
White-sided dolphin	121	Japan	37.6
	56	East United States	50.1
Common dolphin	56	East United States	36.5
	31	New Zealand	0.3–1.5
Harbor porpoise	102	United Kingdom	55.5
Dusky dolphin	121	South of New Zealand	1.4
Hector's dolphin	31	New Zealand	0.4–2.6
Baleen whales	31	New Zealand	0.01–0.02
Minke whales	122	West United States	3.3
Beaked whales	31	New Zealand	0.1–0.5
Baird's beaked whales	123	Japan	3.0
Pilot whale	122	East United States	17
	124	United Kingdom	36.9
	29	New Zealand	0.03–0.93

TABLE 9.4
Summary of Average Dioxin and PCB Concentrations in New Zealand Cetaceans

Analyte	Pilot Whale[29]	Baleen Whales[21]	Oceanic Dolphins[31]	Beaked Whales[31]	Hector's Dolphin[31]
ΣPCBs (ng/g)	311	12.9	833	251	1018
ΣPCDD/F (pg/g)	—	50.01	870.1	281.6	1204
TEQ (pg/g)	39.7	1.9	15.7	12.5	81.4

Note: TEQ based on PCB congeners only.

Available concentration data and the previously derived TRV values for marine mammals were used to calculate hazard quotients (HQs) to assess the possible risks of dioxin-like compounds to marine mammals in New Zealand. HQs were calculated as the quotient of the measured exposure concentration over the TRV concentration. They were calculated based on both blubber residue concentrations (Table 9.5) and on estimated dietary intake (Table 9.6) from concentrations in likely prey items (Table 9.2). The HQs were also calculated based on either total PCB concentrations or on TEQ concentrations to allow assessment of the relative risks of the different compound groups.

TABLE 9.5
Blubber Residue HQs for New Zealand Cetaceans Based on Mean Total PCBs and TEQ

Species	Blubber PCB TRV (ng/g)[a]	Blubber ΣPCB (ng/g)[b]	ΣPCB HQ	Blubber TEQ TRV (pg/g)	Blubber TEQ (pg/g)[b]	TEQ HQ
Pilot whale[c]	13,600	310	0.023	>300	39.7	0.13
Baleen whales	13,600	12.9	0.0001	>300	1.9	0.006
Oceanic dolphins	13,600	833	0.061	>300	15.7	0.052
Beaked whales	13,600	251	0.018	>300	12.5	0.041
Hector's dolphin	13,600	1018	0.0749	>300	81.4	0.271
New Zealand fur seal[d]	13,600	1069	0.08	>300		
Pilot whale (USA)[e]	13,600	17,000	1.25	>300		
Dall's porpoise[f]	13,600	8,600	0.63	>300		
Bottlenose dolphin (USA)[g]	13,600	81,400	6.0	>300		

[a] Converted to ng/g wet weight from 17,000 (ng/g lipid) assuming lipid content of 80%.
[b] Data from Reference 31 unless specified.
[c] Data from Reference 29.
[d] Data from Reference 117.
[e] Data from Reference 121.
[f] Data from Reference 120.
[g] Data from Reference 56.

TABLE 9.6
Dietary HQs for New Zealand Cetaceans Based on Total PCBs and TEQ

	Dietary					
Species	PCB TRV (ng/g)	PCB (ng/g)	PCB HQ	TEQ TRV (pg/g)	TEQ (pg/g)	TEQ HQ
Pilot whale	10–150	0.14–0.73	0.001–0.073	1.6	0.1–1.3	0.063–0.81
Baleen whales	10–150	0.14–0.73	0.001–0.073	1.6		
Oceanic dolphins	10–150	0.47–2.19	0.0031–0.219	1.6	0.1–1.3	0.063–0.81
Beaked whales	10–150	0.47–2.19	0.0031–0.219	1.6	0.1–1.3	0.063–0.81
Hector's dolphin	10–150	0.36–7.2	0.0024–0.72	1.6	0.1–1.3	0.063–0.81
New Zealand fur seal	10–150	0.37–2.19	0.002–0.219			

As HQs are simple ratios of exposure to effect concentrations (TRV), they reflect how close the current exposure concentrations are to those known to cause adverse effects. Interpretation of HQ categories can be facilitated by applying the twofold safety factor to the HQ value, which results in a ratio that generates a presumption of acceptable hazard for all HQ values that are less than 1.0.[6] Any exposure pattern generating an exposure concentration with a HQ greater than 1.0 is determined to have exceeded a level of concern and the potential for adverse effects on organisms is assumed. Because of the inherent conservatism of the approach, a HQ value greater than 1 does not indicate that population or community-level effects would be expected to occur, but rather that there is a sufficient level of concern to warrant additional study. In logical terms, the hypothesis that effects were occurring could not be rejected. Generally, HQ values need to exceed 20 before a very great level of concern would be warranted.

In cases where extrapolation of the TRV is required to predict a TRV in a different species or in a different tissue in the same species, uncertainty factors are commonly applied as multipliers to the TRV to provide a conservative margin of safety. Uncertainty factors were not applied in this assessment as the TRVs used were based on a variety of species and a UF had already been applied in the derivation of the TRV. The species used in the derivation were also similar to the species of interest. Also, use of mink and otter, which are among the most sensitive species to the effects of dioxin-like chemicals, to derive TRVs makes it less likely that other marine mammals would be more sensitive than suggested by the TRV.

All assessments indicated that risks are generally low for open-ocean species if based on total PCB concentrations in food. Risks posed to open-ocean marine mammals by dioxins and PCBs, expressed as TEQ, are also relatively low. However, as has been concluded in other studies of the open-ocean environment, we conclude that current concentrations leave little margin of safety for added exposure to these compounds despite that concentrations are generally below adverse effect concentrations.[114]

Risk estimates were highest for species at higher trophic levels such as the pilot whale and Hector's dolphin (*C. hectori*) which had HQ values of 0.13 and 0.27, respectively (Table 8.5). The HQs for all other New Zealand species were below 0.1, suggesting a low probability of adverse effects to these species. The HQs were greater if the biological potency of the whole mixture was considered by assessing the measured concentrations relative to the TRVs based on TEQ. This suggested that dioxins and furans are the major chemicals of concern to be considered in the risk assessment and also explains the lower risks predicted for open-ocean marine mammals. In contrast to the New Zealand species, HQs for Northern Hemisphere species were closer to or greater than 1, indicating a higher likelihood of adverse effects.

Risk estimates based on dietary intake and TRV estimates (see Table 8.6) were similar to those for the blubber residue-based assessment. For Hector's dolphin and pilot whales, the HQs were similar to those determined using blubber residue concentrations. The similarity of the dietary estimates among species can be explained by the use of the same dietary concentrations for all species. HQs were not calculated for baleen whales because of the limited amount of data available for their prey items. Similarly, diet-based HQ assessments were not made for Northern Hemisphere studies because of the relative lack of suitable prey concentration data.

9.7 CONCLUSIONS

Exposure to POCs in the open ocean is limited compared with inshore regions. This is particularly the case for dioxins that do not appear to travel great distances from the regions of origin and, consequently, are not present at high concentrations in open-ocean biota.[31] In contrast, the more volatile PCBs are distributed globally and accumulate in open-ocean biota. Although contamination in the open ocean is relatively low, increases in contaminant concentrations can be seen in inshore areas. This is presumably due to human activity, even for countries with relatively low human impact such as New Zealand. Therefore, risks posed by PCBs and dioxins are greater in the case of inshore feeding species of marine mammals. This study and those conducted with albatross in the North Pacific demonstrate that there appear to be narrow margins of safety for inshore species based on current environmental concentrations and those known to cause adverse effects in other species. As margins of safety for these chemicals are generally narrow, the world appears to have little remaining assimilative capacity for these contaminants of global concern.

Quantitative assessment of the risks posed to marine mammals by organic contaminants is limited by a lack of controlled toxicological studies from which toxicological reference doses can be derived. In the case of dioxin-like chemicals, uncertainty associated with this lack of data is exacerbated by the wide range of species sensitivities evidenced in the existing data. This range of sensitivity suggests that surrogate species selection is particularly crucial in reducing uncertainty in risk assessments. Clearly, ethical and logistic factors preclude *in vivo* exposure studies on a wide range of marine mammal species. However, there are currently available a range of *in vitro* toxicological assay procedures, using cultured cells or tissues, that could be used to provide comparative data for a range of species.

This screening level risk assessment has demonstrated that the possibility of risks to marine mammals from exposure to POCs cannot be discounted. It suggests the need for further study to refine the risk assessment procedures and assumptions. Studies should initially focus on tissue residue-based assessments because adequate modeling of contaminant accumulation is difficult for such wide-ranging animals. In this context, studies are required to better correlate tissue contaminant concentrations with measures of adverse biological effects. Such studies will permit better estimates of marine mammal–specific tissue residue-based TRVs that can be used in subsequent risk assessment.

ACKNOWLEDGMENTS

The authors wish to acknowledge the thorough and helpful reviews of Dr. Alonso Aguirre of Tufts University and one anonymous reviewer. The significant contributions made by Caren Schroeder and Peter Day for the analysis of New Zealand marine mammal samples are acknowledged as is the Institute for Environmental Science and Research, Wellington, New Zealand, for making that research possible.

All New Zealand marine mammal samples analyzed in these studies were collected from dead stranded animals or from animals killed because they could not be returned to the sea. All decisions and procedures pertaining to euthanasia of New Zealand marine mammals were carried out by Department of Conservation field staff who retain statutory authority over stranded marine mammals.

REFERENCES

1. Simmonds, M., Cetacean mass mortalities and their potential relationship with pollution, in *Whales. Biology — Threats — Conservation — 1991*, Symoens, J.J., Ed., Royal Academy of Overseas Sciences, Brussels, 1991.
2. Aguilar, A. and Borrell, A., Abnormally high polychlorinated biphenyl levels in striped dolphins (*Stenella coeruleoalba*) affected by the 1990–1992 Mediterranean epizootic, *Sci. Total Environ.*, 154, 237, 1994.
3. Reijnders, P.J.H., Reproductive failure in common seals feeding on fish from polluted coastal waters, *Nature*, 324, 456, 1986.
4. Ross, P.S. et al., Suppression of natural killer cell activity in harbour seals (*Phoca vitulina*) fed Baltic Sea herring, *Aquat. Toxicol.*, 34, 71, 1996.
5. Kannan, K. et al., Toxicity reference values for the toxic effects of polychlorinated biphenyls to aquatic mammals, *Hum. Ecol. Risk Assess.*, 6, 181, 2000.
6. U.S. EPA, Guidelines for Ecological Risk Assessment. U.S. EPA/630/R-95/002F, Washington, D.C., 1998.
7. Wania, F. and MacKay, D., Tracking the distribution of persistent organic pollutants, *Environ. Sci. Technol.*, 30, 390, U.S. EPA, Washington, D.C., 1996.
8. Iwata, H. et al., Distribution of persistent organochlorines in the oceanic air and surface seawater and the role of ocean on their global transport and fate, *Environ. Sci. Technol.*, 27, 1080, 1993.
9. Tanabe, S., Iwata, H. and Tatsukawa, R., Global contamination by persistent organochlorines and their ecotoxicological impact on marine mammals, *Sci. Total Environ.*, 154, 163, 1994.

10. Bowles, D., An overview of the concentrations and effects of metals in cetacean species, *J. Cetacean Res. Manage.* (Special Issue 1), 125, 1999.
11. Rawson, A.J., Patton, G.W., Hofmann, S., Pietra, G.G., and Johns, L., Liver abnormalities associated with chronic mercury accumulation in stranded Atlantic bottlenose dolphins, *Ecotoxicol. Environ. Saf.*, 25, 41, 1993.
12. O'Shea, T.J. and Brownell, R.L., Organochlorine and metal contaminants in baleen whales: a review and evaluation of conservation implications, *Sci. Total Environ.*, 154, 179, 1994.
13. Tanabe S. et al., Capacity and mode of PCB metabolism in small cetaceans, *Mar. Mammal Sci.*, 4, 103, 1988.
14. Jones, P.D. et al., Persistent synthetic chlorinated hydrocarbons in albatross tissue samples from Midway Atoll, *Environ. Toxicol. Chem.*, 15, 1793, 1996.
15. Rappe, C. and Kjeller, L.-O., PCDDs and PCDFs in the environment. Historical trends and budget calculations, *Organohalogen Compd.*, 20, 1, 1994.
16. Safe, S., Polychlorinated biphenyls (PCBs), dibenzo-*p*-dioxins (PCDDs), dibenzofurans (PCDFs), and related compounds: environmental and mechanistic considerations which support the development of toxic equivalency factors (TEFs), *Crit. Rev. Toxicol.*, 21, 51, 1990.
17. Okey, A.B., Riddick, D.S., and Harper, P.A., The Ah receptor: mediator of the toxicity of 2,3,7,8-tetrachlorodibenzo-*p*-dioxin (TCDD) and related compounds, *Tox. Lett.*, 1, 1, 1994.
18. Peterson, R.E., Theobald, H.M., and Kimmel, G.L., Developmental and reproductive toxicity of dioxins and related compounds: cross-species comparisons, *Crit. Rev. Toxicol.*, 23, 283, 1993.
19. Tanabe, S. and Tatsukawa, R., Chemical modernization and vulnerability of cetaceans: increasing toxic threat of organochlorine contaminants, in *Persistent Pollutants in Marine Ecosystems*, Walker, C.H., Livingstone, D.R., Lipnick, R.L., and La Point, T.W., Eds, Pergamon Press, Oxford, 1992, 161.
20. Kennedy, S.W. et al., Cytochrome P4501A induction in avian hepatocyte cultures: a promising approach for predicting the sensitivity of avian species to toxic effects of halogenated aromatic hydrocarbons, *Toxicol. Appl. Pharmacol.*, 141, 214, 1996.
21. Giesy, J.P., Ludwig, J.P., and Tillitt, D.E., Deformities in birds of the Great Lakes region. Assigning causality, *Environ. Sci. Technol.*, 28, 128A, 1994.
22. Bowerman, W.W. et al., A review of factors affecting productivity of bald eagles in the Great Lakes region: implications for recovery, *Environ. Health Perspect.*, 103 (Suppl.), 51, 1995.
23. Leonards, P.E.G. et al., Assessment of experimental data on PCB-induced reproduction inhibition in mink, based on an isomer- and congener-specific approach using 2,3,7,8-tetrachlorodibenzo-*p*-dioxin toxic equivalency, *Environ. Toxicol. Chem.*, 14, 639, 1995.
24. De Guise, S. et al., Effects of *in vitro* exposure of beluga whale leukocytes to selected organochlorines, *J. Toxicol. Environ. Health*, 55, 479, 1998.
25. Lahvis, G.P. et al., Decreased lymphocyte responses in free-ranging bottlenose dolphins (*Tursiops truncatus*) are associated with increased concentrations of PCBs and DDT in peripheral blood, *Environ. Health Perspect.*, 103, 67, 1995.
26. de Swart, R.L. et al., Impairment of immune function in harbor seals (*Phoca vitulina*) feeding on fish from polluted waters, *Ambio*, 23, 155, 1994.
27. Tanabe, S. et al., Transplacental transfer of PCBs and chlorinated hydrocarbon pesticides from the pregnant striped dolphin (*Stenella coeruleoalba*) to her fetus, *Agric. Biol. Chem.*, 46, 1249, 1982.

28. Borrell, A., Bloch, D., and Desportes, G., Age trends and reproductive transfer of organochlorine compounds in long-finned pilot whales from the Faroe Islands, *Environ. Pollut.*, 88, 283, 1995.
29. Schröder, C., Levels of Polychlorinated Biphenyls and Life History Parameters in Long-Finned Pilot Whales (*Globicephalus melas*) from New Zealand Strandings, M.Sc. thesis, Victoria University of Wellington, Wellington, New Zealand, 1998.
30. Krowke, R. et al., Transfer of various PCDDs and PCDFs via placenta and mother's milk to marmoset offspring, *Chemosphere*, 20, 1065, 1990.
31. Jones, P.D. et al., Polychlorinated dibenzo-p-dioxins, dibenzofurans and polychlorinated biphenyls in New Zealand cetaceans, *J. Cetacean Res. Manage.* (Special Issue 1), 157, 1999.
32. Buckland, S. et al., Polychlorinated dibenzo-p-dioxins and dibenzofurans in New Zealand's Hector's dolphin, *Chemosphere*, 20, 1035, 1990.
33. Borlakoglu, J.T. et al., Polychlorinated biphenyls (PCBs) in fish-eating sea birds. II. Molecular features of PCB isomers and congeners in adipose tissue of male and female puffins (*Fratercula arctica*), guillemots (*Uria aalga*), shags (*Phalacrocorax aristotelis*) and cormorants (*Phalacrocorax carbo*) of British and Irish coastal waters, *Comp. Biochem. Physiol.*, 97C, 161, 1990.
34. Walker, C.H., Persistent pollutants in fish-eating sea birds — bioaccumulation, metabolism and effects, *Aquat. Toxicol.*, 17, 293, 1990.
35. Solly, S.R.B. and Shanks, V., Organochlorine residues in New Zealand birds and mammals, *N.Z. J. Sci.*, 19, 53, 1976.
36. Bennington, S.L. et al., Patterns of chlorinated hydrocarbon contamination in New Zealand sub-Antarctic and coastal marine birds, *Environ. Pollut.*, 8, 135, 1975.
37. Polischuk, S.C. et al., Relationship between PCB concentration, body burden, and percent body fat in female polar bears while fasting, *Organohalogen Compd.*, 20, 535, 1994.
38. Oehme, M. et al., Concentrations of polychlorinated dibenzo-p-dioxins, dibenzofurans and non-ortho substituted biphenyls in polar bear milk from Svalbard (Norway), *Environ. Pollut.*, 90, 401, 1996.
39. McKim, J.M. and Johnson, K.L., Polychlorinated biphenyls and p,p'-DDE in loggerhead and green postyearling Atlantic sea turtles, *Bull. Environ. Contam. Toxicol.*, 31, 53, 1983.
40. Jones, P.D. et al., Biomagnification of PCBs and 2,3,7,8-substituted polychlorinated dibenzo-p-dioxins and dibenzofurans in New Zealand's Hector's dolphin (*Cephalorhynchus hectori*), *Organohalogen Compd.*, 29, 108, 1996.
41. Auman, H.J. et al., PCBS, DDE, DDT, and TCDD-EQ in two species of albatross on Sand Island, Midway atoll, North Pacific Ocean, *Environ. Toxicol. Chem.*, 16, 498, 1997.
42. Jones, P.D. et al., Biomagnification of bioassay derived 2,3,7,8-tetrachlorodibenzo-p-dioxin equivalents, *Chemosphere*, 26, 1203, 1993.
43. Di Toro, D.M. et al., Technical basis for establishing sediment quality criteria for nonionic organic chemicals using equilibrium partitioning, *Environ. Toxicol. Chem.*, 10, 1541, 1991.
44. Suter, G.W. et al., Ecological risk assessment in a large river-reservoir: 2. Fish community, *Environ. Toxicol. Chem.*, 18, 589, 1999.
45. Leatherwood, S., Reeves, R., and Foster, L., *The Sierra Club Handbook of Whales and Dolphins,* Sierra Club Books, San Francisco, CA, 1983.
46. U.S. EPA, Wildlife Exposure Handbook, U.S. EPA/600/R-93/187, U.S. Environmental Protection Agency, Office of Research and Development, Washington, D.C., 1998.

47. Duce, R.A., Olney, C.E., and Piotrowicz, S.R., Enrichment of heavy metals and organic compounds in the surface microlayer of Narragansett Bay, Rhode Island, *Science,* 176, 161, 1972.
48. Liu, K. and Dickhut, R.M., Surface microlayer enrichment of polycyclic hydrocarbons in southern Chesapeake Bay, *Environ. Sci. Technol.,* 31, 2777, 1997.
49. Beyer, W.N., Heinz, G.H., and Redmond-Norwood, A.W., Eds. *Environmental Contaminants in Wildlife — Interpreting Tissue Concentrations,* Special publication of SETAC Press/CRC Press/Lewis Publishers, New York, 1996.
50. Jarvinen, A.W. and Ankley, G.T., *Linkage of Effects to Tissue Residues: Development of a Comprehensive Database for Aquatic Organisms Exposed to Inorganic and Organic Chemicals.* Society of Environmental Toxicology and Chemistry (SETAC), Pensacola, FL, 1999, 364 pp.
51. Canadian Council of Ministers of the Environment (CCME), Canadian Tissue Residue Guidelines for Polychlorinated Biphenyls for the Protection of Wildlife Consumers of Biota, Guidelines and Standards Division, Science Policy and Environmental Quality Branch, Environment Canada, Hull, Quebec, 1998.
52. Belfroid, A.C., Sijm, D.T.H.M., and Vangestel, C.A.M., Bioavailability and toxicokinetics of hydrophobic aromatic compounds in benthic and terrestrial invertebrates, *Environ. Rev.,* 4, 276, 1996.
53. Colborn, T. and Smolen, M.J., Epidemiological analysis of persistent organochlorine contaminants in cetaceans, *Rev. Environ. Contam. Toxicol.,* 146, 91, 1996.
54. Olsson, M., Karlsson, B., and Ahnland, E., Diseases and environmental contaminants in seals from the Baltic and the Swedish west coast, *Sci. Total Environ.,* 154, 217, 1994.
55. Martineau, D. et al., Levels of organochlorine chemicals in tissues of beluga whale (*Delphinapterus leucas*) from the St. Lawrence estuary, Quebec, Canada, *Arch. Environ. Contam. Toxicol.,* 16, 137, 1987.
56. Kuehl, D.W., Haebler, R., and Potter, C., Chemical residues in dolphin from the U.S. Atlantic coast including Atlantic bottlenose obtained during the 1987/88 mass mortality, *Chemosphere,* 22, 1071, 1991.
57. Lipscomb, T.P. et al., Morbilliviral disease in Atlantic bottlenose dolphins (*Tursiops truncatus*) from the 1987–1988 epizootic, *J. Wildl. Dis.,* 30, 567, 1994.
58. Grachev, M.A. et al., Distemper virus in Baikal seals, *Nature,* 338, 209, 1989.
59. Aguilar, A. and Raga, J.A., The striped dolphin epizootic in the Mediterranean Sea, *Ambio,* 22, 524, 1993.
60. Kannan, K. et al., Isomer-specific analysis and toxic evaluation of polychlorinated biphenyls in striped dolphins affected by an epizootic in the western Mediterranean Sea, *Arch. Environ. Contam. Toxicol.,* 25, 227, 1993.
61. Anderson, D.M. and White, A.W., Toxic dinoflagellates and marine mammal mortalities, 89-3 (CRC-89-6), Woods Hole Oceanographic Institution, Woods Hole, MA, 1989.
62. Lavigne, D.M. and Schmitz, O.J., Global warming and increasing population densities: a prescription for seal plagues, *Mar. Pollut. Bull.,* 21, 280, 1990.
63. Ross, P.S. et al., Contaminant-related suppression of delayed-type hypersensitivity and antibody responses in harbor seals fed herring from the Baltic Sea, *Environ. Health Perspect.,* 103, 162, 1995.
64. Ross, P. et al., Contaminant-induced immunotoxicity in harbor seals: wildlife at risk? *Toxicology,* 112, 157, 1996.
65. Ross, P.S. et al., Impaired cellular immune response in rats exposed perinatally to Baltic Sea herring oil or 2,3,7,8-TCDD, *Arch. Toxicol.,* 17, 563, 1997.

66. Safe, S., Polychlorinated biphenyls (PCBs) — environmental impact, biochemical and toxic responses, and implications for risk assessment, *Crit. Rev. Toxicol.*, 24, 87, 1994.
67. Boon, J.P. et al., The kinetics of individual polychlorinated biphenyl congeners in female harbour seals (*Phoca vitulina*), with evidence for structure-related metabolism, *Aquat. Toxicol.*, 10, 307, 1987.
68. Brouwer, A., Reijnders, P.J.H., and Koeman, J.H., Polychlorinated biphenyl (PCB)-contaminated fish induces vitamin A and thyroid hormone deficiency in the common seal (*Phoca vitulina*), *Aquat. Toxicol.*, 15, 99, 1989.
69. Wren, C.D., Cause-effect linkages between chemicals and populations of mink (*Mustela vison*) and otter (*Lutra canadensis*) in the Great Lakes basin, *J. Toxicol. Environ. Health*, 33, 549, 1991.
70. Olsson, M. and Sandegren, F., Is PCB partly responsible for the decline of the otter in Europe? Proceedings of the Third International Otter Symposium, Strasbourg, November 24–27, 1983, in Proc. V. International Otter Colloquium. Habitat 6, Reuther, C. and Röchert, R., Eds., Hankensbüttel, 1991, 223–227.
71. Leonards, P.E.G. et al., The selective dietary accumulation of planar polychlorinated biphenyls in the otter (*Lutra lutra*), *Environ. Toxicol. Chem.*, 16, 1807, 1997.
72. Mason, C.F. and Macdonald, S.M., Impact of organochlorine pesticide residues and PCBs on otters (*Lutra lutra*) in eastern England, *Sci. Total Environ.*, 138, 147, 1993.
73. Aulerich, R.J. and Ringer, R.K., Current status of PCB toxicity to mink, and effect on their reproduction, *Arch. Environ. Contam. Toxicol.*, 6, 279, 1977.
74. Heaton, S.N. et al., Dietary exposure mink to carp from Saginaw Bay, Michigan. 1. Effects on reproduction and survival, and the potential risks to wild mink populations, *Arch. Environ. Contam. Toxicol.* 28, 334, 1995.
75. Smit, M.D. et al., *Development of Otter-Based Quality Objectives for PCBs*, Institute for Environmental Studies, Vrije Universiteit, Amsterdam, the Netherlands, 1996.
76. Murk, A.J. et al., Application of biomarkers for exposure and effect of polyhalogenated aromatic hydrocarbons in naturally exposed European otters (*Lutra lutra*), *Environ. Toxicol. Pharmacol.*, 6, 91, 1998.
77. Jensen, S. et al., Effects of PCB and DDT on mink (*Mustela vison*) during the reproductive season, *Ambio*, 6, 239, 1977.
78. Tillitt, D.E. et al., Dietary exposure of mink to carp from Saginaw Bay. 3. Characterization of dietary exposure to planar halogenated hydrocarbons, dioxin equivalents, and biomagnification, *Environ. Sci. Technol.*, 30, 283, 1996.
79. Kihlström, J.E. et al., Effects of PCB and different fractions of PCB on the reproduction of the mink (*Mustela vison*), *Ambio*, 21, 563, 1992.
80. Giesy, J.P. et al., Contaminants in fishes from Great Lakes-influenced sections and above dams of three Michigan rivers: II: Implications for health of mink, *Arch. Environ. Contam. Toxicol.*, 27, 213, 1994.
81. Giesy, J.P. and Kannan, K., Dioxin-like and non-dioxin-like toxic effects of polychlorinated biphenyls (PCBs): implications for risk assessment, *Crit. Rev. Toxicol.*, 28, 511, 1998.
82. Ross, P.S., Marine mammals as sentinels in ecological risk assessment, *Hum. Environ. Risk Assess.*, 6, 29, 2000.
83. Gasiewicz, T.A., Nitro compounds and related phenolic pesticides, in *Handbook of Pesticide Toxicology*, Hayes, W.J. and Laws, E.R., Eds., Academic Press, San Diego, 1991, 1191–1270.

84. Storr-Hansen, E. and Spliid, H., Distribution pattern of polychlorinated biphenyl congeners in harbor seal (*Phoca vitulina*) tissues: statistical analysis, *Arch. Environ. Contam. Toxicol.*, 25, 328, 1993.
85. Storr-Hansen, E., Spliid, H., and Boon, J.P., Chlorinated biphenyl congeners in harbour seals (*Phoca vittulina*) and in their food. Statistical comparison of the patterns, *Organohalogen Compd.*, 20, 225, 1994.
86. Elliot, J.E. et al., Biological effects of polychlorinated dibenzo-*p*-dioxins, dibenzofurans, and biphenyls, in bald eagle chicks (*Haliaeetus leucocephalus*) chicks, *Environ. Toxicol. Chem.*, 15, 782, 1996.
87. Van den Berg, M. et al., Toxic equivalency factors (TEFs) for PCBs, PCDDs, PCDFs for humans and wildlife, *Environ. Health Perspect.*, 106, 775, 1998.
88. Ahlborg, U.G. et al., Toxic equivalency factors for dioxin-like PCBS — report on a WHO-ECEH and IPCS consultation — September 1993, *Chemosphere*, 28, 1049, 1994.
89. Watanabe, S. et al., Specific profile of liver microsomal cytochrome P_{450} in dolphin and whales, *Mar. Environ. Res.*, 27, 51, 1989.
90. Boon, J.P. et al., The toxicokinetics of PCBs in marine mammals with special reference to possible interactions of individual congeners with the cytochrome P450-dependent monooxygenase system: an overview, in *Peristent Pollutants in Marine Ecosystems*, Walker, C.H. and Livingstone, D.R., Eds., Pergamon Press, New York, 1992, 119–159.
91. Reijnders, P.J.H., Toxicokinetics of chlorobiphenyls and associated physiological responses in marine mammals, with particular reference to their potential for ecotoxicological risk assessment, *Sci. Total Environ.*, 154, 229, 1994.
92. Murk, A. et al., *In vitro* metabolism of 3,3′,4,4′-tetrachlorobiphenyl in relation to ethoxyresorufin-*O*-deethylase activity in liver microsomes of some wildlife species and rat, *Eur. J. Pharmacol.*, 270, 253, 1994.
93. Goksøyr, A. et al., Cytochrome P450 in seals: monooxygenase activities, immunochemical cross-reactions and response to phenobarbital treatment, *Mar. Environ. Res.*, 34, 113, 1992.
94. Norstrom, R.J. et al., Indications of P450 monoxygenase activities in beluga *Delphinapterus leucas* and narwhal *Monodon monoceros* from patterns of PCB, PCDD and PCDF accumulation, *Mar. Environ. Res.*, 34, 267, 1992.
95. Holcomb, L.C., Reproductive Physiology in Mink (*Mustela vison*). Ph.D. thesis, Department of Zoology, Michigan State University, East Lansing, MI, 1963.
96. Norris, K.S., *Whales, Dolphins and Porpoises*, University of California Press, Berkeley, 1977, 789.
97. Watanabe, M. et al., Congener-specific analysis of polychlorinated biphenyls and organochlorine pesticides, *tris*(4-chlorophenyl)methane and *tris*(4-chlorophenyl)methanol in small cetaceans stranded along Florida coastal waters, USA, *Environ. Toxicol. Chem.*, 19, 1566, 2000.
98. Nakata, H. et al., Accumulation pattern of organochlorine pesticides and polychlorinated biphenyls in southern sea otters (*Enhydra lutris nereis*) found stranded along coastal California, USA, *Environ. Pollut.*, 103, 45, 1998.
99. Lake, C.A. et al., Contaminant levels in harbor seals from the northeastern United States, *Arch. Environ. Contam. Toxicol.*, 29, 128, 1995.
100. Bernhoft, A. and Skaare, J.U., Levels of selected individual polychlorinated biphenyls in different tissues of harbour seals (*Phoca vitulina*) from the southern coast of Norway, *Environ. Pollut.*, 86, 99, 1994.

101. Kannan, K. et al., Persistent organochlorines in harbour porpoises from Puck Bay, Poland, *Mar. Pollut. Bull.*, 26, 162, 1993.
102. Morris, R.J. et al., Metals and organochlorines in dolphins and porpoises of Cardigan Bay, West Wales, *Mar. Pollut. Bull.*, 20, 512, 1989.
103. Troisi, G.M. and Mason, C.F., Cytochromes P450, P420 and mixed-function oxidases as biomarkers of polychlorinated biphenyl (PCB) exposure in harbour seals (*Phoca vitulina*), *Chemosphere*, 35, 1933, 1997.
104. Boon, J.P. et al., A model for the bioaccumulation of chlorobiphenyl congeners in marine mammals, *Eur. J. Pharmacol.*, 270, 237, 1994.
105. U.S. EPA, Use of Uncertainty Factors in Toxicity Extrapolations Involving Terrestrial Wildlife, Technical Basis, Office of Research and Development, U.S. Environmental Protection Agency, Washington, D.C., 1996.
106. Poole, K.G., Elkin, B.T., and Bethke, R.W., Environmental contaminants in wild mink in the Northwest Territories, Canada, *Sci. Total Environ.*, 160, 473, 1995.
107. Wells, D.E. and Echarri, I., Determination of individual chlorobiphenyls (CBs), including non-*ortho*, and mono-*ortho* chloro substituted CBs in marine mammals from Scottish waters, *Int. J. Environ. Anal. Chem.*, 47, 75, 1992.
108. Corsolini, S. et al., Congener profile and toxicity assessment of polychlorinated biphenyls in dolphins, sharks and tuna collected from Italian coastal waters, *Mar. Environ. Res.*, 40, 33, 1995.
109. Macdonald, S.M. and Mason, C.F., Status and conservation needs of the otter (*Lutra lutra*) in the western Palearctic, in *Nature and Environment*, No. 67, Council of Europe, Convention on the Conservation of European Wildlife and Natural Habitats, Strasbourg, 1992.
110. Environment Canada, Canadian Tissue Residue Guidelines for the Polychlorinated Biphenyls for the Protection of Wildlife Consumers of Aquatic Life, November 1998, Final Unpublished draft, Water Quality Guidelines Standards Division, Hull, QC, Canada, 1998, 177 pp plus tables.
111. Harding, L.E. et al., Reproductive and morphological condition of wild mink (*Mustela vison*) and river otters (*Lutra canadensis*) in relation to chlorinated hydrocarbon contamination, *Environ. Health Perspect.*, 107, 141, 1999.
112. Kannan, K. et al., Elevated accumulation of tributyltin and its breakdown products in bottlenose dolphins (*Tursiops truncatus*) found stranded along the U.S. Atlantic and Gulf coasts, *Environ. Sci. Technol.*, 31, 296, 1997.
113. Kannan, K. et al., Interaction of tributyltin with 3,3′,4,4′,5-pentachlorobiphenyl-induced ethoxyresorufin-O-deethylase activity in rat hepatoma cells, *J. Toxicol. Environ. Health*, 55, 373, 1998.
114. Nakata, H. et al., Evaluation of immunotoxic potentials of butyltins and non-*ortho* coplanar PCBs in marine mammals, *Arch. Environ. Contam. Toxicol.*, in press.
115. Bleavins, M.R. et al., Polychlorinated biphenyls (Aroclors 1016 and 1242): effects on survival and reproduction in mink and ferret, *Arch. Environ. Contam. Toxicol.*, 9, 627, 1980.
116. Leonards, P.E.G. et al., Studies of bioaccumulation and biotransformation of PCBs in mustelids based on concentration and congener patterns in predators and preys, *Arch. Environ. Contam. Toxicol.*, 35, 654, 1998.
117. Day, P.J., Bioaccumulation of Persistent Organochlorine Contaminants in a New Zealand Marine Food Chain. M.Sc. thesis, Victoria University of Wellington, Wellington, New Zealand, 1996.

118. Ballschmiter, K. and Wittlinger, R., Interhemisphere exchange of hexachlorocyclohexanes, hexachlorobenzene, polychlorobiphenyls, and 1,1,1-trichloro-2,2-bis(*p*-chlorophenyl)ethane in the lower trophosphere, *Environ. Sci. Technol.*, 25, 1103, 1991.
119. Slooten, E. and Dawson, S.M., Hector's dolphin *Cephalorhynchus hectori* (van Beneden, 1881), in *Handbook of Marine Mammals,* Vol. 5, 1994, chap. 14, 311–333.
120. Jones, P.D., Organochlorine contaminants in albatross from the South Pacific Ocean, Wellington, New Zealand, Department of Conservation, *Conserv. Advisory Sci.*, 226, 1, 1999.
121. Tanabe, S. et al., Global pollution of marine mammals by PCBs, DDTs and HCHs (BHCs), *Chemosphere,* 12, 1269, 1983.
122. Varanasi, U. et al., Contaminant monitoring for N.M.F.S. marine mammal health and stranding response program, *Coastal Zone '93,* 3, 1993, 516–530.
123. Subramanian, A., Tanabe, S., and Tatsukawa, R., Estimating some biological parameters of Baird's beaked whales using PCBs and DDE as tracers, *Mar. Pollut. Bull.,* 19, 284, 1988.
124. Law, R.J., Collaborative U.K. Marine Mammal Research Project: Summary of Data Produced 1988–1992, U.K. Ministry of Agriculture, Lowestoft, Fisheries and Food Directorate of Fisheries Research, 97, 1–42, 1994.

10 Effects of Chronic Stress on Wildlife Populations: A Population Modeling Approach and Case Study

Diane E. Nacci, Timothy R. Gleason, Ruth Gutjahr-Gobell, Marina Huber, and Wayne R. Munns, Jr.

CONTENTS

10.1 Introduction 247
10.2 A Population Matrix Modeling Approach 248
10.3 A Stressor of Ecotoxicological Concern 252
10.4 A Case Study 253
    10.4.1 Toxicological Responses 255
    10.4.2 Matrix Model Projections 257
    10.4.3 Compensatory Mechanisms 259
        10.4.3.1 Life History Shifts:
                Compensatory Demographic Responses 259
        10.4.3.2 Physiological Response Shifts:
                Compensatory Toxicological Responses 261
    10.4.4 The Scale of Evolutionary Effects 263
    10.4.5 Risks of Selection and Adaptation 264
10.5 A Population Modeling Approach and Case Study: Conclusions 265
Acknowledgments 267
References 267

10.1 INTRODUCTION

As a society, we have made commitments to preserve environmental quality, not only for its direct value to humans, but also to support aquatic and other wildlife species. These commitments have become legal mandates in the form of legislation such as the Clean Water Act. In the most general sense, adverse effects on wildlife species caused by human activities, or anthropogenic stress, result in changes to their densities and distributions. Although such changes can be measured at varying levels of biological organization, populations have been defined as a valued unit for wildlife protection and management.

There has been controversy as to whether or not wildlife protection at levels of biological organization higher than the individual truly reflects societal values and whether or not it is effective.[1] However, the impetus toward using the population as the protection unit has occurred for scientific as well as political reasons. Scientifically, some ecologists regard populations as sustainable units, valued for important properties beyond those inherent to individuals (i.e., emergent properties). Others regard this move as a practical response to the increased recognition that environmental management involves choices and costs. In any case, implicit in this approach is the philosophy that the loss of some individuals does not affect population, and therefore species, persistence, except when population sizes are very low (i.e., threatened or endangered species). While a healthy debate on the value of population protection continues, approaches to quantify effects on populations should be developed and evaluated.

This chapter describes a matrix modeling approach to characterize and project risks to wildlife populations subject to chronic stress. Population matrix modeling was used to estimate effects of one class of environmental contaminants, dioxin-like compounds (DLCs), to populations of an ecologically important estuarine fish species, *Fundulus heteroclitus,* or mummichogs. This approach was applied to a case study site highly contaminated with polychlorinated biphenyls (PCBs), including DLCs. Model projections suggested high risks to populations of mummichogs subject to intense DLC exposures. However, field observations of mummichog populations indigenous to this site appeared to be inconsistent with these projections. This apparent disparity provided an opportunity to use the population model structure to develop and test hypotheses on how wildlife populations respond to chronic stress. The directed research that followed has resulted in a more holistic assessment that integrates the perspectives of the contributing toxicologists, biologists, and ecologists.

10.2 A POPULATION MATRIX MODELING APPROACH

An essential component of the analysis phase of risk assessment is the development of a quantitative relationship between the stressor of concern and an ecological response[2,3] (Figure 10.1). To assess risks to populations, this response should reflect some attribute of population health, such as size, growth rate, or probability of persistence. However, direct measurements of population responses are difficult to acquire and often unavailable. Instead, wildlife toxicologists and risk assessors have used laboratory bioassays to develop quantitative relationships between stressors and individual responses. In these bioassays, adverse effects are often defined as reduced reproductive output or increased mortality for individuals exposed to stressors throughout vulnerable portions of their life cycle.[4] However, these measures of

Effects of Chronic Stress on Wildlife Populations

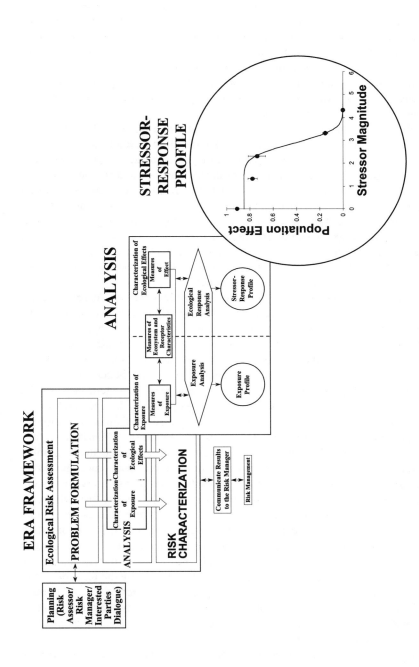

FIGURE 10.1 Stressor–response profile illustrating effects on wildlife populations, shown as one component of the analysis phase of the U.S. EPA's ecological risk assessment (ERA) framework. (Adapted from Reference 2.)

impaired performance or loss of individuals do not provide sufficient information to define a quantitative relationship between stressors and population health.

Life history theory provides the theoretical basis to link performance of individuals and population dynamics that is fundamental for understanding how environmental stressors can affect population regulation and life history evolution.[5,6] In general, the relationship between individual and population traits is not linear, and changes in some traits have a greater impact than others on populations.[7] For example, changes in survival often have a greater impact on populations than changes of a similar magnitude in reproduction.[9]

A matrix modeling approach provides a mathematical mechanism for integrating individual performance traits into estimates of population size and dynamics.[6,9–12] Specifically, the rate of increase per individual (r), is dependent upon vital rates, i.e., survival probabilities, time between reproductive events, and reproductive output. The number of individuals (n_t) is calculated at regular intervals (n_{t+1}) using matrix algebra, and the rate of change for the total number of individuals over this interval is termed population growth rate ($\lambda = e^r$). Population projections reflect population dynamics when conditions remain constant and responses are fixed throughout the period of concern. For example, when population growth rate is projected to be less than 1, the trajectory over time shows decreasing population size, and an increased probability for local extinction.

Population matrix models can be constructed to reflect varying degrees of complexity and specificity. In simple matrix models, the population is assumed to be closed (i.e., no immigration or emigration) and unbounded by carrying capacity (i.e., density independent). Therefore, changes in population size are affected only by initial population size, distribution, and vital rates. Further, the simplest models are deterministic, i.e., demographic or environmental stochasticity is not considered.

In matrix modeling, average rates of vital parameters are determined for the population as a whole or for classes within the population. In age- and stage-structured models, populations are considered as aggregates of linked but discrete classes, permitting the incorporation of class-specific vital rates. Matrix algebra is used to solve one or more difference equations that are used to calculate the number of individuals in each class and, through summation, in the population. This structuring also permits the incorporation of class-specific stressor responses. For example, early life stages are often toxicologically more sensitive stages than adults.

As an illustration, a matrix model can be constructed to represent a species with a four-stage life cycle, e.g., typical of many fish species (Figure 10.2). The numbers of individuals in each stage for any period will be affected by the starting number and the rates for processes by which individuals move in and out of stages. For example, the number of individuals in stage 0 will be affected by the fecundity rates for mature stages 1, 2, and 3 (described by $f_2, f_3,$ and f_4, respectively). The number of stage 0 individuals also will be affected by the relative probabilities that an individual will remain in that stage (survival probability, P_1) or develop into a stage 1–classified individual (transition probability, G_1).

How the model is constructed and parameterized defines the relative contributions of each stage or process to population dynamics. Sensitivity or elasticity analysis can be used to evaluate and rank parameters for their influence on population

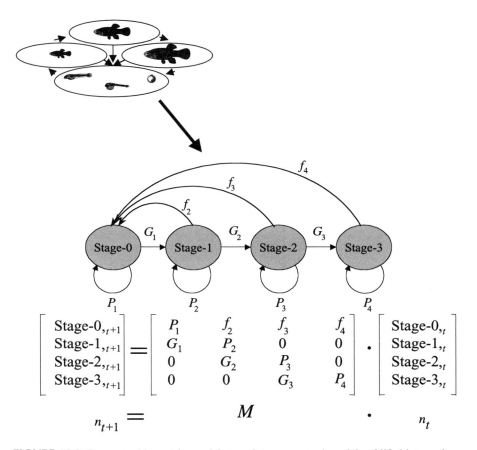

FIGURE 10.2 Demographic matrix models translate conceptual models of life history characteristics into mathematical models (M) that integrate stage-specific rates for survival (P), development (stage transition, G), and reproduction (f) into projections of population size at some time ($n_{t,t+1}$).

dynamics, i.e., the magnitude by which population attributes change in response to small changes in parameter values.[13,14] Processes that strongly influence population effects may be demographically important regulators of populations under stress. For example, for species with high reproductive output, changes in the survival of reproductively mature life stages will have a greater impact on population growth rate than changes of similar magnitude on survival of early life stages.[15]

Demographic data for unstressed populations are often acquired from published studies of populations in field or laboratory conditions. These initial or reference parameter estimates can be replaced or modified to reflect stress responses (Figure 10.3), e.g., as described by stressor–response relationships from laboratory studies.[16] Specifically, population models parameterized with stressor responses can be used to project how population attributes like population growth rate would be affected by a constant or chronic level of stress. By integrating stressor–response

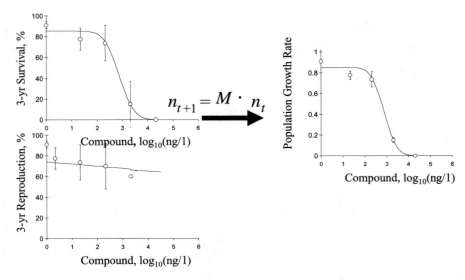

FIGURE 10.3 Ecotoxicological/demographic matrix models integrate mathematically stage- and process-specific stressor–response relationships into projections of population-level effects (i.e., population growth rate) relative to reference or unstressed level.

relationships into a demographic framework, effects on individuals are translated into effects on populations. Therefore, the accuracy of modeling projections is dependent upon the accuracy and completeness of both demographic and toxicological relationships for the populations of concern.

10.3 A STRESSOR OF ECOTOXICOLOGICAL CONCERN

The U.S. Environmental Protection Agency (U.S. EPA) has recognized national concerns about the effects on aquatic and other wildlife species of dioxin and other contaminants that act through similar toxicological mechanisms.[17] These contaminants, classified as dioxin-like compounds, include polychlorinated dibenzo-*p*-dioxin (PCDD), dibenzofuran (PCDF), and certain PCB congeners.[18] Like other persistent bioaccumulative and toxic contaminants, DLCs occur in detectable concentrations in many wildlife populations.[2] Concerns about their effects on fish have been reinforced by results of eco-epidemiological studies demonstrating that DLCs have contributed to the decline of lake trout in the Great Lakes.[19,20] A population matrix modeling approach provided an opportunity to predict risk of DLCs for populations of other fish species.

Although PCB congeners act through several mechanisms, and vary widely in toxic potencies, the mechanism of action for the most potent congeners that resemble dioxins (i.e., DLCs) has been the subject of much study.[18,21] These congeners are non- and mono-*ortho* substituted congeners whose toxic potency can be evaluated relative to 2,3,7,8-tetrachlorodibenzo-*p*-dioxin (TCDD or dioxin) using a toxic-equivalency approach.[18] The toxic effects of DLCs are mediated, in large part, through the aryl hydrocarbon receptor (AhR).[22,24] The DLCs are extremely toxic to the early life stages of fish.[25-27] Specifically, pericardial and yolk sac edema ("blue sac" disease) and subcutaneous hemorrhaging are characteristic pathologies in developing fish exposed to DLCs.[25] Poor growth, "wasting syndrome," and direct or indirect reductions in reproductive output have also been produced by DLC exposures to adult vertebrates, including fish species.[21,27,28]

The AhR signal transduction pathway is activated through binding with xenobiotic ligands that include DLCs, and results in the transcriptional regulation of several proteins.[22,24] Proteins induced by the AhR pathway include a major xenobiotic-metabolizing enzyme, cytochrome P-4501A1 (CYP1A1). The CYP1A1, induced when ligands bind the AhR, is a specific catalyst for ethoxyresorufin o-deethylase (EROD). Thus, elevated EROD activity has been used as a specific indicator of vertebrate exposure and response to AhR ligands, including DLCs.[22,29,30]

10.4 A CASE STUDY

New Bedford Harbor (NBH), Massachusetts was selected as a case study site to evaluate the utility of a matrix modeling approach for the projection of population-level effects associated with DLC exposures (Figure 10.4). Although typical in terms of nutrient overenrichment, habitat loss, and other characteristics of anthropogenic disturbance[31] of many urban estuaries of the northeast coast of the United States, NBH sediment and biota contain extraordinarily high concentrations of PCBs.[31-33] These findings suggest that PCBs, especially DLCs, are toxicologically important stressors in NBH.

According to historical records, PCBs were discharged into the northern or upper harbor as industrial wastes from the 1940s to the 1970s, producing contamination of sufficient magnitude to warrant listing on the U.S. EPA National Priorities List as a Superfund site.[31] Sediment PCBs in the Superfund site have been measured at levels as high as 2100 µg/g dry weight in NBH (total PCBs).[32] This value is four orders of magnitude greater than the sediment guideline value for total PCBs that has been correlated with probable adverse biological effects (180 ng/g dry weight).[34] Consistent with historical records, PCB concentrations in sediments at the Superfund site have been at toxic levels for decades[35] (Figure 10.5). Although the entire harbor is contaminated,[31] there is a steep gradient of PCB concentrations in sediment[36] and biota[37,38] from the northern to southern (Hurricane Barrier) boundaries of the NBH.

Despite high levels of contamination, a few fish species, including mummichogs, exist in great abundance in NBH.[39] Although PCB discharge ceased in 1976,[40] biota sampled from NBH more than 20 years later continue to accumulate PCBs.[33] For example, the mean concentration for total PCBs in livers of mummichogs collected in 1996 from the upper harbor Superfund site was 324 µg/g dry weight.[38] In comparison, the mean concentration of total PCBs in livers of mummichogs from West Island (WI), a reference site outside NBH, was 2.4 µg/g dry weight.[38]

Mummichogs are a nonmigratory fish with no dispersive life stages.[41] Although mummichogs reside in an essentially continuous band along the East Coast of the United States, studies have shown that there is limited gene flow between populations.[42] These findings suggest that mummichogs are subject to the environmental attributes of a limited geographical location throughout their

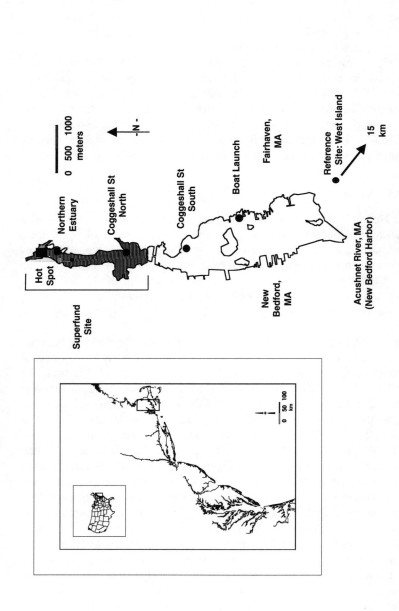

FIGURE 10.4 Case study site, New Bedford, Massachusetts. The northern estuary has been designated a Superfund site by the U.S. EPA because of a high sediment levels of PCBs. A local reference site (West Island, Fairhaven, Massachusetts) is located about 15 km away.

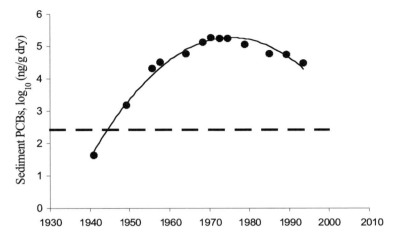

FIGURE 10.5 Sediment contamination at the Superfund site at NBH, as inferred by measurements of PCBs in sediment cores. Dashed line indicates sediment concentrations of PCBs associated with probable ecological effects (180 ng/g).[34] (Courtesy of J. Latimer, 2000.)

life cycle. In addition, local populations of mummichogs exist in dense schools[41,43] of highly fecund and genetically variable individuals.[44,45] For these reasons, mummichogs have been used extensively as a model for evolutionary studies on environmental adaptation.[45,46] These attributes, and their amenability to laboratory conditions, make mummichogs an ideal species for evaluating the chronic effects of environmental stressors.

10.4.1 TOXICOLOGICAL RESPONSES

The occurrence of mummichog populations in varied estuarine environments has promoted their reputation as a hardy species.[47] However, they can be quite sensitive to DLCs. Early life stage toxicity tests for dioxin suggest that the sensitivities of seven freshwater fish species ranged over two orders of magnitude.[48] In comparison to these species, the mummichog has an intermediate sensitivity to dioxin.[49]

Salomon[50] showed that adult mummichogs from reference populations demonstrated reductions in survival when exposed to dietary dioxin under laboratory conditions. Similarly, Black et al.[51] (Figure 10.6A) reported that injections of a mixture of dioxin-like PCB congeners mimicking the mixture and concentration found in NBH mummichogs produced mortalities and reduced egg production in female fish from reference populations.[51] Gutjahr-Gobell et al.[52] found that dietary exposure of DLCs to adult reference mummichogs also reduced their growth, feeding, and survival (Figure 10.6B). Together, these results indicate that exposure to DLCs at concentrations similar to those measured in NBH mummichogs produce toxic effects in reference mummichogs. Similarly, a recent literature review concluded that tissue concentrations as high as those measured in NBH mummichogs increased embryonic and larval mortality and altered neurotransmitter concentrations, hormone metabolism, and gonadal development in many fish species.[28]

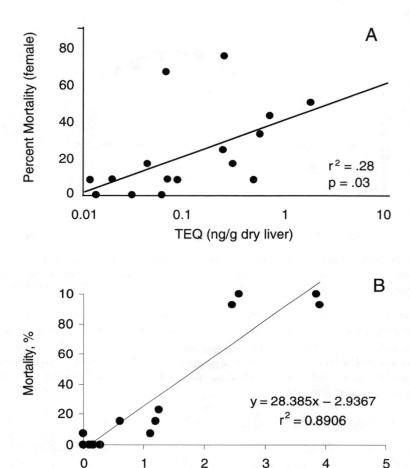

FIGURE 10.6 Regressions of mortalities of female *F. heteroclitus* exposed to a mixture of non-*ortho* and mono-*ortho* PCBs as TEQs of PCBs in liver tissue. Fish were exposed by injection (A) or diet (B). (A, adapted from Black, D.E. et al., *Environ. Toxicol. Chem.*, 17, 1396, 1998. With permission.) (B, adapted from Gutjahr-Gobell, R.E. et al., *Environ. Toxicol. Chem.*, 18, 699, 1999. With permission.)

Consistent with the direct toxic effects of DLCs as assessed using short-term laboratory exposures to reference mummichogs, Black et al.[37] demonstrated increased mortality among 3-year-old NBH mummichogs held in the laboratory during a portion of the summer spawning season. However, the laboratory results for DLC-exposed mummichogs from reference populations and NBH mummichogs were not identical: NBH mummichogs did not exhibit reduced fecundity.[37] In addition, laboratory-held NBH mummichogs produced larvae with unique developmental abnormalities,[37] unlike DLC-exposed mummichogs from reference populations.[51]

10.4.2 Matrix Model Projections

Munns et al.[53] developed a stage-classified matrix model for mummichogs with the explicit intent of producing projections of population effects associated with DLC exposure. This model was constructed as a simple representation of population dynamics (i.e., density independent, and without immigration or emigration) that could incorporate the toxicological responses of key developmental stages (Figure 10.7). Parameters were derived using values for this species available from the literature, such as mean annual adult survival. These values were assumed to represent reference or unstressed populations.[53] Data for survivorship and reproduction, collected in laboratory studies, were used to modify these transition rates. Specifically, data were incorporated from laboratory exposures to dioxin of a single adult life stage of mummichogs collected from a reference site.[50] These projections showed a dose–responsive decline in population growth rates associated with dioxin exposure to naive populations (Figure 10.8A).[53] Consistent with these projections, data from laboratory studies of NBH mummichogs[37] were used to demonstrate correlations between PCB tissue concentrations and reductions in population growth rate, with lowest population growth rates described for mummichogs from the NBH Superfund site[53] (Figure 10.8B). Together, these results suggested

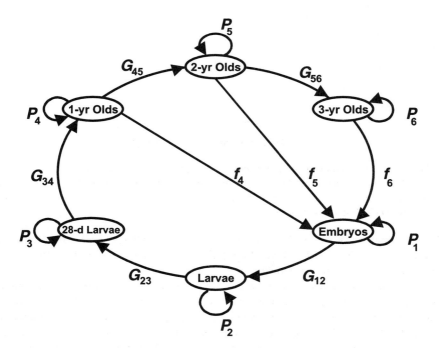

FIGURE 10.7 A stage-classified matrix model developed for *F. heteroclitus* using laboratory- and literature-derived (mean) parameter values for stage-specific P (probability of remaining in stage), G (probability of making the transition to subsequent stage), and f (fecundity). (Adapted from Munns, W.R., Jr. et al., *Environ. Toxicol. Chem.*, 16, 1074, 1997. With permission.)

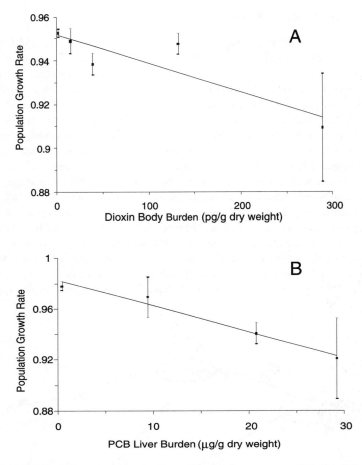

FIGURE 10.8 Population growth rate effects (per 14 day, means ± SE) for *F. heteroclitus* for (A) laboratory exposure to dioxin or (B) PCB mixture measured in NBH fish as a function of their liver concentrations. (Adapted from Munns, W.R., Jr. et al., *Environ. Toxicol. Chem.*, 16, 1074, 1997. With permission.)

that exposure to DLCs at high but environmentally realistic levels could produce adverse population-level effects. Population projections of NBH mummichogs seemed to confirm this conclusion.

However, casual observations at the NBH site revealed an abundance of mummichogs of all size classes, in apparently healthy condition. An earlier field survey that included measures of fish condition, biomass, and individual growth rates also concluded that NBH mummichog populations are not in a degraded condition relative to a population indigenous to a local uncontaminated estuary.[39] This apparent disparity between matrix model projections[53] and field observations suggested that NBH mummichogs might not be characterized accurately by the model projections.

To improve the accuracy and completeness of demographic contributions to model projections, field surveys of mummichog populations have been conducted over a 5-year period. These data will be used to estimate the mean and variance for demographic parameters of mummichogs indigenous to NBH and a local reference site.[54] It is hoped that these analyses will be sufficiently sensitive to elucidate potentially subtle changes in population health, such as change in age-structure, that NBH fish may exhibit relative to an uncontaminated population. However, other laboratory and field studies have been conducted to address specific hypotheses concerning compensatory mechanisms that may permit NBH mummichog populations to persist despite intense, multigenerational exposures to toxic levels of contaminants.

10.4.3 COMPENSATORY MECHANISMS

Using the ecotoxicological population modeling approach as a research framework, we considered that a variety of compensatory mechanisms might enable NBH mummichog populations to persist under conditions of chronic stress. These compensatory responses could be categorized broadly as demographic or toxicological. Demographic mechanisms include those related to life history attributes or population dynamics. In this case, the mummichog matrix model assumptions and parameters[53] might not capture demographic characteristics that could offset high adult mortality rates that were measured in the laboratory study of NBH fish (i.e., older-age-class NBH fish during the summer spawning season).[37] Three specific demographic mechanisms were examined: (1) NBH fish might demonstrate unusually high reproductive output (especially among younger age classes not measured in laboratory studies[37]), (2) overwinter mortality might differ between NBH and reference populations, and (3) mummichog populations from more contaminated sites within the harbor could be supported by immigration from less contaminated harbor sites.

Toxicological compensatory mechanisms also were proposed to explain NBH population persistence. Mummichogs from other contaminated sites had been shown to demonstrate tolerance or insensitivity to local pollutants.[47,55–57] Although the impaired performance of NBH mummichogs under laboratory conditions[37] was somewhat consistent with toxic effects by local DLC exposures, other factors indirectly related or unrelated to the chemical exposures experienced by NBH mummichogs, such as increased susceptibility to infection, could also explain these results. Therefore, we hypothesized that NBH mummichogs might be insensitive to the toxic effects of DLCs at NBH exposure concentrations. Laboratory studies were conducted to test for tolerance to local contaminants in NBH mummichogs, and whether tolerance, if present, reflects acclimation (a temporary response) or adaptation (an inherited response).

10.4.3.1 Life History Shifts: Compensatory Demographic Responses

Laboratory and field studies were conducted to measure demographic parameters in NBH and reference mummichogs. Theoretically, life history shifts can compensate for life stage specific losses.[5] In this case, NBH mummichog populations might

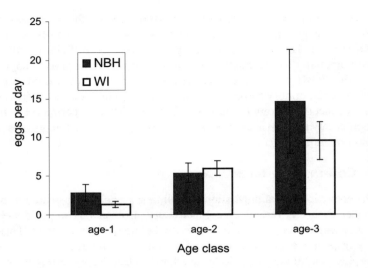

FIGURE 10.9 Fecundity varied by age class, but not between fish populations for *F. heteroclitus* from NBH (dark bars) and a reference site, West Island (WI, light bars). (Courtesy of T.R. Gleason, 2000.)

demonstrate increased reproductive output in younger age classes, an important contributor to population growth, as has been reported for mummichogs indigenous to other highly contaminated sites.[58] This increased reproductive output could offset the losses of older adults as suggested to occur in NBH population.[37] However, laboratory studies showed no significant differences between NBH and reference mummichogs when egg production was compared between females from the three reproductively mature age classes (Figure 10.9).[59] These findings do not support the hypothesis that NBH populations persist via higher reproductive output in younger or older age classes[59] that could compensate for reduced adult survival.[37]

Overwinter survival, especially in the first year, has been shown to be an important regulator of population dynamics for many temperate fish species.[60] Preliminary results of laboratory studies designed to mimic winter conditions have suggested that NBH mummichogs have higher rates of overwinter survival than fish from reference populations.[61] In addition, studies of fish collected before and after the winter season have shown that NBH mummichogs have higher levels of stored fat[61] and relatively similar levels of stored vitamin A[62] compared with fish from a reference site. Together these findings suggest that condition of NBH mummichogs is relatively good and may contribute to a higher rate of overwinter survival than that demonstrated by a local reference population. An overwinter survival advantage could be an important demographic factor contributing to the persistence of NBH mummichog populations that was not captured in the initial matrix model projections developed for this site by Munns et al.[53]

Further analysis of field data will be required to estimate and compare mean annual survival rates for NBH and reference mummichogs. It appears that different factors may regulate age-specific survival rates for mummichogs indigenous to NBH and reference sites. For example, after-winter survival rates may be high for NBH mummichogs. However, during some years in late summer, most NBH mummichogs (nearly 100%) are infested with high numbers and unusual forms of parasites that are not observed in mummichogs from reference sites.[63] Parasite loads may contribute directly or indirectly to increased late-summer mortalities of NBH mummichogs, suggested in preliminary analyses of field data[54] and by some laboratory studies.[37,59]

The extent to which mummichog populations migrate throughout NBH and the relative contribution of immigration to population persistence has not yet been quantified. However, the short-term movement patterns of mummichogs across the Superfund boundary is under investigation currently to evaluate the hypothesis that contaminated populations are replenished by immigration from less-contaminated areas.[64] Based on field observations[43] and molecular genetic techniques,[42] the home range for mummichogs of the mid-Atlantic region has been estimated to be about 2 km of shoreline, about the length of the Superfund site. However, it has also been suggested that mummichogs may not traverse rocky shorelines or deep, fast-moving currents.[41,65] These movement patterns suggest that mummichogs resident to the upper harbor might be isolated from those of the lower, less-contaminated region of the harbor. Preliminary results from tagging and recapture studies conducted over a 1-year period[54] suggest that the bridge abutments at the Superfund border are not complete barriers to mummichog movement. Although NBH mummichogs display high site fidelity, they do not appear to be restricted from moving along continuous shoreline within the harbor. Because the east shore of the Superfund site consists of relatively undeveloped marsh that provides ideal mummichog habitat,[65] this highly contaminated area may actually support the highest density of mummichogs within NBH.

10.4.3.2 Physiological Response Shifts: Compensatory Toxicological Responses

As summarized from studies using many fish species, DLCs disrupt development and reproduction in laboratory-exposed fish, but results from the field are less clear.[28] While there are many reasons results from laboratory studies may differ from field studies, of primary importance are differences in stressor magnitude. These differences in magnitude are related not only to dose and administrative route, but also to duration. Multigenerational stress can result in the selective loss of sensitive individuals, producing populations dominated by resistant individuals. Tolerant populations are characterized by modest stressor–response relationships relative to sensitive populations of the same species.

Recent research has shown that NBH mummichogs do not respond to DLCs like reference mummichogs; i.e., they are dramatically insensitive to the effects of DLCs.[38,66,67] Specific indicators of AhR-mediated effects (i.e., CYP1A concentration or EROD activity) have also shown reduced AhR pathway responsiveness in NBH mummichog embryos[38] and adults.[66,67] For example, DLC-associated increases in *in ovo* EROD fluorescence is a sensitive indicator of AhR-mediated effects during early embryonic development of mummichogs from reference populations.[29] While these responses are also demonstrated in NBH mummichog embryos, DLC exposure concentrations two orders of magnitude greater than those required to elicit effects

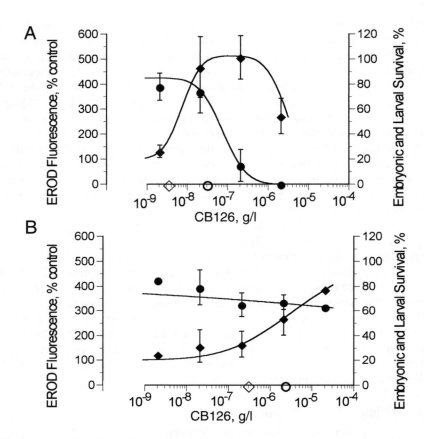

FIGURE 10.10 Survival (●) and EROD (♦) data (means ± SE) and response models for WI (A) and NBH (B) *F. heteroclitus* embryos at estimated exposure concentrations of 3,3′,4,4′,5-pentachlorobiphenyl, CB126. Open symbols indicate concentrations producing differences of 20% from control values for survival (LC_{20}) or EROD fluorescence (EC_{20}). (Adapted from Nacci, D. et al., *Mar. Biol.* 134, 9, 1999. With permission.)

in reference mummichogs are necessary[38] (Figure 10.10A and B). Similarly, mummichog embryos from NBH are profoundly less sensitive to the lethal effects associated with DLC exposures than reference fish[38] (Figure 10.10A and B). Specifically, results of laboratory exposures showed that concentrations of DLCs similar to those measured in NBH mummichog eggs were lethal to reference embryos[38] (Figure 10.11). These results suggest that reference mummichogs could not survive early life stages if they were as contaminated as mummichogs from NBH.

Similar comparisons in sensitivity between laboratory-raised progeny of field-collected mummichogs demonstrated that DLC responsiveness was inherited and independent of maternal contaminant contributions.[38] These findings are consistent with the conclusion that DLC contamination in NBH has contributed to the selection of DLC-adapted fish. Genetic adaptation or evolved tolerance to DLCs may be a critical compensatory mechanism by which fish populations persist in this highly contaminated site.

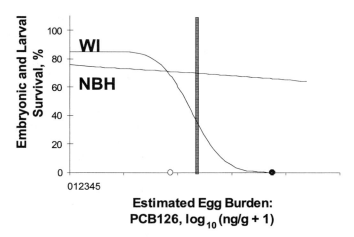

FIGURE 10.11 Response models for survival of WI and NBH *F. heteroclitus* embryos at estimated tissue concentrations of 3,3',4,4',5-pentachlorobiphenyl, PCB126. Circles indicate concentrations producing differences of 20% from control values for survival (LC_{20}) in WI (○) or NBH (●) embryos. Location and width of shaded bar shows concentration of PCB126 (mean ± SD) measured in eggs from NBH *F. heteroclitus*. (Adapted from Nacci, D. et al., *Mar. Biol.*, 134, 9, 1999. With permission.)

This characterization supports the conclusions that long-term effects of chronic stress may include genetic restructuring of the NBH mummichog populations. Although the specific physiological mechanism by which tolerance to DLCs in NBH mummichogs is produced is not yet known, this phenomenon is consistent with changes in the AhR transduction pathway.[24,66–69] Recent research indicating that there are two forms of the AhR in mummichogs may contribute to an understanding of how DLC sensitivity and resistance to the toxic effects of DLCs are affected in mummichogs.[70] As these and other candidate genes of the AhR pathway are sequenced in mummichogs from sensitive and resistant populations,[69,71] a direct linkage to the mechanism of resistance to DLCs in mummichogs may be established. A comprehensive understanding for this single model species may contribute toward a mechanistic basis to extrapolate and predict evolutionary responses to DLCs across wildlife species.

10.4.4 THE SCALE OF EVOLUTIONARY EFFECTS

Indigenous populations of mummichogs that are resistant to the toxicological effects of local contaminants have been characterized at several sites highly contaminated with persistent, toxic- and bioaccumulative contaminants, including DLCs and other AhR agonists.[24,68] In addition to NBH, these sites include Newark, New Jersey, contaminated with dioxins,[55,56,72] and the Atlantic Wood site in the Elizabeth River, Virginia, contaminated with polyaromatic hydrocarbons.[57,73] These occurrences have contributed to the prediction that field conditions that are toxic to reference mummichogs will result in either extinction or adaptation by local populations.

However, predictions based on sediment PCB concentrations (correlated with DLC concentrations) suggest that field conditions that are toxic to mummichogs occur at sites far less contaminated than Superfund sites.[36] To test this hypothesis, 12 mummichog populations were sampled from sites that ranged over five orders of magnitude in sediment PCB concentrations.[74] Laboratory bioassays were conducted on progeny of these field-collected fish, and sensitivity to DLCs was used as an indicator of adaptation. Evidence of DLC insensitivity[36] supports the conclusion that genetic adaptation to DLCs can occur in mummichog populations indigenous to areas that are not extraordinarily contaminated, i.e., at concentrations of sediment PCBs equal to or greater than concentrations correlated with probable ecological effects (i.e., 180 ng/g).[34] If results from this study can be extrapolated to other geographical areas and ecosystems, then the force of contaminants and other anthropogenic stressors as selective agents has been underestimated. These results suggest that exposures to persistent, bioaccumulative, and toxic contaminants, including but not limited to DLCs, may increase risks of genetic or evolutionary effects not previously considered, and on a geographical scale larger than expected.

10.4.5 Risks of Selection and Adaptation

Studies demonstrating inherited changes in NBH mummichogs[36,38] suggest changes in genetic structure in these populations driven by hard selection for DLC tolerance. Recent efforts have been directed toward measuring changes in the genetic structure of chronically stressed fish populations, including those in and around NBH.[63,76–78] These studies address the concern that populations that suffer losses due to chronic stress or rapid and intense selection may demonstrate changes in genetic composition or diversity that present long-term risks to population persistence. For example, Cohen[63] has interpreted from nuclear DNA sequence data that mummichogs from NBH demonstrate significantly different amino acid replacement patterns in proteins associated with immune response (i.e., the major histocompatibility complex) in comparison with reference mummichog populations. These results suggest that immune stressors have also acted as strong selection agents, directing the evolution of populations indigenous to complex sites like NBH. Studies are under way currently to examine whether these genetic changes are correlated with changes in immunoresponsiveness, i.e., increased vulnerability to parasitism and disease, that may be important in the regulation of population dynamics, especially in chronically stressed populations.

Recent studies have reported reduced genetic diversity in populations of aquatic organisms exposed to contaminants.[79,80] However, the relationship between changes in genetic diversity and contaminant exposure is not always clear.[81,82] There is concern, and some evidence, that reduced genetic diversity is detrimental to wildlife populations.[83,84] Many wildlife studies have focused on consequences of reduced genetic diversity in small populations[85] but very little work has been done to understand consequences of genetic loss and homozygosity in species that exist in large populations, like mummichogs. For example, beneficial mutations and selectively adapted genotypes may spread and become fixed more quickly in large populations.[86]

Whether NBH mummichogs show significant reductions in genetic diversity and, if so, what the long-term effects may be is not yet known.

Theoretically, there are potential costs or "trade-offs" associated with selection and adaptation.[87] Ongoing studies are testing whether specific consequences associated with the genetic or biochemical mechanisms of DLC adaptation are realized in adapted mummichog populations.[72,73] Results from these studies will be useful in determining to what extent the short-term benefits of genetic adaptation to the toxic effects of DLCs may increase long-term costs to population persistence.

10.5 A POPULATION MODELING APPROACH AND CASE STUDY: CONCLUSIONS

Population effects of dioxin-like contaminants, an important class of persistent, bioaccumulative, and toxic chemicals, were projected for a common estuarine fish species, the mummichog. Specifically, population effects were quantified through a species-specific matrix model that incorporated results of short-term laboratory studies using a contaminant-naive, reference population exposed to dioxin, the most potent contaminant in this chemical class. Results from these studies conducted using a single reproductively mature stage suggested that DLCs produced mortalities that could be important contributors to reduced population growth rate. Mortality rates under laboratory conditions were also correlated with increased tissue concentrations of DLCs in reproductively mature fish from a population indigenous to a site contaminated for decades with high levels of DLCs. Therefore, projections of population growth for this population subject to high-level, multigenerational exposures were consistent with increased risk for extinction from the direct toxicity of DLC exposures.

However, the projection and characterization of poor population health for mummichogs subject to chronic DLC exposure was inconsistent with observations of abundant, persistent mummichog populations in this highly contaminated site. Therefore, an intense investigation was undertaken to evaluate the accuracy of the characterization of DLC effects on indigenous populations. These efforts were undertaken to reduce uncertainties in the prediction of risks associated with this highly contaminated Superfund site. In addition, the structure of the population model and information derived from this case study were used to develop and test hypotheses on how wildlife populations respond to chronic stress. These exercises reiterated that the accuracy of population modeling projections is dependent upon the accuracy and completeness of demographic and toxicological relationships for the populations and stressors of concern.

Even for well-studied wildlife species, there is a paucity of demographic data available in the literature. Values for important demographic variables and their spatially or temporally specific variation are often unknown. Currently, these deficits limit species-specific model development and contribute to model projection uncertainties. For example, in this case study, patterns of seasonal mortality seem to differ between fish populations indigenous to a reference and contaminated sites. Intensive site-specific field and laboratory studies have demonstrated that fish indigenous to this contaminated urban harbor have high condition indices in autumn[61] and, episodically, high and unusual parasite loads in late summer.[63] We speculate that these factors may contribute, respectively, to decreased overwinter and increased postspawning mortalities. However, we do not understand to what extent local contaminants and other site-specific anthropogenic stressors play a direct or indirect role in mortality patterns. For example, have toxic levels of contaminants reduced species diversity, affecting parasite–host relationships? Has nutrient enrichment contributed to changes in parasite abundance and distribution? Finally, do contaminated populations demonstrate increased susceptibility to parasite infestations? Further studies examining demographic patterns in sites that vary by stressor type and magnitude may help distinguish factors that regulate population dynamics.

Although DLCs are among the most studied environmental contaminants, the sensitivity of fish to DLCs is known for only a few species.[28,48] For this case study, laboratory studies were conducted using mummichogs and demonstrated that DLCs at environmental concentrations reduce reproduction and the survival of young and adult life stages.[38,51,52] Consistently, a population model projected declining growth rates for populations subject to novel exposure to dioxin.[53] Therefore, DLC exposures could result in the local extinction of toxicologically sensitive populations. However, population projections that incorporate fixed stressor–response patterns could not reflect compensatory responses that evolve in populations subject to multigenerational stress. As concluded in this case study, mummichogs indigenous to this highly contaminated site are dramatically insensitive to toxic effects of DLC exposures during early life stages[38] and as adults.[66] These findings suggest that chronic, multigenerational exposures to toxic contaminants can result in adaptive or evolved tolerance, as demonstrated by DLC response patterns that differ profoundly from those of reference populations.[38] More broadly, results from this case study suggest that stressors exerting long-term effects can produce long-term consequences that are not fully understood at this time. The NBH mummichogs may provide an important model to examine evolutionary effects produced by chronic stress.

Although the prediction of local extinction of mummichogs was unmet at this contaminated site, mummichogs may be unusual in their capacity for rapid adaptation. Few other aquatic species have demonstrated genetic adaptation to local contamination, with notable exceptions.[10,88] We do not know to what extent chemical contamination or other anthropogenic stressors may have contributed to the reduced species richness and diversity apparent in this Superfund estuary.[39] Perhaps DLC toxicity has contributed to the loss of species that are more uniformly sensitive or genetically less variable than mummichogs. An "evolutionary eco-toxicological" perspective may provide a basis to determine what genetic, toxicological, life history, and environmental characteristics promote a population trajectory that reflects adaptation rather than extinction.[88]

This chapter describes how a population modeling approach was used to integrate demographic and toxicological information into a projection of potential effects of chronic exposure to toxic chemical contaminants on an estuarine fish population. This approach provided a research framework to evaluate systematically how a chronic stressor can affect important biological processes, and how changes in these processes may be reflected in changes in life stage transition rates that can produce quantifiable changes in population health. This case study provided an opportunity to compare predictions of effects with measurements of individual- and population-level attributes for a population subject to chronic, multigenerational stress. It is neither reasonable nor desirable that intensive investigations, such as the one described for this case study, will be undertaken to assess specific risks associated with every species, stressor, and site of concern. Rather, interpretations of results from these efforts should be used to develop and refine hypotheses predicting effects based on important demographic, toxicological, and ecological attributes specific to the concerns. Modeling efforts and laboratory and field studies have contributed to a more comprehensive understanding of the risks of chronic stress to population health.

ACKNOWLEDGMENTS

The authors appreciate the helpful advice provided by two anonymous reviewers as well as reviewers and technical consultants at the U.S. EPA, National Health and Environmental Effects Research Laboratory, Atlantic Ecology Division (AED), Narragansett, Rhode Island. These reviewers include Dr. Sarah Cohen (NRC/U.S. EPA), Dr. Amy McMillan (NRC/U.S. EPA), Dr. Matt Mitro (U.S. EPA), and Ms. Marguerite Pelletier (U.S. EPA). This manuscript has been reviewed and approved for publication by the U.S. EPA (AED contribution number 00-082). Approval does not signify that the contents necessarily reflect the views and policies of the U.S. EPA. Mention of trade names, products, or services does not convey, and should not be interpreted as conveying, official U.S. EPA approval, endorsement, or recommendation.

REFERENCES

1. Beyer, W.N. and Heinz, G.H., Implications of regulating environmental contaminants on the basis of wildlife populations and communities, *Environ. Toxicol. Chem.*, 19, 1703, 2000.
2. U.S. Environmental Protection Agency, Framework for Ecological Risk Assessment, Risk Assessment Forum, U.S. EPA/630/R-92/001, Washington, D.C., 1992.
3. U.S. Environmental Protection Agency, Guidelines for Application of a Framework for Ecological Risk Assessment, Risk Assessment Forum, U.S. EPA/630/R-92/001, Washington, D.C., 1998.
4. Nacci, D. et al., Biological responses of the sea urchin, *Arbacia punctulata*, to lead contamination for an estuarine ecological risk assessment, *J. Aquat. Ecosyst. Stress Recov.*, 7, 187, 2000.
5. Stearns, S.C., Life-history tactics: a review of ideas, *Q. Rev. Biol.*, 51, 3, 1976.
6. Caswell, H., *Matrix Population Models: Construction, Analysis and Interpretation*, Sinauer Associates, Sunderland, MA, 1989.
7. Forbes, V.E. and Calow, P., Is per capita rate of increase a good measure of population-level effects in ecotoxicology? *Environ. Toxicol. Chem.*, 18, 1544, 1999.
8. Pfister, C.A., Patterns of variance in stage-structured populations: evolutionary predictions and ecological implications, *Proc. Natl. Acad. Sci. U.S.A.*, 95, 213, 1998.

9. Caswell, H., Demography meets ecotoxicology: untangling population level effects of toxic substances, in *Ecotoxicology, A Hierarchical Treatment*, Newman, M.C. and Jagoe, C.H., Eds., Lewis Publishers, Boca Raton, FL, 1996, 255.
10. Walker, C.H. et al., *Principles of Ecotoxicology*, Taylor & Francis, London, 1996.
11. Akcakaya, H.R., Burgman, M.A., and Ginzburg, L.R., *Applied Population Ecology*, Sinauer Associates, Sunderland, MA, 1999.
12. Kammenga, J. and Laskowski, R., *Demography in Ecotoxicology*, John Wiley & Sons, Chichester, U.K., 2000.
13. de Kroon, H. et al., Elasticity: the relative contributions of demographic parameters to population growth rate, *Ecology*, 67, 1427, 1986.
14. Benton, T.G. and Grant, A., Elasticity analysis as an important tool in evolutionary and population ecology, *Trends Ecol. Evol.*, 14, 467, 1999.
15. Calow, P., Sibly, R.M., and Forbes, V., Risk assessment on the basis of simplified life-history scenarios, *Environ. Toxicol. Chem.*, 16, 1983, 1997.
16. Gleason, T.R., Nacci, D.E., and Munns, W.R., Jr., Projecting population-level responses of purple sea urchins to lead contamination for an estuarine ecological risk assessment, *J. Aquat. Ecosyst. Stress Recov.*, 7, 177, 2000.
17. U.S. Environmental Protection Agency, Interim Report on the Assessment of 2,3,7,8-Terachlorodibenzo-*p*-Dioxin Risk to Aquatic Life and Associated Wildlife, U.S. EPA/600/R-93/005, Office of Research and Development, Washington, D.C., 1993.
18. Safe, S., Polychlorinated biphenyls (PCBs), dibenzo-*p*-dioxins (PCDDs), dibenzofurans (PCDFs), and related compounds: environmental and mechanistic considerations which support the development of toxic equivalency factors (TEFs), *CRC Crit. Rev. Toxicol.*, 21, 51, 1990.
19. Cook, P.M., Zabel, E.W., and Peterson, R.E., The TCDD toxicity equivalence approach for characterizing risks for early life stage mortality in trout, in *Chemically-Induced Alterations in the Functional Development and Reproduction of Fishes*, Rolland, R., Gilbertson, M., and Peterson, R., Eds., SETAC Press, Pensacola, FL, 1997, chap. 2.
20. Cook, P.M. et al., Effects of Ah receptor mediated early life stage toxicity on lake trout reproduction and survival in Lake Ontario during the 20th century, manuscript in preparation.
21. Safe, S., Polychlorinated biphenyls (PCBs): environmental impact, biochemical and toxic responses, and implications for risk assessment, *Crit. Rev. Toxicol.*, 24, 87, 1994.
22. Stegeman, J.J. and Hahn, M.E., Biochemistry and molecular biology of monooxygenases: current perspectives of forms, functions, and regulations of cytochrome P450 in aquatic species, in *Aquatic Toxicology: Molecular, Biochemical and Cellular Perspectives*, Malins, D.C. and Ostrander, G.K., Eds., CRC Press, Boca Raton, FL, 1994.
23. Hahn, M.E., Ah receptors and the mechanism of dioxin toxicity: insights from homology and phylogeny, in *Interconnections between Human and Ecosystem Health*, Di Giulio, R. and E. Monosson, E., Eds., Chapman & Hall, London, 1995.
24. Hahn, M.E., Mechanisms of innate and acquired resistance to dioxin-like compounds, *Rev. Toxicol. Ser. B Environ. Toxicol.*, 2, 395, 1998.
25. Walker, M.K. and Peterson, R.E., Potencies of polychlorinated dibenzo-*p*-dioxin, dibenzofuran, and biphenyl congeners, relative to 2,3,7,8-tetrachlorodibenzo-*p*-dioxin, for producing early life stage mortality in rainbow trout (*Oncorhynchus mykiss*), *Aquat. Toxicol.*, 21, 219, 1991.
26. Peterson, R.E., Theobald, H.M., and Kimmel, G.L., Developmental and reproductive toxicity of dioxins and related compounds: cross-species comparisons, *Crit. Rev. Toxicol.*, 23, 283, 1993.

27. Eisler, R. and Belisle, A.A., Planar PCB Hazards to Fish, Wildlife, and Invertebrates: A Synoptic Review, Biological Report 31, National Biological Service, Washington, D.C., 1996.
28. Monosson, E., Reproductive and developmental effects of PCBs in fish: a synthesis of laboratory and field studies, *Rev. Toxicol.*, 3, 25, 1999/2000.
29. Nacci, D. et al., A non-destructive indicator of EROD activity in embryonic fish, *Environ. Toxicol. Chem.*, 17, 2481, 1998.
30. Whyte, J.J. et al., Ethoxyresorufin-*o*-deethylase (EROD) activity in fish as a biomarker of chemical exposure, *Crit. Rev. Toxicol.*, 30, 349, 2000.
31. Nelson, W.G. et al., New Bedford Harbor Long-Term Monitoring Assessment Report: Baseline Sampling, EPA/600/R-96/097, U.S. Environmental Protection Agency, National Health and Environmental Effects Research Laboratory, Atlantic Ecology Division, Narragansett, RI, 1996.
32. Pruell, R.J. et al., Geochemical study of sediment contamination in New Bedford Harbor, Massachusetts, *Mar. Environ. Res.*, 29, 77, 1990.
33. Lake, J.L. et al., Comparisons of patterns of polychlorinated biphenyl congeners in water, sediment, and indigenous organisms from New Bedford Habor, Massachussetts, *Arch. Environ. Contam. Toxicol.*, 29, 207, 1995.
34. Long, E.R. et al., Incidence of adverse biological effects within ranges of chemical concentrations in marine and estuarine sediments, *Environ. Manage.*, 19, 81, 1995.
35. Latimer, J., personal communication, 2000.
36. Nacci, D. et al., Predicting the occurrence of adaptation to contaminants in wild populations of the estuarine fish, manuscript submitted.
37. Black, D.E. et al., Reproduction and polychlorinated biphenyls in *Fundulus heteroclitus* (Linnaeus) from New Bedford Harbor, Massachussetts, USA, *Environ. Toxicol. Chem.*, 17, 1405, 1998.
38. Nacci, D. et al., Adaptation of wild populations of the estuarine fish *Fundulus heteroclitus* to persistent environmental contaminants, *Mar. Biol.*, 134, 9, 1999.
39. Mitchell, E.L. and Oviatt, C.A., Comparison of Fish Biomass, Abundance and Community Structure Between a Stressed Estuary, New Bedford Harbor, MA, and a Less-Stressed Estuary, The Slocums River, MA, University of Rhode Island, Graduate School of Oceanography Report, 1996.
40. www.epa.gov/nbh.
41. Bigelow, H.B. and Schroeder, W.G., Fishes of the Gulf of Maine, Fishery Bulletin of the Fish and Wildlife Service, Vol. 53, U.S. Government Printing Office, Washington, D.C., 1953.
42. Brown, B.L. and Chapman, R.W., Gene flow and mitochondrial DNA variation in the killifish, *Fundulus heteroclitus*, *Evolution*, 45, 1147, 1991.
43. Lotrich, V.A., Summer home range and movements of *Fundulus heteroclitus* (pisces: Cyprinodontidae) in a tidal creek, *Ecology*, 56, 191, 1975.
44. Mitton, J.B. and Koehn, R.K., Genetic organization and adaptive response of allozymes to ecological variables in *Fundulus heteroclitus*, *Genetics*, 79, 97, 1975.
45. Mitton, J.B., *Selection in Natural Populations*, Oxford University Press, New York, 1997.
46. Powers, D.A. and Schulte, P.M., Evolutionary adaptations of gene structure and expression in natural populations in relation to a changing environment: a multidisciplinary approach to address the million-year saga of a small fish, *J. Exp. Zool.*, 282, 71, 1998.
47. Weis, J.S. and Weis, P., Tolerance and stress in a polluted environment, the case of the mummichog, *BioSciences*, 39, 89, 1989.

48. Elonen, G.E. et al., Comparative toxicity of 2,3,7,8-tetrachlorodibenzo-*p*-dioxin to seven fresh water fish species during early life-stage development, *Environ. Toxicol. Chem.*, 17, 472, 1998.
49. Toomey, B.H. et al., TCDD induces apoptotic cell, death and cytochrome P4501A expression in developing *Fundulus heteroclitus* embryos, *Aquat. Toxicol.*, 53, 127, 2001.
50. Salomon, K.S., Dietary Uptake of 2,3,7,8-Tetrachlorodibenzo-*p*-Dioxin and Its Effects on Reproduction in the Common Mummichog (*Fundulus heteroclitus*), Master's thesis, University of Rhode Island, Kingston, 1994.
51. Black, D.E. et al., Effects of a mixture of non-*ortho* and mono-*ortho*-polychlorinated biphenyls on reproduction in *Fundulus heteroclitus* (Linnaeus), *Environ. Toxicol. Chem.*, 17, 1396, 1998.
52. Gutjahr-Gobell, R.E. et al., Feeding the mummichog (*Fundulus heteroclitus*) a diet spiked with non-*ortho* and mono-*ortho*-substituted polychlorinated biphenyls: accumulation and effects, *Environ. Toxicol. Chem.*, 18, 699, 1999.
53. Munns, W.R., Jr. et al., Evaluation of the effects of dioxin and PCBs on *Fundulus heteroclitus* populations using a modeling approach, *Environ. Toxicol. Chem.*, 16, 1074, 1997.
54. Gleason, T.R., personal communication, 2000.
55. Prince, R. and Cooper, K.R., Comparisons of the effects of 2,3,7,8-tetrachlorodibenzo-*p*-dioxin on chemically-impacted and non-impacted subpopulations of *Fundulus heteroclitus*: I. TCDD toxicity, *Environ. Toxicol. Chem.*, 14, 579, 1995.
56. Prince, R. and Cooper, K.R., Comparisons of the effects of 2,3,7,8-tetrachlorodibenzo-*p*-dioxin on chemically-impacted and non-impacted subpopulations of *Fundulus heteroclitus*: II. Metabolic considerations, *Environ. Toxicol. Chem.*, 14, 589, 1995.
57. Vogelbein, W.K. et al., Acute toxicity resistance in a fish population with a high prevalence of cancer, presented at Environmental Toxicology and Chemistry Annual Meeting, Washington, D.C., 1996, Abstr. 330.
58. Weis, J.S., Mugue, N., and Weis, P., Mercury tolerance, population effects, and population genetics in the mummichog, *Fundulus heteroclitus*, in *Genetics and Ecotoxicology*, Forbes, V.E., Ed., Taylor & Francis, London, 1999.
59. Gleason, T.R. and DellaVecchia, K.M., Population-specific reproductive success of mummichogs, *Fundulus heteroclitus*, from a highly PCB contaminated site and a reference site, submitted for publication, 2000.
60. Fullerton, A.H. et al., Overwinter growth and survival of largemouth bass: interactions among size, food, origin, and winter severity, *Trans. Am. Fish. Soc.*, 129, 1, 2000.
61. Huber, M. and Gleason, T.R., Laboratory study on overwintering resource utilization patterns of young-of-the-year *Fundulus heteroclitus* from polychlorinated biphenyl-contaminated and reference sites, manuscript in preparation, 2000.
62. Nacci, D., Jayaraman, S., and Specker, J., Stored retinoids in populations of the estuarine fish *Fundulus heteroclitus* indigenous to PCB-contaminated and reference sites, *Arch. Environ. Contam. Toxicol.*, 40, 511, 2001.
63. Cohen, S., Strong positive selection and habitat specific amino-acid substitution patterns in *Mhc* from an estuarine fish under intense selection pressure, submitted for publication, 2000.
64. Spromberg, J.A., John, B.M., and Landis, W.G., Metapopulation dynamics: indirect effects and multiple distinct outcomes in ecological risk assessment, *Environ. Toxicol. Chem.*, 17, 1640, 1998.

65. Abraham, B.J., Species profiles: life histories and environmental requirements of coastal fishes and invertebrates (mid-Atlantic) — mummichog and striped killifish, U.S. Fish and Wildlife Service Biological Report 82 (11.40), U.S. Army Corps of Engineers, TR EL-82-4, 1985.
66. Bello, S., Characterization of Resistance to Halogenated Aromatic Hydrocarbons in a Population of *Fundulus heteroclitus* from a Marine Superfund Site, Ph.D. thesis, Massachusetts Institute of Technology and the Woods Hole Oceanographic Institution, Woods Hole, MA, 1999.
67. Bello, S.B. et al., Acquired resistance to aryl hydrocarbon receptor agonists in a population of *Fundulus heteroclitus* from a marine superfund site: *in vivo* and *in vitro* studies on the induction of xenobiotic metabolizing enzymes, *Toxicol. Sci.*, 60, 77, 2001.
68. Wirgin, I. and Waldman, J.R., Altered gene expression and genetic damage in North American fish populations, *Mutat. Res.*, 399, 193, 1998.
69. Hahn, M.E. et al., Mechanisms of adaptation to dioxin-like compounds, presented at the Environmental Toxicology and Chemistry Annual Meeting, Nashville, TN, November, 2000, Abstr. 032.
70. Karchner, S.I., Powell, W.H., and Hahn, M.E., Identification and functional characterization of two highly divergent aryl hydrocarbon receptors (AhR1 and AhR2) in the teleost *Fundulus heteroclitus*, *J. Biol. Chem.*, 274, 33814, 1999.
71. Powell, W.H. et al., Functional diversity of vertebrate ARNT proteins: identification of ARNT2 as the predominant form of ARNT in the marine teleost, *Fundulus heteroclitus*, *Arch. Biochem. Biophys.*, 361, 156, 1999.
72. Elskus A.A. et al., Altered CYP1A expression in *Fundulus heteroclitus* adults and larvae: a sign of pollutant resistance? *Aquat. Toxicol.*, 45, 99, 1999.
73. Meyer, J.N. and Di Guilio, R.T., Mechanisms of adaptation and fitness costs in F_1 and F_2 offspring of wild-caught killifish (*Fundulus heteroclitus*) from a contaminated site, presented at the Environmental Toxicology and Chemistry Annual Meeting, Nashville, TN, November 2000, Abstr. 034.
74. Champlin, D. et al., Characterizing populations of the estuarine fish *Fundulus heteroclitus* indigenous to sites with differing environmental quality, presented at the Environmental Toxicology and Chemistry Annual Meeting, Nashville, November 2000, Abstr. PWA119.
75. Nacci, D. et al., Effects of benzo[*a*]pyrene exposure on a fish population resistant to the toxic effects of dioxin-like compounds, *Aquat. Toxicol.*, in press.
76. McMillan, A. et al., Fine-scale genetic differentiation between contaminant-tolerant and contaminant-sensitive fish populations, presented at the Environmental Toxicology and Chemistry Annual Meeting, Nashville, TN, November 2000, Abstr. PWA124.
77. Roark, S.A., Guttman, S.I., and Nacci, D., Allozyme analysis of the relationship among contaminant-tolerant and contaminant-sensitive populations of *Fundulus heteroclitus*, presented at the Environmental Toxicology and Chemistry Annual Meeting, Nashville, TN, November 2000, Abstr. PWA118.
78. Cohen, S., Strong selective signal and high genetic variability at an immune system locus in contaminated and uncontaminated populations of an estuarine fish, presented at the Environmental Toxicology and Chemistry Annual Meeting, Nashville, TN, November 2000, Abstr. PWA116.
79. Street, G.T. et al., Reduced genetic diversity in a meiobenthic copepod exposed to a xenobiotic, *J. Exp. Mar. Biol. Ecol.*, 222, 93, 1998.

80. Krane, D.E., Sternberg, D.C., and Burton, G.A., Randomly amplified polymorphic DNA profile-based measures of genetic diversity in crayfish correlated with environmental impacts, *Environ. Toxicol. Chem.*, 18, 504, 1999.
81. Nadig, S.G., Lee, K.L., and Adams, S.M., Evaluating alterations of genetic diversity in sunfish populations exposed to contaminants using RAPD assay, *Aquat. Toxicol.*, 43, 163, 1998.
82. Bickham, J.W. et al., Effects of chemical contaminants on genetic diversity in natural populations: implications for biomonitoring and ecotoxicology, *Mutat. Res.*, 463, 33, 2000.
83. Smith, T.B. and Wayne, R.K., Eds., *Molecular Genetic Approaches in Conservation*, Oxford University Press, Oxford, 1996.
84. Roelke, M.E., Martenson, J.S., and O'Brien, S.J., The consequences of demographic reduction and genetic depletion in the endangered Florida panther, *Curr. Biol.*, 3, 340, 1993.
85. Vrijenhoek, R.C., Genetic diversity and fitness in small populations, in *Conservation Genetics*, 37, Loeschcke, V., Tomiuk, J., and Jain., S.K., Eds., Birkhauser Verlag, Basel, 1994.
86. Weber, K.E. and Diggins, L.T., Increased selection response in larger populations. II. Selection for ethanol vapor resistance in *Drosophila melanogaster* at two population sizes, *Genetics*, 125, 585, 1990.
87. Coustau, C., Chevillon, C., and French-Constant, R., Resistance to xenobiotics and parasites: can we count the cost? *Trends Ecol. Evol.*, 15, 373, 2000.
88. Levinton, J.S. et al., Running the gauntlet: pollution, evolution and reclamation of an estuarine bay and its significance in understanding the population biology of toxicology and food web transfer, in *Aquatic Life Cycle Strategies*, Whitfield, M., Ed., Marine Biological Association of the United Kingdom, Plymouth, U.K., 1999, 125.

11 Structuring Population-Based Ecological Risk Assessments in a Dynamic Landscape

Christopher E. Mackay, Jenee A. Colton, and Gary Bigham

CONTENTS

11.1 Introduction 273
11.2 Ecological Risk Assessment Model..274
    11.2.1 Risk Model Parameterization..276
        11.2.1.1 Mercury Concentration in Fish $P([C]_i|F_i)$ 276
        11.2.1.2 Heron Exposure Rate $(P(ER))$ 278
        11.2.1.3 Spatial Function..279
        11.2.1.4 Toxicity Response Function 285
11.3 Population-Based Risk Characterization 286
    11.3.1 Population Modeling for the Great Blue Heron.............................286
    11.3.2 Characterization of Population Dynamics......................................291
11.4 Discussion 293
References..296

11.1 INTRODUCTION

Ecological risk assessment is not a purely scientific endeavor. Rather, it is most commonly applied as an exercise in regulatory compliance intended to illustrate objectively demonstrable harm (or lack thereof) as the result of identified human activities. Legal guidance is limited to very broad directives requiring the protection of the environment from harm. Unfortunately, the definition of *harm* can be contentious. Risks to individual plants or animals are the easiest types of impacts to identify. However, within the context of providing protection to the environment, the magnitude of individual impacts may very well represent inconsequential events. Therefore, the characterization of harm may be misrepresented if limited solely to the individual. Simple and transparent (i.e., easy to understand and replicate) methods should be developed for application in a regulatory context to expand the scope of ecological risk assessments to the level of spatially defined subpopulations and populations, in order to determine whether an activity or activities in question actually represent an unacceptable risk within the legal context of environmental protection.

This chapter examines a probabilistic method of population-level ecological risk assessment. It is intended for application within the regulatory context of a remedial investigation under the Comprehensive Environmental Response, Compensation and Liability Act (CERCLA). It describes a risk assessment in terms of impact on population dynamics. In this case, risk was modeled for great blue herons (*Ardea herodias*) exposed to methylmercury from fish taken as prey from an inland lake in the northeastern United States. Although the lake and its surrounding environs were used to parameterize both the risk model and the later-discussed heron population dynamics model, it was necessary to assume higher mercury concentrations in fish and other media than those measured in the lake to demonstrate certain attributes of both models. Therefore, the lake and the results of this assessment are hypothetical. This being the case, the general approach described herein is relevant to marine systems as well as freshwater systems.

To describe the impact of this hypothetical exposure, two separate but interconnected models were developed. The first was a probabilistic risk assessment used to estimate the proportional population impact resulting from methylmercury exposure. The second was a population dynamics model to describe the risk estimates in terms of their impacts on the stability of the exposed heron subpopulation. The goal is to provide an approach that may be applied within a regulatory context to better illustrate the results of an ecological risk assessment in terms that may be quantitatively applied to evaluate environmental protection.

11.2 ECOLOGICAL RISK ASSESSMENT MODEL

The objective of the assessment component of this analysis was to express the impact of individual exposures to a toxic contaminant in terms of the risk to receptor populations. To determine the overall impact of all potential individual responses, it was necessary to quantify the probability of every possible response occurring within the exposed population. The paradigm used to characterize risk is the same one first proposed by the American Society for Testing and Materials (ASTM)[1] where risk (r) is expressed as the ratio of the exposure rate (e) to the expected exposure-dependent response (T):

$$r = \frac{e}{T} \qquad (11.1)$$

In ecological risk assessment, the underlying assumption is that there is no risk of an adverse impact if the rate of exposure of a receptor is less than a defined response threshold. However, if the exposure exceeds the threshold, as indicated by an r value greater than 1, then a risk exists that the response ascribed to T will occur.

If it is assumed that the risk to be characterized is the result of the exposure of a receptor to a toxic contaminant (C) that is found in an environmental medium F, then the risk paradigm can be expressed as a model where the risk of a receptor is estimated as the product of its exposure rate (ER) and the contaminant concentration in the medium of exposure ($[C]_F$), divided by the dose-dependent response threshold for $C(T_C)$. Hence, a generic risk model can then be expressed as follows:

$$r = \frac{\text{ER} \times [C]_F}{T_C} \tag{11.2}$$

If the risk is to be characterized for a group or population of individuals, then neither the exposure rate of the receptor, nor the contaminant concentration of the medium, nor the dose-dependent response of the receptor is an absolute value. Each variable parameter possesses a distribution of possible values within the exposed population, and from observations of the relative frequency of occurrence, a probability function for risk can be discerned. The relation between a parameter value and its probability of occurring is referred to as a probability density function. To represent this, the variables must be expressed as probability density functions (generically denoted as $D(x)$, where x represents the independent variable for the function), which is the integral of potential occurrences of all possible parameter values for the exposed population (generically denoted as $P(x)$, where x represents the parameter for which the probability is expressed). Hence, the generic risk model can be expressed as follows:

$$D(r) = D(\text{ER}) \times D([C]_F) \times D(T_C)^{-1} \tag{11.3}$$

Solving the probability function for risk ($D(r)$) can be accomplished either by convolution or faltung (denoted as $D \times D$) or by simulation using Monte Carlo techniques. When using Monte Carlo techniques, the probability of risk is determined using the pooled estimated probabilities associated with the parameters of the model over a range of potential exposure situations, A, for all members of the exposed population, N, as follows:

$$D(r) = \int_{n-1}^{N} \sum_{a=1}^{A} P\left(\frac{\text{ER}_a \times ([C]_F)_a}{T_C}\right) dn \tag{11.4}$$

The risk is now described for all individuals of population N as a function of the probability for all concentrations of contaminant C within medium F, the probability for all the possible exposure rates of the receptor to medium F within A simultaneous situations, and the probability of all possible dose-dependent responses to C by the receptors. Hence, the risk function is now defined as a probability density function $D(r)$ for the characterization of risk.

11.2.1 RISK MODEL PARAMETERIZATION

The generic risk model derived above was parameterized to predict the risk to a population of great blue herons exposed to methylmercury in fish from a lake. Most of the eastern and southern shore of the lake has been developed for either urban or suburban uses. The western shore remains largely undeveloped and provides habitat to numerous avian and mammalian species. Its shallow sloping banks and moderate bank cover make it excellent heron foraging habitat, and herons are commonly observed during the warm-weather seasons.

The exposure received by a population of receptors is not simply related to the distribution of the contaminant (C) in the entire medium, but rather to the concentration of C in the constituents of medium F that the receptors contact directly. In situations where such distinctions may be made within a medium, the probability for exposure to any concentration C can now be made dependent upon the probability of exposure to a subcomponent of F (F_i of known $[C]$) across the range of all possible exposures (F_i to F_i) as follows:

$$P([C]) = P([C]_i|F_i)_a \qquad (11.5)$$

The relationship $P([C]_i|F_i)$ is the probability of encountering a specific contaminant concentration $[C]$; and is based on the probability of the exposure of the receptor to an identifiable subcomponent of the medium F_i. (See Chapter 4 for further explanation of such conditional probabilities.) This may represent empirical distributions such as time spent in a specific location, or may be used to distinguish between the probability of ingesting specific prey items. Methods for the derivation of these probability density functions can be found in Efron and Tibshirani.[2]

11.2.1.1 Mercury Concentration in Fish $P([C]_i | F_i)$

The modeled medium of exposure (F) for the great blue heron was fish. The distribution of mercury concentrations used in the risk model was based on the likelihood observed for the distribution of mercury concentrations in individual fish (F_i) from the lake, and the likelihood that a fish would be preyed upon by a heron. Likelihood, in this context, is defined as a past probability (i.e., observed) based on reported distributions.

The mercury distribution was determined empirically from data collected from the lake in 1992. The probability that a sampled fish, containing a known concentration of mercury ($P(Hg)$) would be prey for the great blue heron was determined as the product of the likelihood (L) of the heron selecting that size of fish (prey), and the likelihood of that species of fish being available from the lake (available). Therefore, the probability that any individual heron would ingest a fish represented by size and species by a specific sampled fish can be expressed as follows:

$$P([Hg]|F) = ([Hg]|L(\text{prey}) \times L(\text{available})) \qquad (11.6)$$

Structuring Population-Based Ecological Risk Assessments in a Dynamic Landscape

The likelihood that a fish would be prey for the heron was determined from empirical observations reported by Alexander.[3] The great blue heron's predominant prey is fish ranging from 3 to 33 cm in length. Proportional dietary content based on fish size was reported from the survey to be 8, 40, and 52% for fish 3 to 7, 7.1 to 14, and 14.1 to 33 cm, respectively. The sampled fish were ranked according to size. Sampled fish outside the range of 3 to 33 cm were excluded. The remainders were classed into three cohorts based on the above size ranges and $L(\text{prey})$ for each sampled fish was then determined as follows:

$$L(\text{prey}) = \frac{P_c}{n_c} \tag{11.7}$$

P_c is the proportion that the cohort represents in the heron's diet, and n_c is the number of sampled fish in that cohort.

The likelihood that a sampled fish is available as prey for the great blue heron is dependent on the abundance of that species in the lake. Because the fish samples were not random with regard to fish species, abundance within the sample cannot be assumed to be representative. However, overall abundance statistics were available from lake surveys. Therefore, the $L(\text{available})$ for any fish in the sampled group was deemed proportional to the abundance of that species, within each size cohort, throughout the entire lake (A_t). Only fish species typically available to the heron were considered. Deepwater species that are not available for prey were excluded, as were fish determined to be either too large or too small to constitute heron prey. To control for bias in the sample due to disproportional representations of fish species in the sample set, the $L(\text{available})$ was made inversely proportional to the species abundance within the cohort (A_c). Therefore, $L(\text{available})$ can be mathematically defined as follows:

$$L(\text{available}) = \frac{A_t}{A_c} \tag{11.8}$$

Assuming that size selection by the heron was independent of species abundance in the lake, the product of these two likelihoods defines the probability of selection ($P(F_i)$). Therefore, the probability of the heron's exposure to a given concentration of mercury could be derived as follows:

$$P([\text{Hg}]|F_i) = \left([\text{Hg}]\bigg|\left(\frac{P_c \times A_t}{N_c \times A_c}\right)_i\right) \tag{11.9}$$

The estimate of the probability density function across all potential prey fish ($D([\text{Hg}]_i|F_i)$) was determined by bootstrapping (with replacement) the sampled mercury data using the individual $P(F_i)$ values as the metric of probability for selection. Each mercury observation was assigned a probability of occurrence based on the above likelihood. The mercury concentrations were then selected with replication

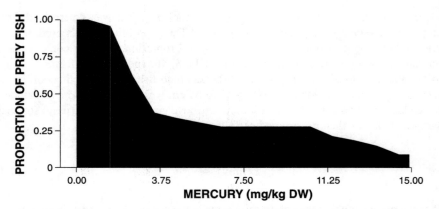

FIGURE 11.1 Reverse cumulative probability density function for mercury exposure concentrations for the great blue heron from lake fish.

based on the probability assigned to derive a probability distribution of potential exposures. A more detailed discussion of this method is available in Chapter 24 of Efron and Tibshirani.[2] The frequency of selection was tracked and used to derive the probability density function ($D([Hg]_i|F_i)$), illustrated in Figure 11.1.

11.2.1.2 Heron Exposure Rate ($P(ER)$)

The dietary intake rate (IR) for the great blue heron may also be described as a probability density function. Unfortunately, there is rarely a sufficient record of empirical observations to develop an adequate distribution for this parameter directly. However, since the dietary requirements of the heron are related to its energy demands, it is possible to model the dietary requirements based upon a metric for which adequate distributions are available. Kushlan[4] developed an allometric equation, specifically for wading birds, by regressing a series of observed dietary intake rates against the paired body masses (BW) for the birds. An estimation for the distribution of body masses for the great blue heron population has been developed by Henning et al.[5] By aggregating reported data on the body masses of great blue herons from the northeastern United States, it was found that the distribution of this parameter conformed to a normal distribution with an average mass (μ_{BW}) of 2300 g with standard deviation (s_{BW}) of 670 g, and a minimum and maximum body mass of 1600 and 3000 g, respectively. The model was truncated to disallow values greater than or less than the minimum and maximum parameters. The variance used remained unchanged. By substituting the allometric relationship, the exposure rate function was parameterized based on the distribution in the heron's body mass as follows:

$$IR(BW) = 0.00925 \times BW^{-1.64}\left(\frac{gDW}{gBW \cdot day}\right) \quad (11.10)$$

$$D(IR) = D(0.00925 \times N(\mu_{BW}, s_{BW})^{-1.64})$$

11.2.1.3 Spatial Function

The great blue herons found in this area are migratory. Although they do nest in this region, they winter in the lower Mississippi Valley, the Gulf Coast, and the Southern Atlantic seaboard.[6] Therefore, herons are not present on the lake for a large part of a year. Within its breeding range, the great blue heron may be either colonial or solitary, depending on its location and situation. Herons in this area tend to be solitary nesters, and will establish and defend a nesting territory.[6] Great blue herons are most likely to be found hunting near their nesting sites, but may range as far as 24 km during daily feeding forays.[7,8]

Because there are two considerations affecting the foraging behavior of the exposed heron population, one for migration and one for local foraging use, two spatial functions had to be developed to control for the heron's feed locations. The first function was modeled as a temporal parameter ($D(t)$) that was used to describe the heron's location in its migratory cycle where $P(t) = 1$ represents 100% residence around the lake, and $P(t) = 0$ represents 100% residence somewhere else along the migration route. The second distribution was a spatially explicit parameter ($D(a)$) that was used to describe the probable foraging patterns for the great blue heron subpopulation that relies on the lake as part of its food source while in residence around it. This also was parameterized in a similar manner where $P(a) = 1$ represented complete dietary reliance on the lake and $P(a) = 0$ represented complete dietary reliance on other locations within the defined foraging range. The generic risk model was therefore structured as follows:

$$D(r) = \frac{\int_{n-1}^{N} \sum_{t=0}^{T} P(t) \times \left(\sum_{a=0}^{A} P(a)_t \times \left(\sum_{i=1}^{I} P(\text{ER}_i \times ([C]_F)_i)_{ta} \right) \right) dn}{D(T_C)} \quad (11.11)$$

The temporal probability, $P(t)$, was parameterized based upon regional observations. Observations indicate that herons arrive in the area around the lake between days 46 and 90, and depart on winter migration between days 258 and 273. With lack of data to the contrary, it was assumed that the dates of arrival and departure of any given individual were independent. For illustration purposes, the distributions for both arrival and departure dates were assumed to be represented by a skewed triangular distribution with the mode defined at the earliest and latest 10th and 90th percentiles for arrival and departure, respectively. By solving for the residency time (departure date minus arrival date) using Monte Carlo techniques, the resulting residence time was found best to fit a beta distribution of alpha 38.72, beta 3.93, scaled to 227 days and truncated at 168 days (Figure 11.2).

The habitat use function, $D(a)$, was modeled based on the great blue heron's bioenergetics. Flight to and from any location within the foraging area would require an energy output proportional to its distance from the nest or colonial roost. Associated with this expenditure is also a loss in potential foraging time equivalent to the time en route. This relationship can be expressed as follows:

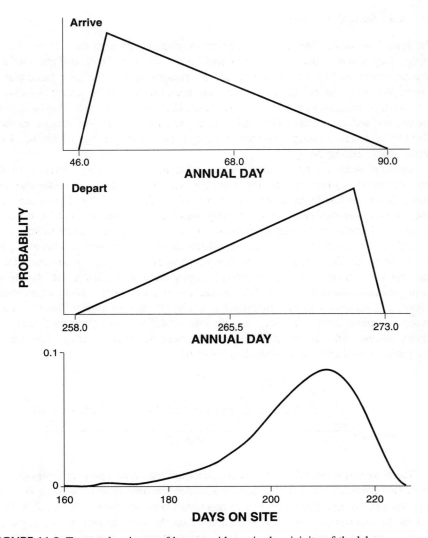

FIGURE 11.2 Temporal estimate of heron residence in the vicinity of the lake.

$$\Delta E_{(x, y)} = (W_{(x, y)} - W_{(0, 0)}) \times (t_{(0, 0)} - t_{\text{Flight}}) - (W_{\text{Flight}} \times t_{\text{Flight}}) \quad (11.12)$$

The net benefit to the heron by foraging at any given location (denoted as (x, y)) is expressed as the net energy availability at (x, y), ($\Delta E_{(x, y)}$), and is proportional to the total power (i.e., energy per unit time foraging) that may be derived at the location ($W_{(x, y)}$) relative to the power derived if the heron had not traveled, but foraged in the immediate vicinity of the roost or nest ($W_{(0, 0)}$). The amount of time available to the heron to forage at location (x, y) is equal to the amount of foraging time available at location (0, 0), minus the time necessary to commute to and from (x, y) (t_{Flight}). It is also necessary to consider the energy expended in commuting between (0, 0) and (x, y). This is determined as the product of the power requirement for flight (W_{Flight}) and the duration of the commute, t_{Flight}. This is subtracted from the net energy difference between $(0, 0)$ and (x, y) to derive the net energy available.

Since the ultimate goal is to develop a probability density function based on the relative benefit of one location over another, another energy term is defined (E_H) that relates the net energy benefit at any given point (x, y) relative to that at $(0, 0)$ ($E_{(0, 0)}$). The probability density function may now be expressed as a proportional function of the relative energy availability at any point (x, y) with its differential radial distance from $(0, 0)$ (dr) as follows:

$$D(E_H) = \int_{r=0}^{23\ km} \frac{\Delta E_{(x, y)}}{E_{(0, 0)}} dr \qquad (11.13)$$

The distribution of this function is illustrated in Figure 11.3.

The probability density function $D(E_H)$ is now a spatially explicit metric that describes the probability of a heron being present at any point (x, y), based solely on the bioenergetic advantage relative to location $(0, 0)$. This may now be applied to a measure of habitat quality at (x, y) relative to $(0, 0)$ to predict the likelihood of a heron's presence based on the overall advantage that a heron would derive by foraging at that location.

Habitat quality is a site-specific parameter and dependent upon estimates of the quantity of available forage fish and the quality of the local environment as adequate foraging habitat. The great blue heron is a wading bird that can utilize a variety of freshwater and marine habitats. It is found in areas of shallow water that have firm substrates and high concentrations of small fish.[6] Great blue herons forage in lakes, rivers, brackish marshes, lagoons, coastal wetlands, tidal flats, and sandbars, as well as wet meadows and pastures.[4,6] For the purposes of this assessment, potential heron habitat was defined as any shoreline or riverbank within 24 km of the lake, with a

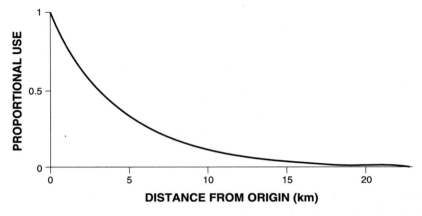

FIGURE 11.3 The function of the relative energetic benefit $EH_{(x, y)}$ with distance from the origin $(0, 0)$. $E_{(0, 0)}$ and $E_{(x, y)}$ are assessed as uniform within the radius of 24 km.

wading depth less than 50 cm for any water body whose minimum dimension was greater than 2 m. Habitat quality was also assumed to be proportional to relative habitat density. Habitat within urban areas was deemed unsuitable. Prey availability was determined from available state surveys. For habitats where survey results were unavailable, prey abundance was estimated based on comparisons with similar surveyed water bodies based on size and location.

To determine the relative habitat quality, all potential foraging locations were identified and mapped relative to location (0, 0) to a radius of 24 km. These were then grouped into octants and segmented into $\frac{1}{20}$ths of the total foraging radius (Figure 11.4). Habitat quality ($H_{(x, y)}$) was expressed as the product of the density of appropriate habitat ($d_{H(x, y)}$) within the segment and the average prey abundance ($a_{(x, y)}$) relative to the prey abundance at point (0, 0) ($a_{(0, 0)}$) as follows:

$$H_{(x, y)} = \frac{d_{H(x, y)}}{d_{H(0, 0)}} \times \frac{a_{(x, y)}}{a_{(0, 0)}} \quad (11.14)$$

The distribution of habitat density relative to the requirements of the great blue heron is provided in Figure 11.5.

The bioenergetics model describes the probability of a heron being at location (x, y) based on the potential advantage of commuting from (0, 0) to (x, y). The spatial habitat quality model describes the potential advantage of a heron being at location (x, y) based on the availability of prey and the quality of the habitat. When these two are combined as follows, the product is a measure of probability that describes the likelihood of a heron foraging at any location within the prescribed foraging range ($P(q_{(x, y)})$):

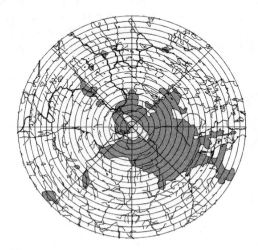

FIGURE 11.4 Spatial distribution of great blue heron habitat in the vicinity of the lake. Circle represents the 24-km radius assumed for potential habitat use. Shaded areas represent urban regions excluded as potential habitat. Segments represent octants segmented radially in $\frac{1}{20}$th of the total radius.

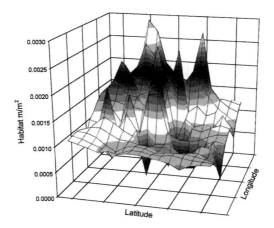

FIGURE 11.5 Habitat density, expressed as meters of usable shoreline per meter square of area, for the environment surrounding the lake to a distance of 24 km.

$$P(q_{(x, y)}) = \left(\frac{H_{(x, y)}}{H_{(0, 0)}}\right) \times P(E_H)_{(x, y)} \qquad (11.15)$$

The estimation of overall area use based on habitat quality for the foraging area, centered on the lake, is illustrated in Figure 11.6.

The final step in the parameterization of habitat use, $D(a)$, required the expression of the heron's foraging behavior relative to the lake. To accomplish this, the probability density function for habitat utilization, illustrated in Figure 11.6, was applied to determine the probability of any heron foraging on the lake ($P(a) = 1$) vs. foraging at any other location ($P(a) = 0$) by first quantifying the proportion of the lake shoreline within each of the sections defined in the habitat map ($A_{site(x, y)}$). First, the three-dimensional distribution of $P(q_{(x, y)})$ was then collapsed to a two-dimensional probability $P(q)$ by grouping all locations relative to the proportional area of the lake contained within each segment. Then the relative area of the lake was then bootstrapped (with replacement) and the results tracked to develop the probability density function for area use by the heron subpopulation relative to the proportion of the lake to which they will be exposed. The algorithm used was as follows:

$$D(a) = \int_{A_{\text{Site}} = 0}^{1} (A_{\text{Site}} | P(q)) dA_{\text{Site}} \qquad (11.16)$$

The probability density function for this relation is illustrated in Figure 11.7 and was used as the distribution for $P(a)$ in the risk model.

Insufficient data were available to model mercury exposures outside of the lake using the $P([Hg]|F_i)$ method detailed above. Therefore, mercury concentrations for exposures that occurred outside of the lake, but within the state, were generalized

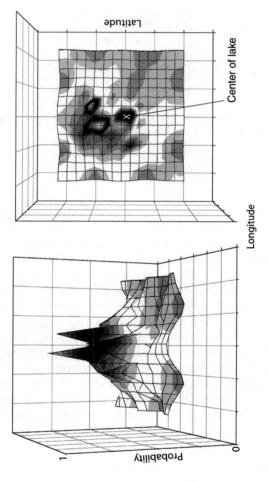

FIGURE 11.6 Relative probability of area use based on available habitat and bioenergetic necessity of the great blue heron.

Structuring Population-Based Ecological Risk Assessments in a Dynamic Landscape

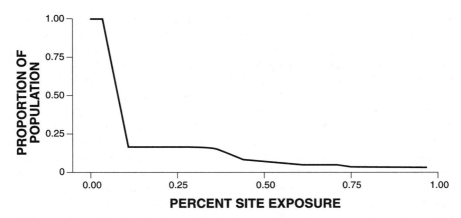

FIGURE 11.7 Reverse cumulative probability density function of heron utilization of foraging habitat expressed as percent exposure to the lake.

to the state averages for freshwater fish.[10] Exposures that were assumed to occur outside of the state were assigned methylmercury concentrations equivalent to the U.S. national average for freshwater fish.[11]

11.2.1.4 Toxicity Response Function

Mercury is a naturally occurring element that exists in the environment in different chemical forms. Methylmercury (CH_3Hg^+; $(CH_3)_2Hg$) generally represents less than 1% of the total mercury in the water column, but is of major ecotoxicological importance because of its high toxicity and natural tendency to bioaccumulate. For this reason, it is the predominant mercury species found in fish tissue and, therefore, the primary form of mercury to which piscivorous birds are exposed. Methylmercury in birds has been demonstrated to affect various organ systems with toxic effects including altered behavior, hepatic lesions, ataxia, weakness, muscular atrophy, and death, as well as reduced fecundity manifested primarily as a decline in fledging rates.[12,13] In this study, it was assumed that methylmercury represented the only toxicant for assessment and that the only significant source was through the ingestion of contaminated fish. Toxicological end points to be assessed here were both chronic lethal toxicity and chronic reductions in adult reproduction rates.

The toxicity response function ($f(T)$) is the most difficult parameter to express as a probability density function. The current understanding of wildlife responses to methylmercury is not sufficient to establish an adequate dose–response curve. Attempts to estimate quantitatively the probability density function for the response to mercury exposure would be highly uncertain. Consequently, the toxicological response was described as a constant toxicity reference value (TRV).

The TRV used to evaluate the reproductive effects of methylmercury in the great blue heron was based on a three-generation study by Heinz[14] in mallards. Mallard ducks were exposed to dietary concentrations of methylmercury ranging from 0.5 to 3.0 mg/kg dry weight for two generations, with the third generation exposed to

0.5 mg/kg dry weight/day. The initial test birds showed no behavioral or reproductive effects at the lowest methylmercury concentration. However, the first-generation ducklings demonstrated a 30% reduction in 1-week survival rates at methylmercury concentrations of 0.5 mg/kg methylmercury.[15] Based on a food intake rate of 138 g dry weight/kg body weight for the adult females, this represents a lowest observed adverse-effect level (LOAEL) of 0.069 mg/kg body weight/day.

The TRV used to evaluate the lethal effects of methylmercury in the great blue heron is based on a study by Bouton.[16] That study used great egrets as a test species. Fledged juveniles exposed to 0.5 mg/kg methylmercury demonstrated significant adverse effects on activities including shade preference and motivation to forage. Since methylmercury is a cumulative toxicant, and the aberrant behaviors manifested would be expected to affect individual survival in the wild, this was assumed to represent a lethal concentration. Based on an intake rate of 0.97 kg fresh weight/kg body weight/day (as reported by Bouton[16]), a lethal TRV of 0.098 mg/kg body weight was derived.

11.3 POPULATION-BASED RISK CHARACTERIZATION

The risk model, parameterized above, yielded a probability density function that described the distribution of exposure concentrations relative to the TRV. This was expressed as a hazard quotient, which is the ratio of the exposure concentration over the TRV. Any proportion of the heron population associated with the lake whose exposure rate exceeded the TRV would manifest a hazard quotient greater than 1, and thus was assumed to have received an exposure sufficient to incur adverse toxicological effects.

At this stage of the risk characterization, the risk was evaluated separately for reproductive effects and lethality end points (Figure 11.8). The evaluation indicated that 9.4% of the population would be at risk for reproductive effects, and 4.4% of the same population would be at risk for mortality. Because determinant values were used to represent the toxicological response of the heron as the result of exposure to mercury, these values were treated as absolute thresholds for later population dynamics modeling. The proportions of the population that exceeded a hazard quotient of 1 were assumed to experience either complete reproductive failure or mercury-induced mortality.

11.3.1 POPULATION MODELING FOR THE GREAT BLUE HERON

Population modeling for the great blue heron was performed using a cohort demographics model, solved over consecutive time steps using Monte Carlo techniques to represent temporal variability. The model was executed under four different scenarios:

1. No impact of methylmercury exposure on the resident heron subpopulation (control);
2. Methylmercury impact to the exposed subpopulation equivalent to those modeled, but only with regard to effects on the herons' reproduction;

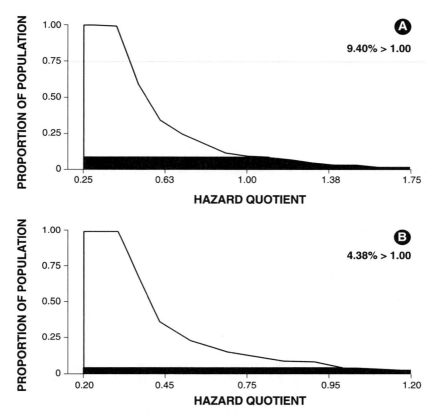

FIGURE 11.8 Reverse cumulative probability density function of risk from methylmercury exposure for the great blue heron based on the disruption of reproduction (A) and lethality to adult receptors (B).

3. Methylmercury impact to the exposed subpopulation equivalent to those modeled, but only with regard to effects on the herons' mortality; and
4. Methylmercury impact to the exposed subpopulation equivalent to those modeled with regard to effects on both heron mortality and reproduction.

The underlying hypothesis for the population model is that all herons in the vicinity of the lake can be characterized into two distinct, but interacting subpopulations (Figure 11.9). First, there is the exposed subpopulation that is composed of those herons that, to a greater or lesser extent, obtain prey from the lake. The second group is the regional subpopulation that consists of all herons that do not receive prey from the lake, but that do interact directly with the exposed subpopulation. Three types of interactions are defined. The first is immigration, where members of the regional subpopulation shift to the exposed subpopulation to fulfill available but unused carrying capacity. The second is emigration from the exposed subpopulation that occurs when recruitment within the exposed subpopulation exceeds the available carrying capacity. A third type of

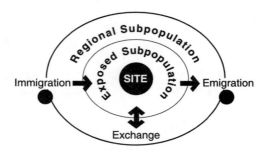

FIGURE 11.9 Conceptual model for the interaction between the population of herons exposed to a site and a great blue heron population that is not exposed, but interacts freely with the exposed population.

interaction, exchange, involves the movement of individuals between the exposed and non-exposed subpopulations independent of carrying capacity. To simplify the model, it was assumed that individuals that moved interchangeably between the populations in this model would not receive sufficient exposures to bias the population impact assumptions.

The life-cycle model used to define the state variables for the population dynamics model is illustrated in Figure 11.10. The great blue herons that comprise the exposed subpopulation were considered to consist of three cohorts. The first cohort is the reproductive adults (P_1). These individuals are responsible for the annual reproduction of the species and experience a cohort-specific mortality rate (M_A). Quantification of P_1 in terms of the number of individuals would require knowledge of the absolute carrying capacity. Unfortunately, this is one of the most difficult aspects to quantify. Carrying capacity is defined as the number of individuals a defined area can support based on available resources necessary for the survival and successful reproduction of a receptor. The evaluation of resources relative to any species' particular needs requires high-resolution habitat characterization, is fraught with uncertainty as to resource availability, and is complicated by a high degree of temporal variability. Therefore, in this population dynamics model, the number of individuals (n) within each cohort was standardized based on the proportion of

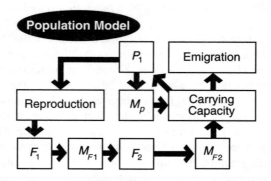

FIGURE 11.10 Conceptual model for the individual life cycle of the great blue heron.

carrying capacity (K) fulfilled. Because the availability of nesting sites is considered the primary limitation on heron population densities,[17] the regional carrying capacity was expressed as the carrying capacity for the P_1 cohort (K_p) as follows:

$$\frac{X}{K_p}, \quad \text{where } X \in \{P_1, F_1, F_2, e\} \tag{11.17}$$

In this model, P_1 represents the population of the reproductive adult cohort, F_1 is the fledged juveniles hatched in the same year, F_2 represents the immature non-reproductive adults, and e is the relative emigration rate from the exposed subpopulation to the regional non-exposed subpopulation.

To permit this type of standardization, an assumption must be made that the mortality and reproduction rates of the receptor are independent of population density, while population density remains dependent on carrying capacity. This assumption may be problematic for some communal species; however, it was considered reasonable for great blue herons, which, although they are ubiquitous within the environment, are found at relatively low population densities.

Progeny of the breeding adult cohort were defined as the juvenile cohort (F_1) and were modeled as follows:

$$F_{1(t)} = r \times P_{1(t)} \tag{11.18}$$

The size of F_1 cohort was the product of the reproductive efficiency of the adults (r) and the size of the reproductive adult population. The reproductive efficiency, in this case, includes not only the fecundity of the adults, but also mortality of the nestlings. Therefore, F_1 represents those individuals that have survived past the fledgling stage.

The juvenile cohort experiences a specific annual mortality rate (M_{F1}) at the end of each iteration, with surviving individuals passed to the next iteration's immature adults cohort (F_2). The immature adults are 1-year-old great blue herons that are not reproductively active and were modeled as follows:

$$F_{2(t)} = F_{(1(t-1)} \times (1 - M_{1(t-1)}) \tag{11.19}$$

The immature adults also experience a cohort-specific mortality rate (M_{F2}) at the end of each iteration, with surviving individuals joining the reproductive adult pool (P_0). The reproductive adult pool represents all reproductive adults that have survived through the past iteration, plus recruitment from the surviving individuals of the past F_2 cohort, and was modeled as follows:

$$P_{0(t)} = (P_{1(t-1)} \times (1 - M_{p(t-1)}) + (F_{2(t-1)} \times (1 - M_{F2(t-1)})) \tag{11.20}$$

By definition, P_1 cannot exceed K_p. Therefore, if P_0 exceeds K_p, then individuals will be forced to emigrate from the exposed subpopulation to the non-exposed regional subpopulation. However, if P_0 is insufficient to fulfill the carrying capacity K_p, then the population dynamics model responded in one of two ways. If the exposed subpopulation was assumed to be open, then individuals would be expected to immigrate into the exposed subpopulation from the regional subpopulation. The magnitude of allowable immigration was limited to the expected surplus recruitment capacity for the non-exposed population $((P_0 - K_p)_c)$. If the exposed population was assumed to be closed, then a deficit $P_{1(t)}$ (i.e., $P_{1(t)} < K_p$) would occur. The algorithm used to model these rules was as follows:

$$P_{t1} = P_{0t} + e_t \qquad (11.21)$$

where

$$e_t = P_{0t} - K_p$$
$$\text{Open: } e_t \in \{\infty > e_t > (P_0 - K_p)_c\}$$
$$\text{Closed: } e_t \in \{e_t \geq 0\}$$

The parameters necessary to fulfill this population model are limited to the distributions for reproductive rate and cohort-specific mortality. Estimates for these parameters were taken from a survey reported by Henny.[12] Parameter variation was accounted for in each generation from the probability density functions describing the reproductive and mortality parameters. State variables included the size of each individual cohort as well as the immigration/emigration between the regional population and the exposed subpopulation (all expressed as a proportion of the P_1 carrying capacity). The population dynamics model was run in yearly time steps with potential impact determined on 60-year runs to ensure steady state. Evaluation of emigration (Equation 21) was determined simultaneously within each of the time-step determinations. Runs were started with the assumptions that P_1 equaled K_p and that F_1 and F_2 equaled 0. Equilibrium between the cohorts was achieved within five time steps.

Results from the probabilistic risk assessment were applied directly to the parameter estimates of mortality and reproduction, after equilibration, to determine the effect of the impact on the population dynamics.

Because no dose–response function could be resolved based on the current understanding of methylmercury toxicity to great blue herons, assumptions had to be made regarding the absolute magnitude of the effect resulting from exposures that exceeded the TRV. For illustrative purposes, it was assumed that exceedance of the reproductive TRV would result in complete reproductive failure. The result of the proportional reproductive impact (I_r) was modeled as follows:

$$F_{1t} = r \times \min((P_{1t} - (P_{1t} \times I_r)), K_p) \qquad (11.22)$$

Similarly, exceedance of the lethal TRV was assumed to result in individual mortality within the same iteration cycle. Since the lethal impact (I_L) was expressed in terms of the entire population, the proportional impact between cohorts was determined based on their relative sizes. This was modeled as follows:

Structuring Population-Based Ecological Risk Assessments in a Dynamic Landscape

$$X_{(t)} = X_{(t-1)} \times (1 - M_x) - \left(I_L \times \frac{X_{(t-1)}}{\sum X_{(t-1)}}\right); \text{ where } X \in \{P_1, F_1, F_2\} \quad (11.23)$$

11.3.2 Characterization of Population Dynamics

Figure 11.11 illustrates the comparative impact on recruitment productivity for the exposed subpopulation of great blue heron as a result of exposure to mercury in the fish of the lake. Productivity values greater than 1 indicate a recruitment productivity in excess of that necessary to maintain the P_1 cohort at the carrying capacity without immigration. Under the control scenario, it was estimated that the median recruitment productivity of the exposed population would result in an excess capacity of 7.5% of the carrying capacity for the P_1 population. When the percentage of herons determined to be at risk for reproductive failure was removed from reproduction, the median excess recruitment productivity fell to 5.0%. When the percentage of herons determined to be at risk from lethal impact was modeled individually, the median excess recruitment productivity fell to 4.1%. When both impacts were combined, it was estimated that the median excess productivity of the exposed population would only be approximately 1.4% of the carrying capacity for the P_1 cohort. This could also be expressed as a 20% probability that the combined impacts would result in a level of productivity that could not sustain the current P_1 population at the carrying capacity. This result demonstrates the importance of the consideration of multiple impacts in the characterization of potential risk. Because both reproductive failure and lethal impact affect different parameters within the population dynamics model, it cannot be assumed that the results would be additive or that one impact would overshadow the other.

By using the risk model to project potential lethal population effects relative to reproductive impacts, it is possible to develop a population response curve for the expected emigration from the exposed population at varying levels of population

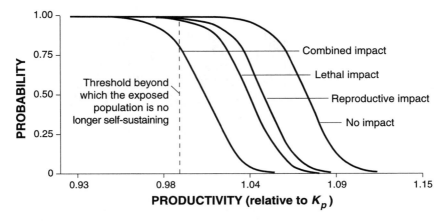

FIGURE 11.11 Reverse cumulative probability distributions of expected recruitment for the exposed great blue heron population on the lake.

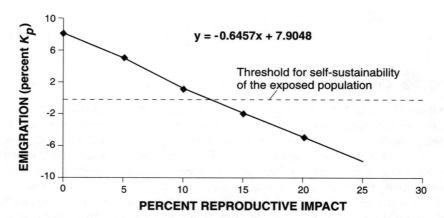

FIGURE 11.12 Estimated emigration capacity for the exposed population at various levels of impact resulting from exposure to mercury from the lake. Impact is expressed in levels of percent reproductive failure and includes the impact of lethal effects.

impact. When this was performed, it was found that a reproductive failure rate of 11.8% (which corresponds to a lethal impact of 5.3%) would result in the exposed subpopulation no longer being self-sustaining and therefore requiring immigration from the regional non-exposed subpopulation in order to fulfill the carrying capacity (Figure 11.12). Distributions for these estimates are illustrated in Figure 11.13.

When the model was executed in an open format, immigration from the regional subpopulation ensured that the P_1 cohort remained at the carrying capacity, K_p, and there was no threat of extinction at any level of impact (exceedance of the excess production capacity of the regional subpopulation was never observed). However,

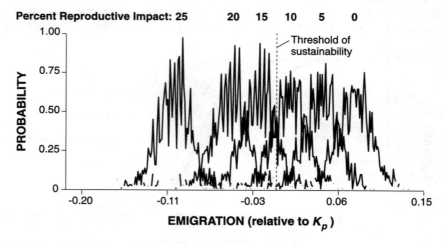

FIGURE 11.13 Frequency distribution of the P_1 cohort for the estimate of emigration from the exposed population (expressed as a proportion of K_p) at various levels of impact resulting from mercury exposure on the lake. Impact is expressed in percent reproductive impact, but includes the effects of lethal impact as well.

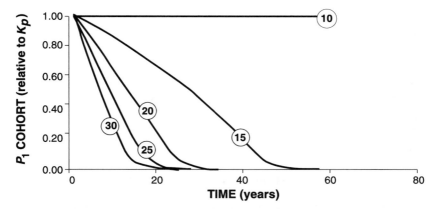

FIGURE 11.14 Time-course extinction curves for the exposed population of great blue herons at various levels of impact (expressed as proportional reproductive failure). Population is expressed as a proportion of K_p. Impacts include both reproductive and lethal effects as determined by the risk model used to characterize the effects of exposure to mercury in fish from the lake.

when the model was executed under the closed format where immigration was not permitted, impacts to the exposed population resulted in levels of recruitment productivity that could not sustain the P_1 cohort at K_p, and therefore resulted in population declines over time. Relative effects at various levels of impact on the exposed subpopulation are illustrated in Figure 11.14. Results from the closed-model analyses indicate that there is no threat of extinction, so long as the proportional impact is less than 11.8% reproductive failure and, correspondingly, 5.3% lethality. The model indicated that the resulting decline in the P_1 cohort eventually resulted in the extinction of the exposed population. An interesting observation in modeling this particular species, which possesses a relatively long lifespan, was the slow rate at which extinction would be expected to occur. For example, if it were assumed that extinction occurred when the P_1 cohort fell below 10% of the carrying capacity, then an impact level equivalent to 15% reproductive failure (7.0% lethality) would result in extinction of the exposed great blue heron population in 44 years of sustained impact (Figure 11.15). Similarly, at 30% reproductive impact (14.0% lethality), great blue heron extinction would be expected to occur within 12 years of sustained impact. This strongly indicates the importance of considerations for the temporal dynamics not only of the receptor, but also of the modeled stress, when attempting to predict the likelihood of subpopulation extinctions.

11.4 DISCUSSION

Species currently present within the environment represent ecological successes. As populations, they have managed to withstand natural fluctuations in reproduction and mortality to ensure the propagation of their species. Different species have evolved different methods to survive transient or localized stresses. Some survive based on excess reproductive capacity. Some survive based on the longevity of the

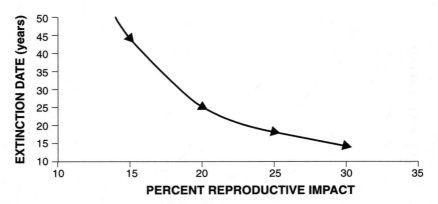

FIGURE 11.15 Relationship between proportional sustained impact (expressed as percent reproductive impact) and the time necessary for the exposed population P_1 cohort to fall to 10% of the K_p. Impact includes the lethal as well as the reproductive response as determined by the risk model used to characterize the impact of exposure to mercury on the lake.

reproductive adults. Some survive by spreading local variations over very broad spatial populations. Such differences in survival strategies make it very difficult to generalize about how a particular population is likely to respond to a given survival pressure. The modeling of population dynamics, in this situation, has a very specific purpose. It is intended solely to provide a context within which the results of the probabilistic risk assessment may be evaluated for a specific species. The great blue heron possesses a relatively low reproductive rate. However, it is relatively long-lived and highly mobile, with populations consisting of a large number of interacting subpopulations. In the example provided here, it was demonstrated that for the great blue heron, an impact greater than 11.8% reproductive failure and 5.3% lethality would result in a situation where an exposed subpopulation would no longer possess the productivity necessary to maintain the adult cohort at carrying capacity. However, the mobility of this species ensures that deficits in recruitment may be overcome by immigration from non-exposed subpopulations. Furthermore, even if available immigration were not sufficient to maintain the reproductive adult cohort, it would require prolonged impact over many reproductive cycles before the subpopulation would approach extinction.

The observation made here for the great blue heron cannot be generalized to other species. Species with different relative rates of reproduction and/or mortality and/or mobility would show differing population dynamics at the same levels of impact. Species with higher reproductive rates, faster maturation cycles, and/or lower mortality rates would be expected to incur a greater level of impact before demonstrating an inability to sustain the exposed subpopulation without immigration. Furthermore, differences in these parameters resulting from differing survival strategies would also affect the rate at which a species in decline would be expected to reach extinction if immigration rates were not sufficient to overcome deficits in local recruitment.

Reliance on immigration from a greater regional population need not necessarily be considered an adverse impact. Species with the mobility of the great blue heron are able to ameliorate localized impacts by spreading the effects over large interacting populations. This is a very common survival strategy among birds where the resistance to immigration (i.e., the expected rate at which individuals from a regional population can infiltrate into a suppressed subpopulation) is insignificant. This may not be the case with some highly territorial avian species or certain terrestrial species where the resistance to immigration could be very important. For example, if an impact were sufficient to reduce recruitment to the point where the rate of necessary immigration could not be met due to resistance to immigration from unimpacted subpopulations, then it is possible that localized extinctions may occur, even in an open system. Such considerations must be taken into account to determine whether a specific level of induced population impairment (reproductive or otherwise), as indicated by the result of an ecological risk assessment, truly does represent a survival threat to a local receptor population.

Currently, the most obvious drawback, which is also the greatest source of uncertainty in this risk model, is the basis for the estimate of toxic response. No study has ever provided a suitable estimate of the relationship between exposure dose and proportional probability of impact in any vertebrate species. As a result, the toxicity has to be evaluated based solely on a single threshold value. This introduces a great deal of unpredictable uncertainty into the assessment regardless of whether it is performed deterministically or stochastically. The intention in this case was to develop a model that could be used in conjunction with a probabilistic dose–response when such information becomes available.

The evolution of tools in the characterization of ecological risk has permitted better predictions and expressions for estimates of potential adverse impact. As this evolution continues, risk managers can expect better evaluations of threats and hazards to wildlife in terms that are easier to conceptualize and that can be quantitatively applied directly to management goals. These improvements in risk communication will allow direct management at the population level, which in turn will permit risk managers to better allocate limited resources to environmental protection and restoration.

The model used to simulate population dynamics is relatively simple. It was structured this way intentionally, in order to provide the transparency necessary for application within a regulatory context. As a result, certain generalizations were necessary that will be addressed in future versions of this model, for example, the assumption of constant carrying capacity. This precluded interactions between the characterized impact and any other transient event that may affect the reproduction or longevity of the receptors. Since all the state variables are standardized based on carrying capacity, this may cause highly significant fluctuations in receptor populations that are currently hidden in the results of the model. Methods for developing a population dynamics model that is dependent on carrying capacity, but does not require it to be constant, are being investigated. Another area of development currently under investigation is the incorporation of aspects of contaminant environmental fate and transport as well as remediation activities to estimate the long-term impacts of contaminant exposure on subpopulations. This will not only permit the estimation of past and current impacts, but also allow the opportunity to predict recoveries, thus providing even more complete information to the risk managers responsible for protecting the environment.

REFERENCES

1. Kenaga, E.E., Ed., *Avian and Mammalian Wildlife Toxicology*, ASTM Special Technical Publication 693, American Society for Testing and Materials, Philadelphia, 1978.
2. Efron, B. and Tibshirani, R.J., *An Introduction to the Bootstrap*, Chapman & Hall, New York, 1993.
3. Alexander, G.R., Food of vertebrate predators on trout waters in north central lower Michigan, *Mich. Acad.*, 10, 181, 1977.
4. Kushlan, J.A., Feeding ecology of wading birds, in *Wading Birds*, Sprunt, A., Ogden, J., and Winckler, S., Eds., *Natl. Audubon Soc. Res. Rep.*, 7, 249, 1978.
5. Henning, M.H. et al., Distributions for key exposure factors controlling the uptake of xenobiotic chemicals by great blue herons (*Ardea herodius*) through ingestion of fish, *Hum. Ecol. Risk Assess.*, 5, 125, 1999.
6. Peifer, R.W., Great blue herons foraging for small mammals, *Wilson Bull.*, 91, 630, 1979.
7. Gibbs, J.P. et al., Determinants of great blue heron colony distribution in coastal Maine, *Auk*, 104, 38, 1987.
8. Gibbs, J.P., Spatial relationships between nesting colonies and foraging areas of great blue herons, *Auk*, 108, 764, 1991.
9. Dowd, E.M. and Flake, L.D., Foraging habitats and movements of nesting great blue herons in a prairie river ecosystem, South Dakota, *J. Field Ornithol.*, 56, 379, 1985.
10. NYSDEC, Unpublished fish data collected by the New York State Department of Environmental Conservation in 1994, New York State Department of Environmental Conservation, Albany, 1995.
11. U.S. EPA, Mercury Study Report to Congress, U.S. EPA-452-R-96-001c, U.S. Environmental Protection Agency, Washington, D.C., 1996.
12. Chlorine Institute, Environmental Fate and Toxicity of Mercury, Final Report, Alliance 5-051-001, The Chlorine Institute, Washington, D.C., 1992.
13. Eisler, R., Mercury Hazards to Fish, Wildlife, and Invertebrates: A Synoptic Review, Contaminant Hazards Review Report No. 10, U.S. Department of the Interior, Fish and Wildlife Service, Laurel, MD, 1987.
14. Heinz, G.H., Methylmercury: reproductive and behavioral effects on three generations of mallard ducks, *J. Wildl. Manage.*, 43, 394, 1979.
15. Heinz, G.H., Methylmercury: second-year feeding effects on mallard reproduction and duckling behavior, *J. Wildl. Manage.*, 40, 82, 1976.
16. Bouton, S.N., Effects of chronic, low concentrations of dietary methylmercury on the behavior of juvenile great egrets, *Environ. Toxicol. Chem.*, 18, 1934, 1999.
17. U.S. FWS, Habitat Suitability Index Models: Great Blue Heron, Biological Report 82(10.99), U.S. Fish and Wildlife Service, U.S. Department of the Interior, Washington, D.C., 1985.
18. Henny, C.J., An Analysis of the Population Dynamics of Selected Avian Species, U.S. Department of the Interior, Fish and Wildlife Service, Washington, D.C., 1972.

12 Incremental Chemical Risks and Damages in Urban Estuaries: Spatial and Historical Ecosystem Analysis

Dave F. Ludwig and Timothy J. Iannuzzi

CONTENTS

12.1 Introduction 297
12.2 Risk and Damage Assessment: Foundations for Urban Ecosystems..........298
    12.2.1 The Problem: Unique Conditions in Urban
           Estuary Environments 299
           12.2.1.1 Ecosystem Conditions: Organisms and Habitats 300
           12.2.1.2 Chemicals: Where and When..300
    12.2.2 The Solution: Ecological Coincidence Analysis 301
    12.2.3 ECA in Practice: Application Examples 301
           12.2.3.1 Birds in an Urban Estuary... 302
           12.2.3.2 Habitat Analysis in a Wisconsin Lacustuary 304
           12.2.3.3 Newark Bay Estuary Historical Baseline........................ 307
12.3 Conclusions 321
References...324

12.1 INTRODUCTION

Urbanized estuaries may be the most abused environments on Earth. After centuries of shoreline development, wetland "reclamation," watershed alteration, physical disturbances from such activities as dredging, shipping, mosquito control, and garbage disposal, biotic communities have endured substantial habitat loss and degradation. For more than 150 years, urban waterways have been subjected to varying degrees of chemical pollution from industrial and municipal sources. Over time, the habitats that support estuarine-dependent organisms in urban areas have decreased in size and become spatially fragmented. Water and sediment quality is so degraded (at least seasonally) in some urban systems that many organisms are excluded from portions of the estuary. Consequently, despite their adaptive flexibility, many estuarine-dependent organisms have been constrained to "patchy" use of the urban environment.

Our ability to evaluate incremental risks and damages from various chemical groups in urban waterways depends on many interrelated factors bridging a number of scientific disciplines. These include the ecology of the system, the form, mode of action, and toxicity of the chemicals, and the physiology of the organisms that may be sensitive to the effects of exposure (i.e., at risk). Effective chemical risk assessment should be accurate: it should neither underestimate nor overestimate risk. To begin the process of conducting an accurate chemical risk assessment, two key factors must be addressed. The first is the influence of nonchemical impacts to the system, or the "baseline" environmental conditions that would exist in the absence of the contamination. The second is the spatial extent of chemical concordance with the habitats that organisms use, given the fragmentation of the ecosystem. This overlap determines the potential for exposure. These factors can be addressed through a combination of site-specific historical/ecological research, and quantification of the findings using Geographic Information System (GIS) analyses.

12.2 RISK AND DAMAGE ASSESSMENT: FOUNDATIONS FOR URBAN ECOSYSTEMS

The risk assessment process is inherently an exercise in causal analysis. This is because the ultimate use of risk assessment information is risk management. As the Presidential/Congressional Commission on Risk Assessment and Risk Management[1] states:

> [I]t is time to modify the traditional approaches to assessing and reducing risks that have relied on a chemical-by-chemical, medium-by-medium, risk-by-risk strategy. While risk assessment has been growing more complex and sophisticated, the output of risk assessment for the regulatory process often seems too focused on refining assumption-laden mathematical estimates of small risks associated with exposure to individual chemicals rather than on the overall goal—risk reduction.

The commission's concept is perhaps most critically important for estuarine risk assessment. Estuaries are at once very open and highly integrated.[2] Exchanges of matter and energy with adjacent lands, with upstream waters, and with downstream coastal marine systems drive many of the overall physical attributes of the estuary. At the same time, the tightly integrated nature of biogeochemical processes within the estuary[3,4] can greatly magnify or dampen the impact of forcing parameters.

This is the critical challenge for risk assessment in the estuarine context, and most particularly in urbanized estuaries. Environmental management necessarily focuses on specific sources of degradation and impact. To support management in complex estuarine environments and to render management actions as effective as possible, risk assessment is a fundamental decision-making tool. The U.S. EPA guidelines for ecological risk assessment[5] make this clear:

> Risk assessments provide a basis for comparing, ranking, and prioritizing risks. The results can also be used in cost-benefit and cost-effectiveness analyses that offer additional interpretation of the effects of alternative management options.

In other words, once the causes of environmental degradation have been identified, risk assessment is the tool by which their importance and management priorities are characterized. Similarly, natural resource damage assessment (NRDA), the regulatory process by which incremental damages from oil spills or other chemical releases are quantified and assessed a compensatory value in terms of monetary or equivalent resource currency, often rely on the risk assessment framework as a primary assessment tool.

The generic risk assessment process[6] as applied most intensively in the regulatory context[7] is neither inherently nor necessarily fully effective in urban estuarine environments. For the risk assessment framework to be effective in urban estuaries, basic ecology, as an integrative discipline by which effects can be characterized and causality evaluated, must be emphasized. Techniques for implementing this focus are only now being developed and integrated into the risk assessment framework. The objective of this chapter is to identify some of the techniques for quantifying and integrating the ecological components of risk assessments for urban estuaries. As the examples make clear, these methods are equally applicable for environmental remediation/cleanup and damage assessment/restoration.

12.2.1 The Problem: Unique Conditions in Urban Estuary Environments

There may no longer be any pristine or undisturbed ecosystems on Earth, if the source of disturbance is considered human influence.[8,9] From the perspective of the ecologist as well as the environmentalist, human interactions with estuaries are usually perceived as highly negative perturbations. Indeed, one excellent estuarine ecology text[10] titles its chapter on people and estuaries "Human Impact in Estuaries," and provides a detailed classification and discussion of the many sources and kinds of human "impacts." But the Manichaean view of human interactions as clearly negative forces in an otherwise "positive" world is grossly simplistic and is in any case counterproductive. As Ludwig[11] wrote:

> The view that one system state is "better" than another, that we humans in our "bad" way push ecosystems away from initial "good" states, and if we push too hard, things won't get "good" again, is not relevant. Ecosystems operate on a contingent, not a value, basis. Parameter states have no intrinsic "goodness" or "badness." Human technology now controls the state of the entire biosphere. We "manage" the biosphere, primarily by default. To manage effectively, we must determine what values we desire in the ecosystem ... identify parameter states that yield those values, and manage to achieve those parameter states.

For estuarine ecosystems, this means that human influence is assumed to be an unavoidable constant, and that its magnitude will only increase into the foreseeable future. Risk assessments must, of necessity, take current conditions as the baseline. Management decisions and management actions must build from the present patterns and processes of our admittedly highly disturbed estuaries. So we must apply our assessment tools to the unique conditions of modern estuarine ecosystems.

12.2.1.1 Ecosystem Conditions: Organisms and Habitats

Before European colonization, native Americans impacted watersheds (and water quality) by farming and burning.[12] The most fundamental fact of estuarine ecology after nearly 200 years of industrial development is habitat alteration. In practice, estuarine habitat alteration began thousands of years ago, with the berms that controlled inundation of agriculture and aquaculture sites. Since that time, tidal waters were dammed to power mills, and wetlands were diked for land "reclamation." European settlement shocked the ecology of the Western Hemisphere.[13] Since the beginning of the industrial age, dredging has usurped large areas of natural bottoms,[14] and shorelines and intertidal wetlands have been replaced wholesale by anthropogenic land and structures.[15]

The major effect of habitat alteration on the biotic components of the ecosystem has been community fragmentation. Where once large areas of marshes, shallow flats, or oyster reefs might have stretched unbroken across suitable portions of estuaries, there are now habitat patchworks and parcels.[16] The conversion of human landscapes to patchworks is a well-studied phenomenon,[17] but such conversion of "seascapes" has received less attention, despite being a critically important problem for risk assessment.

Simply stated, risk assessment is an analytical process by which probability of exposure to a stressor is evaluated, in the context of the known severity (effect) of a particular level of exposure to a particular stressor. For chemical risk assessment in estuarine ecosystems, habitat patchiness means that receptor organisms are not always evenly distributed within an area. They are distributed where available or appropriate habitats exist, and can only be exposed to chemicals and chemical concentrations present in those areas.

12.2.1.2 Chemicals: Where and When

The physical, hydrological, and geological conditions in estuaries are complex and heterogeneous.[18] Even the prehuman, natural distribution of chemical concentrations must have varied considerably in the spatial context of estuarine waters and sediments. However, the extensive and intensive modification of estuaries in industrial times has enhanced spatial heterogeneity, and the variety of chemicals present has increased concomitantly. Sediment conditions, in particular, affect chemical concentrations, bioavailability, and thus potential exposure. Sediment heterogeneity is reflected in highly heterogeneous exposure assessment outcomes.[19] The distribution of chemicals, like the distribution of biota, is patchy in modern estuaries.[20]

12.2.2 THE SOLUTION: ECOLOGICAL COINCIDENCE ANALYSIS

Chemical risk assessment in aquatic ecosystems is essentially analysis of the overlap of bioaccessible and bioavailable chemicals with susceptible receptor organisms and quantification of the effects at this overlap. It is the co-occurrence of chemicals and biota that drives ecological risks:[21]

> Distributional analyses of measured exposures can consider both spatial and temporal distributions of environmental concentrations ... [T]he probability of cooccurrence of the sensitive organisms and the greater concentrations of a stressor may, in fact, be small ... [C]oincidence of dominance and greater exposure concentrations at a particular location could ... increase risk in some situations but reduce it in others.

We term the suite of tools used to quantify the co-occurrence of chemicals and receptor organisms ecological coincidence analysis (ECA). Similar techniques have been used (at much larger spatial and temporal scales) for land-use planning for many years. The concept of coincidence analysis was pioneered by geographers, and popularized as a planning tool in the laboratories of urban land-use specialists.[22] In risk assessment, an initial quantitative application of ECA was published for a terrestrial site with multiple contaminants and receptors[23] and ECA has been applied in other studies.[24,25] Analyses similar to ECA are integral to the modern risk assessment process.[5] But the complexity and difficulty of urban estuarine risk analysis remains a challenge to these tools.

Implementing ECA for urban estuarine risk assessment requires detailed characterization and quantitative understanding of two sets of parameters:

1. Habitat suitability and receptor distribution
2. Chemical distribution

The first depends on heterogeneity in parameters that control the presence and abundance of organisms, such as currents, tides, sediment type, vegetation, and bottom and shoreline structures. The second depends on parameters controlling the bioaccessibility and bioavailability of specific chemicals. Chemical behavior and sediment conditions are particularly important parameters. For example, sediments high in organic matter might sequester high concentrations of hydrocarbon contaminants, but little or none of the hydrocarbons might be bioavailable. Conversely, sands with low organic content may have low concentrations of nonpolar hydrocarbons, but the molecules present may be highly bioavailable.

12.2.3 ECA IN PRACTICE: APPLICATION EXAMPLES

The following sections provide three practical illustrations of the application of ECA to real-world risk analysis problems. The examples vary in concepts addressed, level of detail, and completeness.

12.2.3.1 Birds in an Urban Estuary

The Passaic River in northeastern New Jersey flows into the New York/New Jersey (NY/NJ) Harbor Estuary, a quintessential example of a complex urbanized estuary. Water and sediments in the estuary are contaminated with a wide variety of chemicals arising from a large number of municipal and industrial sources and as non-point input from the highly developed watershed.[26] Our observations indicated that the tidal portion of the Passaic River is extremely heterogeneous relative to habitat. Depositional areas (represented by intertidal mud flats) are interspersed with erosional areas (many adjacent to vertical bulkheads and seawalls). Riparian habitat is limited primarily to mudflats with little or no associated vegetation. The shorelines are highly developed and dominated by bulkheads, riprap, buildings, parking lots, roadways, and other structures. However, there are areas of narrow riparian weedlots and even some widely dispersed small groves of *Ailanthus* trees and other ruderal vegetation. We are applying ECA to the tidal portion of this river to determine where birds are found as a first step for quantitative risk and damage assessment.

The critical questions are as follows:

1. Is bird use of this highly urbanized river, relative to their use of surrounding waterways, high enough to drive substantive risks or damages?
2. Can co-occurrence of birds and chemicals can be quantified and analyzed to ascertain incremental chemical exposure risk?
3. Are particular habitats favored by birds in this river that could be the focus of restoration activities?

As a first step in the ECA process, bird distribution was evaluated in detail relative to temporal and spatial parameters.

Temporal parameters were investigated at an annual scale by conducting four intensive seasonal surveys. Bird use of estuarine habitats in this area is seasonal (for examples, see Figure 12.1). Exposure to chemicals is, therefore, time dependent — exposure will be higher during periods when species-specific abundance is highest. Absolute abundance of all waterbird species (shorebirds, waders, waterfowl, gulls, and terns) physically present on mudflats is compared seasonally in Figure 12.1. As expected for this Atlantic flyway waterway, autumn is the time of peak abundance.[27] Winter bird use of the estuary is very low, and chemical exposure is expected to be correspondingly low. These findings, when data analyses are completed, will provide seasonal exposure information (as time-dependent differential site-specific doses) for quantitative ECA.

Temporal and spatial exposure of birds also varies on smaller scales. Daily use of particular habitats is determined by tidal exposure (of flats, for example) and by time of day. Activity peaks vary by species, but many estuarine birds are crepuscular. To characterize habitat use, we conducted spring and autumn surveys over two periods each: once during a period when low tide corresponded to midday (testing whether tide was a stronger driver of bird use than time of day); and once when low tide corresponded to morning and evening. In both periods,

Incremental Chemical Risks and Damages in Urban Estuaries 303

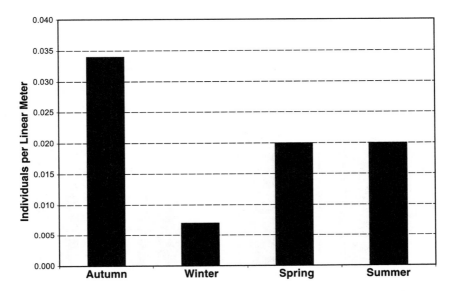

FIGURE 12.1 Seasonal comparison of mudflat use by birds (all species combined).

we surveyed the identical stretch of estuary intensively in morning, at midday, and in the evening. Low tide morning/evening survey periods included three, thrice-daily surveys each. Low tide midday surveys were one thrice-daily survey per period. Figure 12.2 shows an example of the data generated by this intensive sampling effort, which included observations of birds actually using mud flats and those using other structures or shoreline types (e.g., bulkheads, weedy banks). Clearly, mud flats exposed at low tide are the focus for bird use in this river, dominating the relative abundance. During high tides, birds use whatever portion of the flats remain available or are forced into adjacent riparian shoreline areas or out of the river altogether.

Our observations and preliminary data suggested that bird populations are very low in this urban waterway compared with those in similar, nearby waterways, with much less development and more substantial and diverse habitats. Further analyses will consider the home range of the birds using this river, and a quantitative assessment of the likely habitat use within this home range. The objectives are to ascertain the incremental chemical exposure represented by this river, as well as to assess what habitat restoration may be most effective for increasing bird populations in this area. Results to date suggest that the tidal Passaic River likely represents only a very small portion of the overall area used by bird populations in the NY/NJ Harbor estuary because of limited areas of habitat available (confined generally to mud flats and ruderal riparian vegetation strips). For quantitative ECA-based risk analysis, this means that (1) overall bird exposure is relatively low in this particular river; (2) what exposure there is arises from feeding and occupying mudflats as preferred habitat; and (3) restoration of more diverse habitat would likely contribute substantially to bird use of the system.

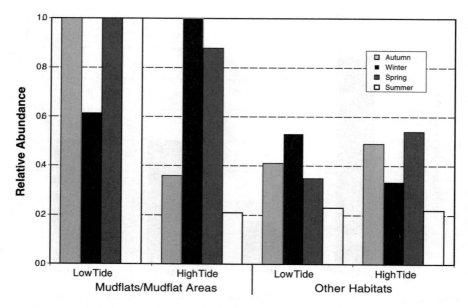

FIGURE 12.2 Relative abundance of birds on mudflats and in adjoining riparian habitat relative to tide level in spring and autumn surveys.

12.2.3.2 Habitat Analysis in a Wisconsin Lacustuary

The Fox River in northeastern Wisconsin flows into southern Green Bay, and is an active lacustuary of Lake Michigan. Upper reaches of the watershed are agricultural lands; lower reaches are highly urbanized (the river flows directly through the city of Green Bay). Many chemicals have been found in Fox River sediments, and the lower river and southern portions of the bay are the subject of ongoing risk assessment and NRDA.

As a component of an ECA for fish, bird, and mammal receptors in the lower Fox River system, we conducted a detailed habitat characterization. The objective was to help quantify the co-occurrence of receptors and chemical concentrations for detailed risk analyses. Aquatic and shoreline habitats were characterized by key parameters controlling the distribution of fish and invertebrates (as critical components of the aquatic food webs, and links between sediment contaminants and birds and mammals). Key characterization parameters included water depth, presence or absence of in-stream cover, bottom substrate type, in-stream structures, shoreline structure, and detailed habitat characterization/classification of adjacent land areas. These parameters were characterized by the application of sidescan sonar throughout the study area, coupled with a complete videotape record of bank condition and ecological surveys along both shores of the lacustuary. Shoreline types present include natural shoreline and wetlands, riprap and bulkheads, and pilings. Figure 12.3 shows an example of sidescan sonar output with features indicated, and demonstrate how the sonar analysis supported the shoreline characterization (in conjunction with the videotape evaluation).

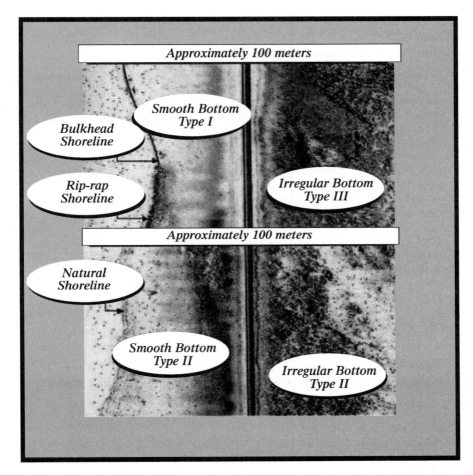

FIGURE 12.3 Substrate characterization showing sonar signal of shoreline features in addition to bottom type.

For quantitative ECA, habitat parameters were ranked on a categorical relative scale of value for aquatic organisms (Table 12.1). These qualitative ranks were converted by professional judgment to relative scores for each habitat component so they could be incorporated in quantitative analysis of co-occurrence of high-quality habitat and chemical concentrations (Table 12.2). The integration of these components for the aquatic habitats is illustrated in Figure 12.4, which shows the fundamental GIS application for ECA. The GIS overlays are prepared sequentially for each key habitat component (water depth, substrate type, presence of submerged aquatic vegetation). These habitat components are then overlaid on Thiessen polygons derived from chemical concentration measurements in the river sediments. In this way, habitat areas of different quality can be quantified relative to extrapolated chemical concentrations. This provides the fundamental basis for exposure characterization, incorporating realistic estimates of habitat heterogeneity and thus receptor use of areas of differing chemical concentrations.

TABLE 12.1
Value Ranking of Aquatic Habitat Parameters

Habitat	Function	Relative Value
Surface water depth		
Shallow (<3.0 m)	Access to shoreline/wetlands, high productivity, key foraging habitat, spawning and nursery	High
Deep (>3.0 m)	Relatively featureless corridor to shallows	Medium–Low
Substrate		
Rock/cobble	Spawning substrate, high diversity	High–Medium
Soft silt	Relatively featureless	Low
Shoreline		
Natural and wetlands	Key foraging habitat, cover and refuge, spawning and nursery	High
Riprap	Cover and refuge, spawning and nursery	Medium
Bulkhead	Relatively featureless	Low
In-stream cover		
SAV[a]	Key foraging habitat, abundant prey, cover and refuge, spawning and nursery	High–Medium

[a] SAV = submerged aquatic vegetation.

TABLE 12.2
Example of Aquatic Habitat Scoring Derived from Qualitative Ranking

Habitat	Relative Value	Score
Surface water depth		
Shallow (<3.0 m)	High	0.7
Deep (>3.0 m)	Medium–Low	0.3
Substrate		
Rock/cobble	High–Medium	0.9
Soft silt	Low	0.1
Shoreline		
Natural and wetlands	High	0.6
Riprap	Medium	0.3
Bulkhead	Low	0.5
No shoreline	Low	0.5
In-stream cover		
With SAV[a]	High	0.7
Without SAV[a]	Medium–Low	0.3

[a] SAV = submerged aquatic vegetation.

The results of the shoreline habitat characterization/ecological surveys were used to map and quantify available habitats for mink and birds that could be exposed to chemicals via ingestion of contaminated prey (i.e., fish and other aquatic organisms) in the system. A habitat ranking and scoring system was used to classify the shoreline areas into good,

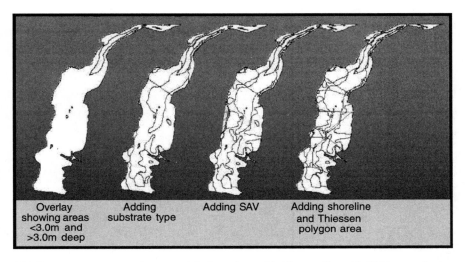

FIGURE 12.4 Example of integrated GIS overlays of habitat quality with Thiessen polygons of chemical concentration.

moderate, marginal, or poor habitat categories for mink, and to identify primary habitats for a number of birds including bald eagle, terns, and cormorants. To do this, a detailed assessment of the ecology of each of these organisms (based on existing literature) was performed, yielding a comprehensive habitat suitability profile for each, complete with a list and rank of key habitat requirements. A GIS analysis was then performed to delineate, classify, and map suitable habitats for the birds (e.g., Figure 12.5), and to classify and map the shoreline in terms of mink habitat by category (e.g., Figure 12.6). The results of these analyses are being used to assess exposure risks of these organisms in the system. Applications include calculating the number of organisms that the habitats can support (and that could therefore be at risk), and incremental estimation of risk and damages based on chemical distribution within or adjacent to the various habitat types.

12.2.3.3 Newark Bay Estuary Historical Baseline

For NRDA, the incremental injury associated with a particular chemical release is the key issue.[28] In urban ecosystems, of course, it is very difficult to discern such injury in the context of the many other alterations and insults the estuaries have suffered (Figure 12.7). Overall, in the historical context, the effect of a particular chemical is likely to be small, and such effects can only be quantified if the total historical service impairments are understood.

In the Newark Bay sub-basin of the NY/NJ Harbor Estuary, we have reconstructed the historical "baseline" of post-colonial estuarine alterations (Figure 12.8). Figure 12.9 summarizes critical baseline impacts, aggregating categories of change and overlaying those changes on a timeline showing quantitative habitat losses at the same temporal scale. As the analysis demonstrates, habitat losses have been very high on both relative and absolute scales — nearly all of the wetlands, riparian corridors, tributaries, and productive bottoms are gone from this estuary, and have been lost for decades. Nearly half of all productive habitat has been gone for more than a century.

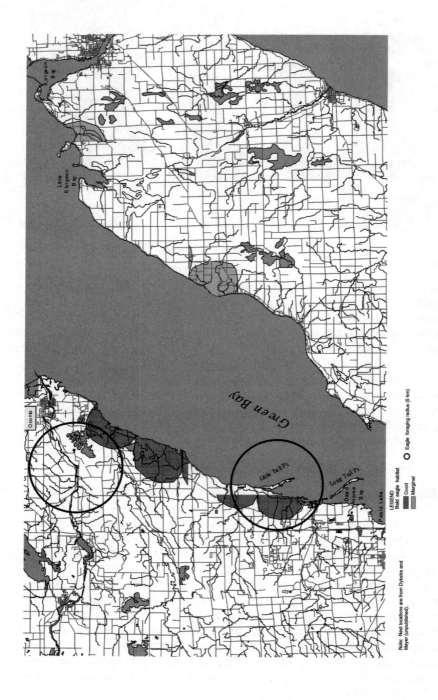

FIGURE 12.5 Bald eagle nests and estimated foraging range adjacent to Green Bay.

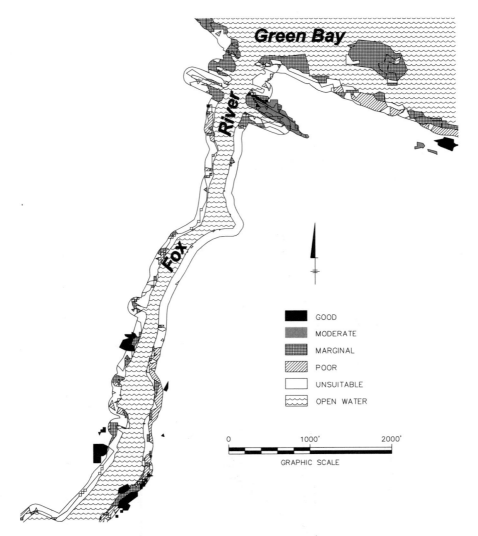

FIGURE 12.6 Mink habitat suitability along the lower Fox River in Green Bay.

As explained earlier in this chapter, habitat alteration is an overriding factor in estuarine degradation. Quantifying such alteration is an exercise in ecological examination and historical analysis. Excellent examples are available at a regional scale and for primarily terrestrial habitats.[29] The tools of historical analysis for urban estuaries are only now being developed. However, they are critical to ECA, and ECA is fundamental to accurate estuarine risk and damage assessment. In the following paragraphs, we illustrate some of the historical data gathering that can be used to develop quantitative understanding of ecological baseline in urban estuaries.

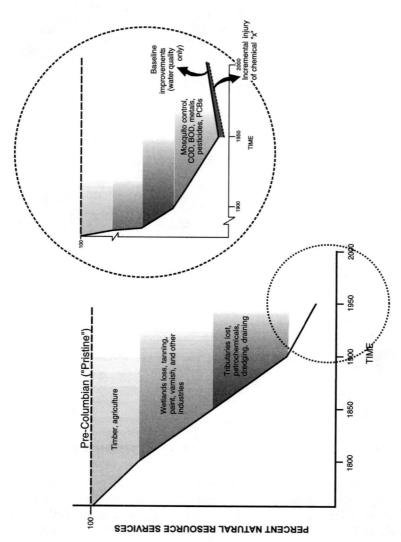

FIGURE 12.7 Historical effects of human presence on natural resource services in urban estuaries.

Incremental Chemical Risks and Damages in Urban Estuaries 311

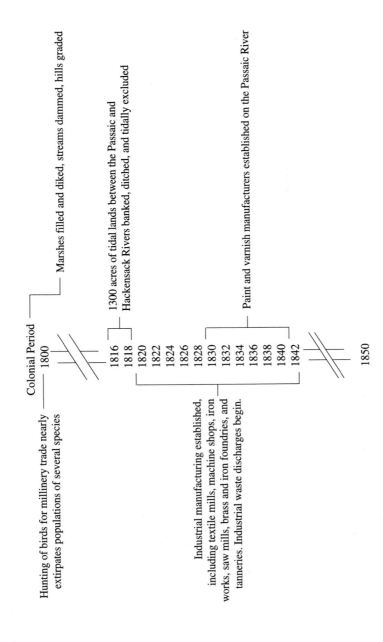

FIGURE 12.8 Historical timeline of post-colonial alterations of the Newark Bay Estuary.

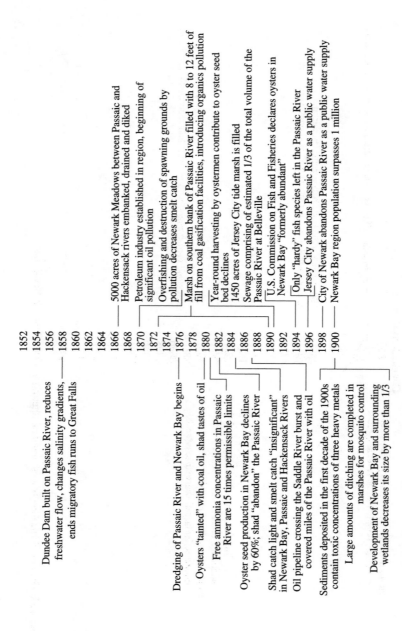

FIGURE 12.8 (CONTINUED) Historical timeline of post-colonial alterations of the Newark Bay Estuary.

Incremental Chemical Risks and Damages in Urban Estuaries

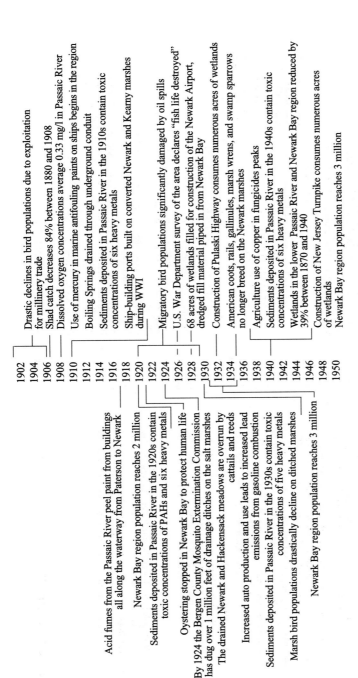

FIGURE 12.8 (CONTINUED) Historical timeline of post-colonial alterations of the Newark Bay Estuary.

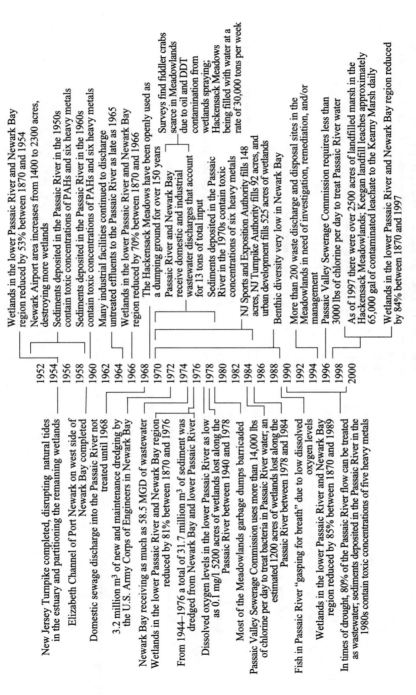

FIGURE 12.8 (CONTINUED) Historical timeline of post-colonial alterations of the Newark Bay Estuary.

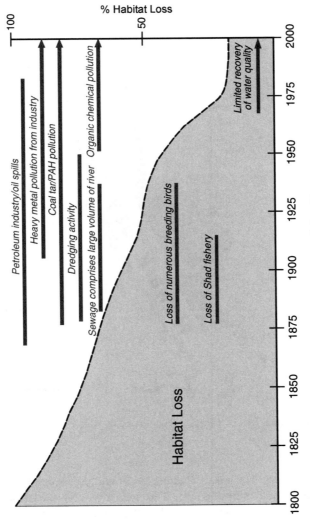

FIGURE 12.9 Timeline of aggregated impacts and habitat losses in the Newark Bay Estuary.

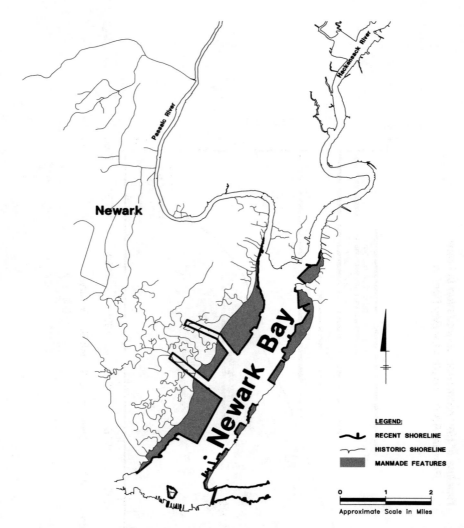

FIGURE 12.10 Historical loss of aquatic habitat to make land in the Newark Bay estuary, 1700 to present day.

Figure 12.10 shows that aquatic habitat was rapidly converted to fast land in Newark Bay during the industrial age. This map was compiled from numerous historical sources, primarily maps prepared or published from the early 1700s to the present day, and from aerial photographs dating from the 1930s.

Before European settlement, many tributaries defined the drainage of the lower Passaic River and Newark Bay systems (Figure 12.11). These small waterways provided critically important habitat for fish, shellfish, birds, and furbearing mammals. These tributaries are all gone with the exception of a short length of the Second

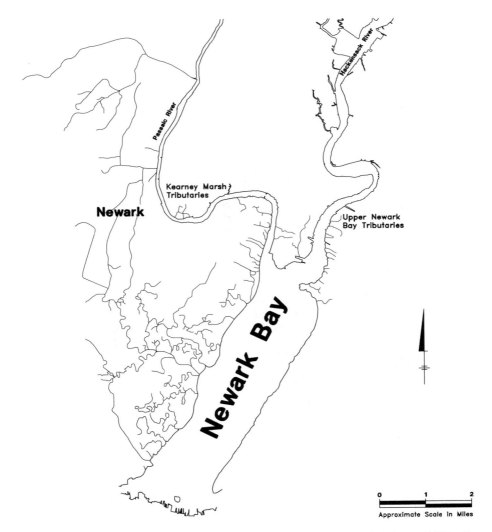

FIGURE 12.11 Historical tributaries and drainages of the lower Passaic River and Newark Bay.

River. They were lost directly to development and are represented now, if at all, by storm sewer culverts.

Wetlands, among the most ecologically valuable components of the estuarine complex, have been largely lost in many urban ecosystems. Figure 12.12 illustrates our reconstruction of wetland losses from the mid-1800s to the mid-1900s in the Newark Bay estuary. The devastating effect of these losses on a critical natural resource service (bird population support) is illustrated in Figure 12.13.[30] These estimates are calculated from equations relating bird production to wetland area.

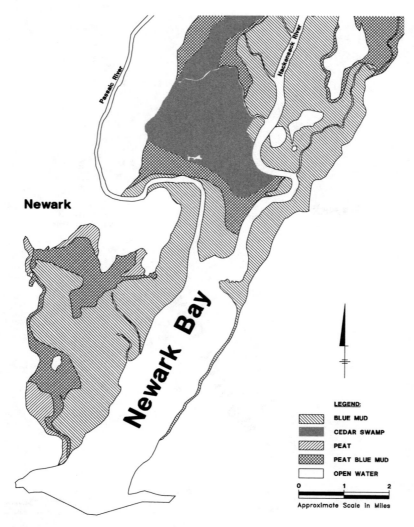

FIGURE 12.12A Wetland losses in the Newark Bay estuary, 1857.

Although they are only rough approximations, they indicate the substantive impact of wetland losses in urbanized estuarine ecosystems.

Finally, there is once again the issue of chemical contamination of estuarine sediments. Using radio-dated sediment cores, concentrations of chemicals measured at depth in the sediment can be related to the time of deposition and their relative exceedance of toxicological benchmarks. A timeline showing exceedance factors for a number of metals and polycyclic aromatic hydrocarbons (PAHs) based on highly conservative sediment assessment benchmarks is presented in Figure 12.14. The rise of industrial inputs can be seen in this figure, as chemical concentrations peak in

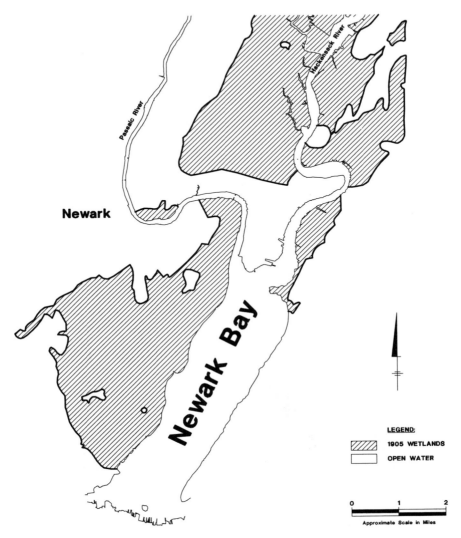

FIGURE 12.12B Wetland losses in the Newark Bay estuary, 1905.

the 1950s and 1960s, and the beneficial effects of technology and water quality regulation in the decline of exceedances into the 1990s. This pattern has also been seen for a number of chemical compounds (and their by-products) that were produced and released into the estuary in the post-World War II era, including pesticides such as DDT and its metabolites, polychlorinated biphenyls (PCBs), and polychlorinated dibenzo-*p*-dioxins and dibenzofurans (PCDD/Fs). Overall, however, it is clear that sediments throughout the estuary have exceeded benchmark sediment quality thresholds for many chemicals for decades.

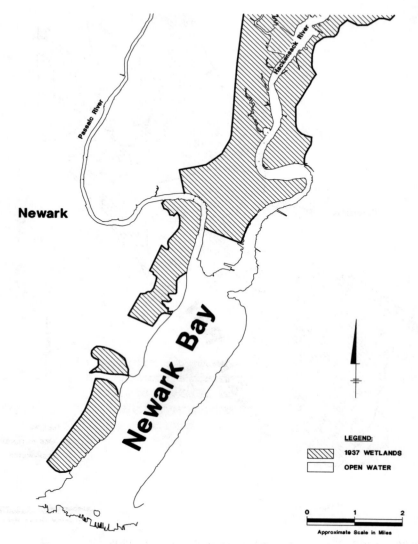

FIGURE 12.12C Wetland losses in the Newark Bay estuary, 1937.

In summary, ECA allows ecological risk to be characterized in the context of both past and present environmental conditions. An ECA is conducted by analyzing and quantifying the spatial and temporal distributions (at appropriate scales) of stressors and receptors. The resulting overlap provides the foundation for determining the incremental contribution of specific stressors to total system risk to particular receptors.

Incremental Chemical Risks and Damages in Urban Estuaries 321

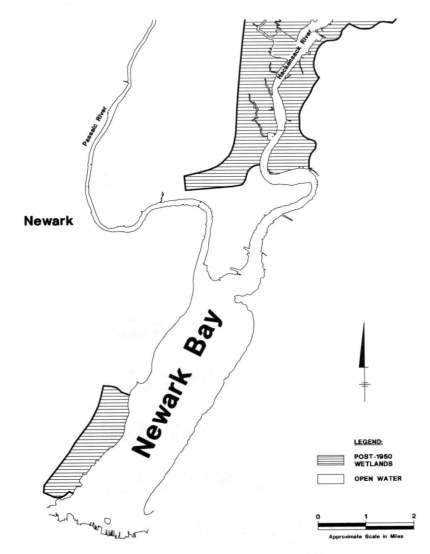

FIGURE 12.12D Wetland losses in the Newark Bay estuary, 1950.

12.3 CONCLUSIONS

The ecological risk assessment paradigm is a useful framework for decision making in urbanized estuaries only if supplemented by detailed ecological coincidence analysis. Depending on the questions being asked or the hypotheses being tested, ECA can be conducted on existing conditions or in a historical framework. Because

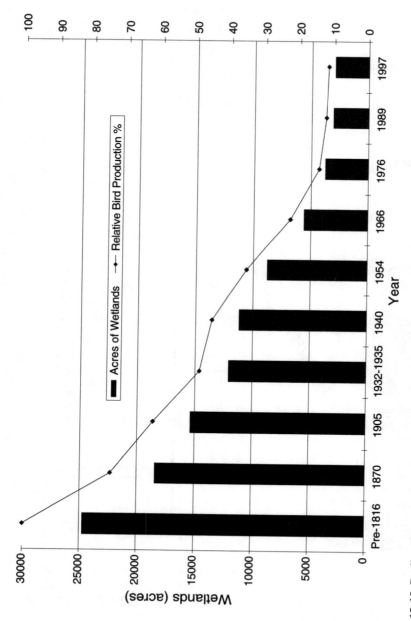

FIGURE 12.13 Declines in production of estuarine-dependent birds with wetland losses.

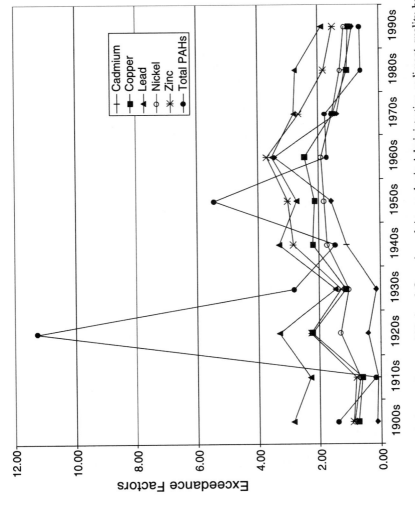

FIGURE 12.14 Timeline of historical exceedances of National Oceanic and Atmospheric Administration sediment quality benchmarks in Newark Bay estuary sediments.

of the complex issues, alterations, and impacts in urban estuaries, management decisions cannot be made simply for single impact drivers or for single components of the ecosystem. Effective decision making in these highly dynamic and most complicated ecosystems requires detailed baseline analysis and quantitative understanding of the temporal and spatial relationships between the resources and the sources of impact.

REFERENCES

1. Presidential/Congressional Commission on Risk Assessment and Risk Management, Risk Assessment and Risk Management in Regulatory Decision-Making. Final Report, Vol. 2, Washington, D.C., 1977.
2. Odum, H.T., *Systems Ecology: An Introduction*, John Wiley & Sons, New York, 1983.
3. Valiela, I., *Marine Ecological Processes*, Springer-Verlag, New York, 1995.
4. Hopkinson, C.S. and Day, J.W., Jr., A model of the Barataria Bay salt marsh ecosystem, in *Ecosystem Modeling in Theory and Practice*, Hall, C.S. and Day, J.W., Jr., Eds., John Wiley & Sons, New York, 1977, 235–265.
5. U.S. EPA, Guidelines for Ecological Risk Assessment, 63 FR, May 14, 1998/Notices.
6. NRC (National Research Council), Risk Assessment in the Federal Government: Managing the Process, National Academy Press, Washington, D.C., 1983.
7. U.S. EPA, Ecological Risk Assessment Guidance for Superfund: Process for Designing and Conducting Ecological Risk Assessments, EPA 540-R-97-OCS, U.S. Environmental Protection Agency, Washington, D.C., 1997.
8. Ludwig, D.F., The final frontier, *Ecol. Soc. Am. Bull.*, 66, 332, 1985.
9. Ludwig, D.F., Anthropic ecosystems, *Ecol. Soc. Am. Bull.*, 70, 12, 1989.
10. Day, J.W., Jr., et al., *Estuarine Ecology*, John Wiley & Sons, New York, 1989.
11. Ludwig, D.F., Economic growth and environmental policy, *Science*, 268, 1550 (letter), 1995.
12. Krech, S., III, *The Ecological Indian*, W.W. Norton and Company, New York, 1999.
13. Crosby, A.W., Jr., *The Columbian Exchange — Biological and Cultural Consequences of 1492*, Greenwood Press, Westport, CT, 1972.
14. Kennish, M.J., *Ecology of Estuaries: Anthropogenic Effects*, CRC Press, Boca Raton, FL, 1992.
15. NRC (National Research Council), *Restoration of Aquatic Ecosystems*, National Academy Press, Washington, D.C., 1992.
16. Weinstein, J.E., Anthropogenic impacts on salt marshes — a review, in *Sustainable Development in the Southeastern Coastal Zone*, Vernberg, F.J., Vernberg, W.B., and Siewicki, T., Eds., University of South Carolina Press, Columbia, 1996, 135–170.
17. Turner, M.G. and Gardner, R.H., Eds, *Quantitative Methods in Landscape Ecology*, Springer-Verlag, New York, 1991.
18. Pomeroy, L.R. and Imberger, J., The physical and chemical environment, in *The Ecology of a Salt Marsh*, Pomeroy, L.R. and Wiegert, R.G., Eds., Springer-Verlag, New York, 1981, 21–36.
19. Hall, L.W., Jr. and Giddings, J.M., The need for multiple lines of evidence for predicting site-specific ecological effects, *Hum. Ecol. Risk Assess.*, 6, 679, 2000.
20. Lick, W., Modeling the transport and fate of hydrophobic contaminations, in *Ecological Risk Assessment of Contaminated Sediments*, Ingersoll, C.G., Dillon, T., and Biddinger, G.R., Eds., SETAC Press, Pensacola, FL, 1997, 239–253.

21. Giesy, J., Jr. et al., Chlorpyrifos: ecological risk assessment in North American aquatic environments, *Rev. Environ. Contam. Toxicol.,* 160, 1, 1999.
22. McHarg, I., *Design with Nature*, Natural History Press, Garden City, NY, 1969.
23. Clifford, P.A. et al., An approach to quantifying spatial components of exposure for ecological risk assessment, *Environ. Toxicol. Chem.,* 14, 895, 1995.
24. Henriques, W.D. and Dixon, K.R., Estimating spatial distribution of exposure by integrating radiotelemetry, computer simulation, and geographic information systems (GIS) techniques, *Hum. Ecol. Risk Assess.,* 2, 527, 1996.
25. Sample, B.E., Hinzman, R., Jackson, B.L., and Baron, L.A., Preliminary Assessment of the Ecological Risks to Wide-Ranging Wildlife Species on the Oak Ridge Reservation: 1996 Update, DOE/OR/01, 1407&D2, Oak Ridge National Laboratory, Oak Ridge, TN, 1996.
26. Brydon, N.F., *The Passaic River's Past, Present, Future*, Rutgers University Press, New Brunswick, NJ, 1974.
27. Leck, C.F., *The Status and Distribution of New Jersey's Birds*, Rutgers University Press, New Brunswick, NJ, 1984.
28. Stewart, R.B., Ed., *Natural Resource Damages: A Legal, Economic, and Policy Analysis*, National Legal Center for the Public Interest, Washington, D.C., 1995.
29. Cronon, W., *Changes in the Land — Indians, Colonists, and the Ecology of New England*, Hill and Wang, New York, 1983.
30. Custer, T.W. and Osborn, R.G., Wading Birds and Biological Indicators: 1975 Colony Survey, U.S. Fish and Wildlife Service, Washington, D.C., 1975.

13 Ecological Risk Assessment in Coastal and Estuarine Environments

Michael C. Newman, Robert C. Hale, and Morris H. Roberts, Jr.

CONTENTS

13.1 Introduction 327
13.2 Chapter Contributions to Coastal and Estuarine Risk Assessment.............329
13.3 Conclusion...335
References..336

13.1 INTRODUCTION

> Hearing the rising tide, I think how it is pressing also against other shores I know ...
>
> — **Rachel Carson**[1]

When Rachel Carson published *Silent Spring*,[2] she had already established her reputation by authoring two popular books, *Under the Sea Wind*[3] and *The Edge of the Sea*.[1] This being the case, it is puzzling that *Silent Spring*, a book literally changing how we view our relationship with our environment, contained so little material about marine pollution. Such inconsistencies about marine environments were commonplace at that time. At the same time that the last two books were published, the first author spent many indolent hours as a child on a particular Long Island Sound beach, alternately watching for dolphins on the horizon and, at the sand's edge, watching rats scurrying between the riprap in search of edible garbage. Much more effort was spent scanning for surfacing dolphins than watching the rats compete for garbage. Tar balls and rusty aerosol cans were as plentiful in the drift zone as were skate egg cases and strings of whelk eggs. At one end of the beach was a picturesque New England lighthouse silhouetted against plumes of smoke rising above the Bridgeport city dump. Coastal pollution was as obvious as that in nearby streams, lakes, and lands, but, for the coastal and estuarine habitats, the eye was drawn more to attractive, not degraded, seaside features. The tide of contamination was similarly rising in other coastal environs but was viewed only peripherally. We focused on the aesthetic and recreational pleasures of the coast. Although much less apparent today, remnants of this tendency to ignore the evidence of degradation exist in our activities.

What are the roots of this incongruity? A review of current U.S. environmental legislation[4,5] indicates that a lack of legal mandates was not the reason. Ample legislation included the overarching National Environmental Protection Act (1969); Marine Mammal Protection Act (1972); Coastal Zone Management Act (1972); Clean Water Act (1972); Federal Water Pollution Control Act (1977); Marine Protection Research and Sanctuaries Act (1972), especially Chapter 27 and amendments in the Ocean Dumping Ban (1988); the Comprehensive Environmental Response, Compensation and Liability Act (1980) as amended by the Superfund Amendments and Reauthorization Act (1986), which established a risk assessment context; the Organotin Antifouling Paint Control Act (1988); the Ocean Dumping Ban (1988); and the Oil Pollution Act (1990). The need was equally obvious and legislative initiatives were nearly as timely for stewardship for marine as for freshwater and land resources.

Full consideration of coastal and estuarine pollution was delayed by cultural biases. There were contrasting romantic and pragmatic delusions about the oceans that delayed action as incisive as that made for freshwater and land stewardship. Deeply embedded attitudes and complex use patterns regarding marine resources made marine regulation more difficult. For example, consequences of complex, traditional use patterns coupled with efficient modern fishing gear are manifest in the current worldwide decline of marine fisheries.[6] In contrast to a complex of mores surrounding land ownership and obligations associated with terrestrial landscapes, tradition associated with coastal resources emerged more from the concept of the commons. Laissez-faire mores for commons use resulted in a higher risk of degradation with coastal resources compared with terrestrial resources.

Perhaps more importantly, our culture has consistently portrayed the ocean as larger than humankind, a relationship reflected in innumerable literary works. Humans might be overpowered by the seas, but not vice versa. This sense that the ocean was too vast to be adversely impacted delayed rejection of "the solution to pollution is dilution" paradigm for marine systems. One piece of evidence for this mind-set was the relatively late passage of the Ocean Dumping Ban (1988) amendment to the Marine Protection Research and Sanctuaries Act (1972).

Despite an initial delay, the capacity of coastal and estuarine environments to absorb the cumulative loading of contaminants is now the focus of much insightful research and effective action. Key aspects of marine pollution are being synthesized; for example, see Reference 7. An entire issue of the leading journal, *Environmental Toxicology and Chemistry* (Vol. 20, No. 1, 2001) was recently devoted to toxicant chemistry, effect, and risk assessment in marine and coastal systems. (However, only one article in the journal issue addressed risk assessment.) In the lead editorial, Scott[8] emphasized the importance of studying contaminants in coastal systems:

... [The coastal] zone represents 8% of the planet's surface but is the source of 26% of the world's primary productivity. More than 76% of all commercially and recreationally important fish and shellfish species are estuarine-dependent. Only recently have these coastal areas been recognized as resource bases of national significance and also among the nation's most highly stressed natural systems.

... Historically, the focus of environmental toxicology and risk assessment has been on rivers and lakes. The available dilution in near-coastal areas was thought to mitigate adverse effects of anthropogenic contaminants. However, this perception has changed because of constantly increasing population densities and scientific evidence of environmental degradation. Approximately 44% of U.S. estuaries, assessed in 1998 for environmental quality, were impaired.

The complexity of attitudes, uses, and stressors impinging on coastal and estuarine systems is now being confronted directly and fully. A recent issue of *Limnology and Oceanography* (Vol. 44, No. 3, 1999) was dedicated to multiple stressors in freshwater and marine ecosystems. Marine ecosystems discussed in that issue range widely from southeastern U.S. estuaries and bays, northeastern U.S. harbors, and coral reefs (Florida Keys reef system and Great Barrier Reef). Integrated coastal planning and assessment activities are becoming more prominent (e.g., References 9 and 10), including those for acutely endangered systems such as coral reefs (e.g., References 11 and 12).

Despite recent progress, implementation of the ecological risk assessment paradigm to coastal and estuarine ecosystems still lags behind that for freshwater and terrestrial systems. As an example, the excellent treatments of ecological risk assessment written by Suter and colleagues[13,14] provide limited coverage of assessments of coastal and estuarine systems. The U.S. EPA Guidelines for Ecological Risk Assessment[15] requires considerable augmentation prior to optimal use for coastal and estuarine systems. Enrichment of the ecological risk assessment paradigm is the next crucial step in eliminating the differences in effective environmental stewardship for marine, freshwater, and terrestrial environments. The chapters contained in this edited book were developed with that goal in mind.

13.2 CHAPTER CONTRIBUTIONS TO COASTAL AND ESTUARINE RISK ASSESSMENT

In this brief summary chapter, the preceding chapters will be placed into the context of the current ecological risk assessment paradigm (see Chapter 1). The framework for discussion will be Figure 1.1 (reproduced here) from Chapter 1 in which the steps of ecological risk assessments are diagrammed.

The intent of Chapter 1 was to provide a context for ecological risk assessment in estuarine and coastal regions, to describe major sources of information applicable to risk assessments in these environments, and to highlight areas needing special attention during assessments in marine systems. Strong and dynamic gradients, e.g., salinity and hydraulic flow, dominate the ecology and chemistry of these systems, and their influence must be considered in estuarine risk assessments more than in

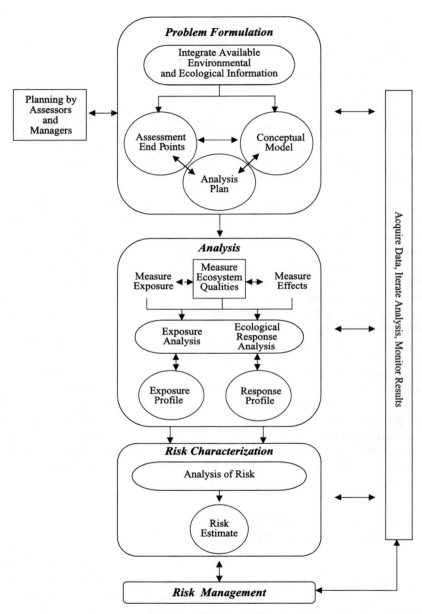

The ecological risk assessment paradigm as presented by the U.S. EPA (see Chapter 1 for further detail).

other environments. By their nature, coastal and estuarine environments are transition zones within landscapes. Risk assessments must fully consider relevant landscape features, e.g., the contaminant loadings in river discharge into an estuary or the land use around the estuary proper. Further, as is the case for all ecotones, crucial features of coastal environments cannot be captured solely by envisioning them as ecosystems existing between adjacent ecosystems. The ecotone framework may be equally, or more, important than the ecosystem context underpinning problem formulation in most ecological risk assessments.

Currently, several general methods are applied for identifying hazards in coastal and estuarine systems, and these methods generate important information applicable to risk assessments. Methods currently include water quality criteria generation, sediment quality guidelines, and toxics characterization methods. While the criteria and guidelines have utility for management, the tools of probabilistic risk assessment can improve our ability to protect marine and other environmental resources.

The European, or more specifically European Union (EU), context for coastal and estuarine risk assessments was explored in Chapter 2. Consequently, this chapter covered all aspects of the risk assessment paradigm diagrammed in the figure, but from a European perspective. After outlining the political institutions of the EU, the authors discussed how environmental legislation is promulgated, a process distinct from that in the United States. Under existing and developing legislation, both prospective and retrospective risk assessment are primary tools used in the management of hazardous chemicals. In both cases, the focus is the hazard quotient approach, although probabilistic risk assessment approaches are being evaluated for incorporation into future directives. At present, difficulties arising from incomplete information regarding existing and new chemicals are acknowledged as the reason for delay in actual management of many chemicals even when impacts of their use are apparent. In the EU, no action is taken to regulate a material until the risk assessment is complete. Many people conclude from the resulting conundrum that the EU should apply the controversial Precautionary Principle, that is, to take regulatory action to prevent present or possible impacts while the risk assessment proceeds.

Reflecting the theme discussed in the introduction to this chapter, the predominant freshwater focus was acknowledged for risk assessment and discussed in the context of using the more abundant data for freshwater species to predict consequences to saltwater species. The EU approach to saltwater assessments is described in detail, including current shortcomings. The use of freshwater species data to predict risks to saltwater species is a significant shortcoming that applies not only to the EU but also to the United States (see Chapter 1). To change the current practice effectively will require substantially more data for marine species.

The authors of Chapter 3 addressed emerging contaminants of concern and provided examples of several classes. They discussed chemicals underemphasized in current research, legislation, and risk assessments and those that may elicit inadequately studied effects, such as endocrine disruption discussed in Chapter 8. Their treatment here contributes directly to the "problem formulation" and "exposure analysis" components of the risk assessment paradigm (see the figure). Emphasis was on assuring that these important contaminants are more fully considered in assessments of multiple stressors in freshwater, estuarine, and coastal environments.

Emerging toxicants are those recently introduced in significant amounts, those historically ignored in environmental regulation and risk assessment, and those being applied now in new ways that require more careful scrutiny. Some priority pollutants such as polychlorinated biphenyls (PCBs) are considered here as well because new knowledge has changed the context for considering their potential effects. Other emergent toxicants included brominated fire retardants, natural and synthetic estrogens, alkylphenol polyethoxylates and related compounds, pharmaceutical agents, antimicrobial agents, and chemicals in personal care products.

As one studies emerging toxicants, one is struck by the complexity of chemical mixtures that exist in various environments. Chemical mixtures may have synergistic or antagonistic effects that are inadequately considered in risk assessment because of our poor understanding of interactions. This observation calls into question the traditional chemical-by-chemical evaluation of environmental risk, also discussed in Chapters 1 and 2. Biosolids, sewage sludge destined for land application, are another interesting example of a complex stressor. Current regulation of these in the United States is based on incomplete risk assessments that consider only a subset of the contaminants present. For example, the nonylphenols and brominated diphenyl ethers, although present in biosolids in milligram per kilogram quantities, are not considered. Different objectives in assessing resulting risks in the EU vs. the United States also produce different conclusions. In the former, a "do no harm" approach is taken toward contaminant accumulation in receiving soils and effects on soil organisms. In the United States, the maximum allowable toxicant concentration (MATC) for the protection of humans and livestock is the end point of concern. This may relate to differing views of the environment with the perception in the United States that resources are less limited, akin to our attitude toward the oceans expressed earlier in this chapter.

Chapter 4 discussed the causal assessment at the center of every risk assessment. It described inherent errors that emerge in informal methods used to generate preliminary and definitive statements of cause and effect. Qualitative methods of assigning causality were demonstrated with the example of notionally polycyclic aromatic hydrocarbon (PAH)-induced hepatic cancers in fish from contaminated coastal environments. The chapter then provided details of a formal, Bayesian approach that minimizes the likelihood of a mistake in the assessment of causality. An example was provided of quantitatively estimating the likelihood of a fish kill given the presence of *Pfiesteria piscicida* or a related complex of marine dinoflagellates. This chapter contributes especially to the problem formulation and risk characterization stages of the risk assessment paradigm, but also to all the other components that depend on identifying causal linkages between toxicants and effects.

Three chapters addressed key issues relevant to the "analysis phase" of the risk assessment paradigm. Each chapter focused on a specific group of chemicals in the context of speciation, bioavailability, and effects.

Chapter 5 provided details for both bioavailability and effects essential to the analysis phase of risk assessment for organic compounds present in the marine environment. Bioavailability from water, sediment, and food was discussed, including models for quantifying uptake. Mechanisms of xenobiotic metabolism and elimination were described. Considerable discussion was devoted to the mechanisms of contaminant transformation to readily excreted metabolites and the adverse consequences occurring if a metabolite is more toxic or carcinogenic than the parent compound.

Chapter 6 emphasized bioavailability and bioaccumulation of metals, especially those of ionic mercury and methylmercury. It contributed directly to exposure analysis (see the figure). The role of food chain transfer of metals from water to the target organism was discussed specifically as it pertains to mercury, but the author also discussed the role of this route for other metals. The author included a brief discussion of how dissolved organic material, particulate organic material, and acid volatile sulfides affect availability and uptake from water and sediment. The discussion reviewed environmental speciation of metals and the role that speciation plays in bioaccumulation and trophic transfer.

Chapter 7 expanded considerably on the theme of metal exposure and effects to estuarine and coastal organisms. The theme of chemical speciation and its relationship to accumulation of metals was extended in this chapter to a consideration of how one uses this information in risk assessment to meet regulatory requirements. This included discussion of applying equilibrium models such as the Dynamic Multipathway Bioaccumulation Model (DYMBAM) to incorporate dissolved and dietary sources of metals into predictions of bioaccumulation. This chapter addressed many exposure and effects characterization issues of the analysis phase of a risk assessment.

As discussed briefly in Chapter 3, the role of xenobiotics as endocrine-disrupting agents is becoming a major area of effects research for marine organisms. Chapter 8 dealt primarily with the ecological effects aspects of the analysis phase of the risk assessment paradigm for endocrine-disrupting agents, an increasingly important class of xenobiotics. The authors first summarized information about modes of action and effects in marine fish and invertebrates. Many modes of action and effects that were discussed for endocrine disruption extended discussions in previous chapters that focused on biotransformation and clearance of chemicals from tissues (Chapter 5) and emergent contaminants (Chapter 3). The authors then extended the review to describe methods for generating a coherent risk assessment strategy for endocrine disruptors with particular reference to marine organisms. This is a new application of the risk assessment paradigm that is likely to bring significant new issues to the discussion.

The focused assessment of risk to marine mammals was the topic of Chapter 9. This chapter raised the issue of how to describe exposure (a major element in the analysis phase of the risk assessments process) for cetaceans, pinnipeds, and mustelids. Most of these charismatic animals are not amenable to laboratory experimentation and legal issues often preclude controlled experimental exposures. Much discussion focused on determining toxicity reference values (TRVs) for persistent organochlorine (POC) compounds. The ultimate intent was to determine exposure dose to animals that cannot be readily sampled or subjected to experimental exposure. The authors then reviewed literature related to effects on marine mammals, noting that most studies were hampered by the sample size and life mode of these species. Finally, data from New Zealand marine mammals were used as a case study for a risk assessment. This risk analysis employed a hazard quotient approach that used the exposure and effects parameters as estimated by the methods described earlier in the chapter. The chapter thus addressed problem formulation and risk characterization in addition to effects and exposure assessments, resulting in a general risk assessment for marine mammals exposed to POC compounds. However, end points of concern other than those elicited by dioxin-like chemicals may be important, e.g., the endocrine-disrupting agents discussed in Chapter 8.

The remaining chapters shifted focus from effects at the suborganismal and individual levels to those at higher levels of biological organization. These three chapters looked to a more comprehensive risk analysis addressing populations, incorporating analyses of spatial and temporal heterogeneity.

Chapter 10 presented a matrix model for predicting demographic consequences of chronic exposure to a pollutant. Data for a New Bedford Harbor (Massachusetts) fish population (*Fundulus heteroclitus*) chronically exposed to PCBs were used as a case study for model application. Interestingly, demographic projections from toxicological evaluations of fish naive to PCBs predicted poor population status. This prediction contrasted with the apparent robustness of the New Bedford Harbor population. Compensatory shifts including those related to life history and physiological adjustments were described that lead to increased potential for evolutionary effects. The authors suggested that a paucity of demographic data for the well-studied fish used in the case study contributed to the difficulty in deriving a retrospective model that accurately predicted the condition of the population. The authors proposed research at additional sites subject to different types of stressors to evaluate further the applicability of the model and to seek improvements in its use. The disparity between prediction and real populations raised the issue of dealing with risk assessment in a multigenerational context to account for adaptation of resident populations to the gradual introduction of contaminants that occurred historically.

Chapter 11 considered population-based risk assessment in a complex landscape. As mentioned above, such a landscape context can be essential in adequately defining risk to coastal and estuarine species. Although the modeled heron population described was lacustrine, the approach taken for this common coastal species is directly applicable to landscapes encompassing marine features and pollution. In this study, heron exposure to mercury was through their diet. A set of well-known models for uptake by various routes was integrated with demographic analysis. A metapopulation context for this analysis was developed using a Geographical Information System (GIS) analysis of habitat and then applied to predicting population consequences of exposure. Risk was estimated and placed into the context of different metapopulation scenarios including the presence of significant migration to compensate for excess mortality or depressed reproductive success. Important features of this population-level study were its spatially and demographically explicit structure.

There was a similar landscape context for Chapter 12 that included an explicit recognition of the ultimate use of any risk assessment, i.e., risk management to reduce risk as suggested in Chapters 1 and 2. The potential or known incremental chemical risks for a highly modified urban estuary were emphasized in this chapter. The authors discussed the implications that apply to any implementation of risk management; one must understand that the human activities producing environmental risk, especially in urban areas, are an unchangeable and persistent feature of the landscape that must be addressed in any risk assessment. This issue was alluded to in several earlier chapters.

The analysis stage of risk assessment was developed in a context described by the authors as an "ecological coincidence analysis." This approach is one that assumes cause-and-effect relationships rather than attempting to demonstrate them. The distribution of stressors was compared to that of receptors to determine the extent of co-occurrence. Ecological coincidence was the central theme in their analysis and was based on the notion that there can be no effect if there is no co-occurrence. To demonstrate this principle, the authors used three U.S. examples including the Passaic River estuary in New Jersey, the Fox River lacustuary of Lake Michigan (Wisconsin), and the New York Harbor. Here, coincidence in both time and space was essential to the analysis. The ultimate result of such analyses was an estimate of time-dependent and site-specific dose for the exposure analysis. In the examples, specific estuarine considerations are included such as restriction of avian foraging on tidal flats to periods of low tide. Thus, differential exposure in time is not only a matter of historical (an issue raised in Chapter 10) and seasonal differences in spatial distribution of a receptor, but also differences on tidal scales. In the analysis, the potential for behaviorally mediated reduction in exposure was discussed. As in the preceding chapter, GIS methodologies played an important role in the analysis phase of the risk assessment process. Although the importance of stressor and receptor coincidence in time and space is not new, the use of the concept as a central element in specific estuarine landscape risk assessments brought the implications of the concept into sharper focus.

13.3 CONCLUSION

The intent in developing this book was to accelerate the application of ecological risk assessment in coastal and estuarine systems. Despite the need for such assessments, the development of methods applicable to these systems has lagged behind those for freshwater and terrestrial systems. Coastal and estuarine environments have features unique or extreme relative to other environments; therefore, this special attention to risk assessment in these habitats is necessary. These tend to be environments with high risk of adverse impacts stemming from increasing and complex uses by humans. Assessment is particularly difficult due to the inherent complexity of these landscape features and the complex of human activities occurring within them.

Each chapter provides information about or examples of applying the current ecological risk assessment paradigm to these unique and vulnerable environments. Collectively, the authors suggest future directions that assessors can and should pursue.

The authors in this book did not address risks to certain, highly valued habitats such as coral reefs or areas of low human population density such as coastal agricultural areas. Also, little attention was been given to the effects of contaminants on primary producers except relative to toxicant entry into the food chain. In this book, attention to estuarine or marine communities was limited. As we look to the future, these and other issues not included here must be given careful consideration in problem formulation, analysis, and, ultimately, risk characterization. Our risk assessments are never definitive, and the benefits of our risk management are always uncertain without consideration of the full diversity of biological organization and stressor exposure.

This may lend additional support for the Precautionary Principle, already gaining acceptance in the EU (Chapter 2). As we have learned in recent decades, the old paradigm is flawed that the sea exceeds our ability to decimate it. Even as we hasten to develop and implement new paradigms for environmental management to benefit the oceans, our industrial society is changing the environment without adequate attention to the preservation of vital resources.

For our children and grandchildren to also enjoy indolent moments along estuarine and coastal shores, either appropriate risk-based paradigms for reasoned action must be formulated or we must apply precautionary measures. It is our hope that the contents of this book will help in some small but meaningful way to convey the need for solutions to the problems emerging from our use of the coastal resource.

REFERENCES

1. Carson, R., *The Edge of the Sea*, Houghton Mifflin, Boston, 1955.
2. Carson, R., *Silent Spring*, Houghton Mifflin, Boston, 1962.
3. Carson, R., *Under the Sea Wind*, Oxford University Press, New York, 1941.
4. Rand, G.M. and Carriger, J.F., U.S. environmental law statutes in coastal zone protection, *Environ. Toxicol. Chem.*, 20, 115, 2001.
5. WEST Group, *Federal Environmental Laws*, 2000 Edition, WEST Group, Eagan, MN.
6. Sissenwine, M.P., Marine fisheries at a critical juncture, *Fisheries*, 18, 6, 1993.
7. Kennish, M.J., *Practical Handbook of Estuarine and Marine Pollution*, CRC Press, Boca Raton, FL, 1997.
8. Scott, G., Marine and estuarine toxicology and chemistry, *Environ. Toxicol. Chem.*, 20, 3, 2001.
9. Rodríguez, G.R., Breddia, C.A., and Pérez-Martell, E., Eds., *Environmental Coastal Regions III*, WIT Press, Boston, 2000.
10. Pernetta, J. and Elder, D., *Cross-sectoral, Integrated Coastal Area Planning (CICAP): Guidelines and Principles for Coastal Area Development*, IUCN (International Union for Conservation of Nature and Natural Resources), Gland, Switzerland, 1993.
11. Anonymous, *Reefs at Risk*, IUCN (International Union for Conservation of Nature and Natural Resources), Gland, Switzerland, 1993.
12. Gustavson, K., Huber, R.M., and Ruitenbeek, J., Eds., *Integrated Coastal Zone Management of Coral Reefs: Decision Support Modeling*, The International Bank for Restoration and Development/The World Bank, Washington, D.C., 2000.
13. Suter, G.W., Jr., *Ecological Risk Assessment*, CRC Press/Lewis Publishers, Boca Raton, FL, 1993.
14. Suter, G.W., Jr., Efroymson, R.A., Sample, B.E., and Jones, D.S., Eds., *Ecological Risk Assessment of Contaminated Sites*, Lewis Publishers, Boca Raton, FL, 2000.
15. U.S. EPA, Guidelines for Ecological Risk Assessment, U.S. EPA/630/R-95/002F, April 1998, Final, U.S. Environmental Protection Agency, Washington, D.C.

Index

A

Abductive inference 80, 93
Accession States 17
Accumulation (*see* Bioaccumulation)
Acid volatile sulfides (AVS) 134, 138–139, 144, 167, 169, 331
Acipenser sp. (sturgeon) 173
Acquiescence 76
Adaptation (of populations) 262–265, 332
Adverse drug reaction 87–90
AhR-active compounds 106–108, 225, 227–228, 231
Alar 91
Albatross
    Laysan (see *Diomedea immutabilis*)
    Black-footed (see *Diomedea nigripes*)
Alderfly (see *Sialas valeta*)
Algae 166, 169, 179
    cell cytoplasm 162, 171
    growth inhibition 19
Alkene oxides 107
Alkylphenols (*see also* Endocrine disrupting agents) 191–192
    monoethoxylates 192
    polyethoxylates 330
Allometry 278
American plaice (see *Hippoglossoides platessoides*)
Amino acids 163, 165
Ammonia 26, 156
Amodiaquine 87
Amphipod (*see also* Crustacea and specific species) 99, 104, 117, 135–136, 138, 162, 165–166, 171, 176–179
    survival and growth 35
Anadonta grandis (Unionid clam) 169
Anas sp. (ducks) 285–286
Anchoring 75
Annelid (*see also* specific annelid species) 98, 107, 111
Antibiotics 57–58
Antimalarial drug (*see also* Chloroquine and Amodiaquine) 87
Antimicrobial agents 58
Anthracene 117
Arctic Ocean 30
Arctocephalus forsteri (fur seal, New Zealand fur seal) 232–233, 235–236
Ardea herodias (great blue heron) 274–295, 332
Arene oxides 107
Arenicola marina (lugworm) 33, 164–165
Arsenic (As) 127–129, 133–134, 142, 144, 158, 168
Artemia salinas (brine shrimp) 168
Aryl hydrocarbon receptor (*see also* AhR-active compounds) 220, 252, 261, 263
Assimilation efficiency 104–105, 133, 162, 170–173, 179
Atmospheric transportation 219, 232
ATPase (Na^+/K^+ -) 157
AVS (*see* Acid volatile sulfides)
Avoidance (*see* Behavior, avoidance)

B

Bacon (Francis) 75–76
Bacteria 100, 161–162, 167
Baikal seal (see *Phoca sibirica*) 225
Baird's beaked whale (see *Berardius bairdi*) 234
Balaenoptera acutorostrata (minke whale) 233–234
Balaenoptera musculus (blue whale) 233
Balanus amphitrite (acorn barnacle) 195
Bald eagle (see *Haliaeetus leucocephalus*)
Baltic gray seal (see *Halichoerus grypus*) 225
Baltic Sea 30, 226
Barnacles 195
Baseline 298, 300, 307, 309, 323
Bayes (Reverend Thomas) 82
Bayesian 84, 330
Bayes's theorem 81, 84, 85, 90–91, 93
Behavior 285–286
    avoidance 179
    feeding, 104, 221
    foraging (of herons) 279
    predator avoidance 180
Belief
    based on repetition 75
    level/degrees of 81, 83–84
    a posteriori 86
    a priori 86
Beluga whale (see *Delphinapterus leucas*)

Benchmark concentrations 225
Benthic invertebrates (see also specific species) 134–136, 140, 144
Benzo(a)pyrene (BaP) 99, 101–102, 104, 106, 109–110, 113, 115, 117
Berardius bairdi (Baird beaked whale) 234
Bias toward easy representation 76
Bioaccessibility 301
Bioaccumulation 97–98, 118, 129, 131, 133, 136–138, 140–143, 331
Bioaccumulation factor (BAF) 105, 128–129, 131, 135, 137–140, 169
  normalized (BCF) 105
Bioaccumulative 127
  persistent 252
Bioavailability 97–100, 104, 118, 128, 134–136, 140, 144, 161, 165, 179, 300–301, 330, 331
Biocides 19
Bioconcentration 103, 129, 173
  factor (BCF) 103, 105, 169
Biodegradation rate 23
Bioenergetic models 133, 170–175, 181–182, 279–282, 284
BIOENV 35
Biomagnification 143
  factor (BMF) 231
  ratio 233
Biomarker 193, 200, 205, 207
Biosolids (see Sewage sludge)
Biota-sediment accumulation factor (BSAF) 106
Biotic Ligand Model 156–158
Biotransformation (see also Xenobiotics, metabolism) 98, 106–110
Birds 285–286, 302–303, 307, 316–317
Bivalves (see also Mollusk and specific species) 104–105, 160–163, 167, 172–174
  glandular digestion 164–165
Blue crab (see *Callinectes sapidus*)
Blue whale (see *Balaenoptera musculus*)
Bootstrap 202, 277, 283
Bottlenose dolphin (see *Tursiops truncatus*)
Brain 111
Brominated diphenyl ethers (BDEs) 43–48, 61–63, 330
  concentrations in
    biosolids 61–63
    humans 48
    sediment 48
    wildlife 46–48
  degradation 48
  selective bioaccumulation 47–48
  uses of 43–45
  toxicity of 45–46
Brominated flame retardants 43–48, 330

Bromobenzene 115
Bulkheads 302–304
Butyltin (see also Monobutyltin, Dibutyltin, and Tributyltin) 231

C

Caddisfly (Trichopteran insect larvae) 176, 178
Cadmium (Cd) 26–27, 127–129, 132–134, 138–140, 142, 144, 154, 157, 159–162, 164, 167–173, 175–176, 179–181, 195
  chloride (Cl_2) 131
Calcium (Ca) 132
Callinectes sapidus (blue crab) 112, 114, 116, 141
Cancer 78
  liver 79–80, 93, 106, 330
  lung 80
Capelin (fish; see *Mallotus villosus*)
Caperea marginata (pygmy right whale) 233
Capitella sp. (polychaete worm) 167
Carassius auratus (goldfish) 118
Carbon tetrachloride 30
Carcinogen 26
Carcinus maenas (green crab) 168
Carson (Rachel) 18, 325
Causal
  hypothesis 80, 89
  relationship 92
Causality assessment 73–74, 91, 93, 330
Cephalorhynchus hectori (Hector's dolphin) 233–236
Ceriodaphnia dubia (water flea) 169
Cetaceans (see also Whales and Dolphins) 219, 229
Channelization 141
Chaoborus punctipennis (phantom midge) 168–169
Chaoborus sp. (midge) 176
Characterization
  effects 1
  exposure 1
  risk 1,3
Chemical
  monitoring strategies 42
  of Potential Concern 74
  risk assessment 298, 300–301
  persistent, bioaccumulative, toxic (PBT) 42–43, 264
  speciation 132, 331
  tolerance 259, 261–262
Chesapeake Bay 203
Chionectes opolio (snow crab) 176
Chironomid (non-biting midge) 177, 178
Chlorobenzene 115

Index

Chlorobenzoic acid 110, 115
Chlordane 79
Chlorodinitrobenzene 112, 114
Chloroquine 87
Chromium (Cr) 26, 139, 158, 160, 162, 164, 165, 171, 173, 175, 177
Cladocera 166, 169, 176, 179
Clams (bivalve mollusks) 194
Classification
    packaging and labeling of dangerous substances 19
Clean Water Act 248, 326
Climatic change 91
Coastal Zone Management Act 326
Cobalt (Co) 157, 164, 167, 171
Cognitive
    biases 77, 93
    error 75
    overconfidence 77
Combined Monitoring- and Modelling-based Priority Setting (COMMPS) 23
Common dolphin (see *Delphinus delphis*)
Community
    ecology 76
    effects 236
Compensatory mechanisms (*see also* Mechanisms) 259–263
Complexation 140
Complexes 136
Comprehensive Environmental Response, Compensation and Liability Act (CERCLA) 274, 326
Conceptual model 218, 223
Confirmation bias 75
Congener specific analysis (PCB) 222, 225, 228
Convention/Protection of the Marine Environment
    of the North East Atlantic 21
Copepod (*see also* Crustacea and specific species) 104, 132, 163, 171–172, 174, 176–177, 179
    survival and reproduction 35
Copper (Cu) 132–133, 138–139, 154, 157, 164–165, 168
Coral reefs 327
Corophium volutator (amphipod) 33, 195
Council of Ministers 17
Crab (*see also* Crustaceans or specific species) 98, 168, 176
Crassostrea gigas (Pacific oyster) 33, 195, 198
Crassostrea virginica (Atlantic oyster)
Critical body burden (*see also* Toxicity reference values) 24
Crustacea (*see also* specific species, Amphipod, Crab) 98, 107, 109–111

Cryptochironomus sp. (non-biting midge) 168
cumulative (probability) density function 278, 291
Cyanides 26
Cyprindon variegates (sheepshead minnow) 193
Cytochrome P-450 monooxygenase complex 106, 109, 111–112, 229
Cytochrome P-4501A1 (CYP1A1) 80, 253

D

Dall's porpoise (see *Phocoenidae dalli*) 234–235
Daphnia (water flea) 201
    magna 19, 194–195
Database
    AQUIRE 25
    ECETOC Aquatic Toxicity 24
    Quality of Shellfish Waters (QSW) 30
DDT, DDE (*see* Dichlorodiphenyltrichloroethane or Dichlorodiphenyldichloroethylene)
Deduction 75
Delphinapterus leucas (beluga whale) 227
Delphinus delphis (common dolphin) 234
Depuration 102, 106, 172, 178
    rate constant (*see* Elimination, rate constant)
Detoxifying 133
    tissues 134
Diatom 161, 173, 181
Dibutyltin 117
Dichloroaniline 115
Dichlorobenzene 115
Dichlorobenzoic acid 115–116
Dichlorodiphenyldichloroethylene (DDE) 112
Dichlorodiphenyltrichloroethane (DDT) 30, 78–79, 225, 233, 319
Dieldrin 79
Diet 105, 114, 116, 133, 172, 175–176, 179, 181, 221–223, 226, 231, 236
    mercury in heron diet 273–295
Diffusion 130
Digestive
    fluids/juices 99, 118, 163–164
    gland 98, 111, 165
    mechanisms 163–165
Dihydrodiols 107
2,4-Dinitrochlorobenzene 113, 115
Dinoflagellate (see also *Pfiesteria piscicida*) 90, 93
Dinophilus gyrociliatus (polychaete worm) 195
Diomedea immutabilis (Laycans albatross) 233, 237
Diomedea nigripes (Black-footed albatross) 221, 233, 237

Dioxin 105, 220, 232, 252
    effects on fish 253–254, 257, 263, 265–266
Dioxin-like compounds (DLCs) 225, 248, 332
Diporeia spp. (amphipod) 104
Directive
    dangerous substances 26
    existing chemicals 20
    Plant Protection Products (PPPD) 19
Dissolved organic carbon (DOC) 99, 131–132, 144, 157, 331
    complexes
    Hg 132
    methylHg (MMHg) 132
Distribution coefficient (*see* K_D)
DNA 107, 115
DNA adduct 79, 80, 106, 115
Dogwhelk (see *Nucella lapillus*)
Dolphins (*see also* individual dolphin species) 229, 234
    oceanic 234–236
Duck (see *Anas* sp.) 285–286
Dusky dolphin (see *Lagenorhynchus obscurus*)
Dynamic Multipathway Bioaccumulation Model (DYMBAM) 170, 172–173, 175, 331

E

Ecological
    coincidence analysis (ECA) 301–305, 309, 320, 323, 333
    risk assessment framework (ERA) 218, 249, 328
Ecological Committee on FIFRA Risk Assessment Methods 25
Ecotone framework 329
Egret, Greater (see *Egretta alba*)
Egretta alba 286
Elasticity analysis 250
Electrophiles 109
    reactive 107
Elimination 97–98, 103, 115–118
    rate constant 103–105, 117, 171
Elliptio complanata (Unionid clam) 195
Emerging contaminants 41–64, 329, 331
Endocrine disrupting agents (*see also* Alkylphenol, Estradiol, Ethinylestradial, Nonylphenol, Natural and synthetic estrogens) 331–332
Endoplasmic reticulum (ER) 109, 112, 114
Endowment effect 76
English sole (see *Parophrys vetula*)
Environmental Quality Standard (EQS) 202
Epoxide hydrase 107

Equilibrium
    constants 136
    partitioning 221
Estradiol 49–52
Estrogens (*see also* Endocrine disrupting agents, Natural and synthetic estrogens) 330
Ethinylestradiol 50–51
Ethoxyresorufin *o*-deethylase (EROD) 253, 262
European
    Atomic Energy Community 16
    coal and steel community 16–17
    Commission 17
    Commissioner 17
    Community 16
    Court 17
    Directives 17
    Directorates 17
    Economic Community 16
    member state 17
    Parliament 17
    Union (EU) 16–17, 329
    Union System for the Evaluation of Substances (EUSES) 20
Excretion rate 105–106
Expert opinion 74–75, 80, 87, 90, 93
Exposure 298
    assessment 97, 221–225
    from air 223
    from sediment 104–106
    intensity 221
    internal 221
    long term 26
    pathway 221
    profile 222
    spatial extent of 221
    temporal aspects of 222
External tissues 134
Extinction 266, 293
Extrapolation from freshwater data 24

F

Factor
    dilution 23
    safety 23
Fathead minnow (see *Pimephales promelas*)
Federal Pollution Control Act (FPCA) 326
Feeding behavior 104, 221
Fenitrothion 110
Ferret, domestic (see *Mustela furo*) 232
Fish 102–103, 105, 107, 109–111, 118, 141–142
    gill 141–142
    gut 141
        lining 142

Index   341

tissues 144
kill 90–94, 330
Flounder (see *Platichthys flesus*) 191–193
Fluorene 115
Fluorides 26
Food 118, 136, 138, 142
Food chain 129, 223, 333
  length 133
  transfer 331
Food web 106
Foraging 280
Fox River 304, 333
Fox's rules of ecoepidemiology 77, 79–80, 93
Framing 76
Free ion activity model (FIAM) 169
Free radicals 80
Frequentist 82, 84
Fugacity 101–102
  capacity 101
Fundulus heteroclitus 248, 253, 332
Fur seal or New Zealand fur seal (see
  Arctocephalus forsteri)
Furans (*see also* Polychlorinated dibenzofurans) 220, 232

G

Gastropod (snails) 166
Genetic
  adaptation 259, 262, 264, 266
  diversity 264
Geographic Information System (GIS) 298, 305, 307, 332–333
Gill 99, 101–102, 111–112, 118, 170
  membranes 156–158, 179
Global introspection (*see also* Expert opinion) 74, 86, 93
Global redistribution processes 219, 232
Globicephala melaena (pilot whale) 232–236
Glucose 108, 110
Glucuronic acid 108, 110
Glutathione (GSH, g-glutamylcysteinylglycine) 108–110
Glutathione-S-transferase 109–112
Glycosylation 110
Goldfish (see *Carassius auratus*)
Golgi network 112
Gonadal Somatic Index (GSI) 192
Gradients 3–4
Grass shrimp (see *Palaemonetes* sp., *P. pugio*)
Great Lakes 220
Green Bay 304
Green crab (see *Carcinus maenas*)
Green gland 112

Growth 179
  rate constant 105, 170
  dilution 106
Gut 163–164

H

Habitat
  availability 284
  area use 283
  density 283
  heterogeneity 305
  quality 281–282
  suitability 301
Haliaeetus leucocephalus (bald eagle) 228
Halichoerus grypus (Baltic gray seal) 225
Harbor seal (*Phoca vitulina*) 225–227, 231–232
Harbour porpoise (see *Phocoena phocoena*)
Hard clam (see *Mercenaria mercenaria*)
Hazard
  assessment 76
  concentration 76
  labeling 19
  quotient 25, 29, 74, 234–237, 286
Hector's dolphin (see *Cephalorhynchus hectori*)
Hepatitis 87
Hepatopancreas 98, 111–112, 114–117
  E-, F-, R- and B-cells of 112, 115
Heron, Great blue (see *Ardea herodias*)
Heteromastis filiformis (polychaete) 167
Hexachlorocyclohexane 30
Hexachlorobiphenyl (*see also* PCB) 112–114
High production volume (HPV) chemicals 45
Hill's Rules of Disease Association 77–80, 92
Hinia reticulata (netted dog whelk) 194
Hippoglossoides platessoides (American plaice) 176
Hoki (blue grenadier, see *Macruronus novaezelandiae*)
Holothurian 163
Home range 222
Humic acids 141, 161
Humin-kerogen polymers (in sediments) 99
Hyalella azteca (amphipod) 179
Hydrophobicity 101
3-hydroxybenzo(*a*)pyrene 106
Hypotheticodeductive method 75, 81

I

Idol of certainty 77
Idola quantitatis 74
Idols of the tribe 75

Immune system 225, 231, 264
Immunocompetence 220
Immunosuppression 226
Imposex 195, 198–199, 205
Incremental
    chemical exposure 303
    damage 298–299, 307
    risk 298, 307
Individual tolerance concept 77
Induction 75
Industrial inputs 319
Ingestion 140, 162–163, 170, 223–224, 231
In-stream cover 304
Intersex 191, 195, 205
Intersexuality 198
Intestinal
    fluid 136
    tissue 141
Invertebrates (*see* particular species) 142, 194
Ion regulation 156–157
Ionic strength 170
Iron (Fe) 166
Iron oxyhydroxides 161, 169, 171

K

Kidney 111–112
Kit size (mink) 230
K_D, (partition coefficient) 159–160, 167, 170, 172, 175, 181

L

Lactation 220
Lacustuary 304
Lagenorhynchus obscurus (dusky dolphin) 234
Lagenorhynchus sp. (white-sided dolphin) 234
Laissez-faire mores 326
Lake Michigan 304
Landscape 328, 333
Lead (Pb) 26, 127, 129, 134, 139, 142, 144, 160
Lepsiella vinosa (wide-mouthed Lepsiella) 199
Leptocheirus plumulosus (amphipod) 162, 165, 176–178
Life cycle model 288
Life history 219, 250, 259, 332
Ligand 141
Likelihood of the data 86
Likelihood ratio 86, 91
Limnesia maculata (predatory mite) 176, 178
Limnodrilus hoffmeisteri (oligochaetes) 179
Lipid 114–115, 118, 164
    content 101–102, 118

Litter size 230
Littorina littorea (common periwinkle) 195
Liver 112, 230
    disease 88, 111
    function 87
    necrolysis 87
Long Island Sound 172
Low risk testing 75
Lowest observed adverse effect level (LOAEL) 226–227, 229–230, 286
Lugworm (see *Arenicola marina*)
Lumbriculus variegates (oligochaete) 104
Lutra lutra (otter) 227, 229–230
Lymphocyte proliferation 226–227

M

Mackerel (see *Scombrus* sp.)
Macoma balthica (Baltic telling clam) 160–162, 164–165, 167, 171, 176–178
Macoma nasuta (bent-nosed macoma clam) 104–105, 167
Macruronus novaezelandiae (Hoki or blue grenadier) 233
Mallotus villosus (capelin) 192
Marine fisheries decline 326
Marine mammals 217–245, 331
Marine Mammal Protection Act 326
Marine Protection Research and Sanctuaries Act 326
Maximum Allowable Toxicant Concentration (MATC) 226, 229–330
Mechanisms
    compensatory 259, 266
    demographic 258–259, 265
    evolutionary 263–264, 266
    of accumulation 129
    toxicological 259, 261
Medical diagnosis 86–90
Mediterranean Sea 30
Mediterranean striped dolphin (see *Stenella coeruleoalba*)
Membrane
    accumulation rates 141
    gut 141
    transfer 143
    transport 140, 142, 144
    perfusion 144
Mercenaria mercenaria (hard clam) 101–102, 118
Mercury (Hg) 26–27, 127–133, 136–143, 158, 179, 219, 332
    behavior 144
    inorganic 127, 130, 135, 141, 331
    methyl 273–295

Index

monomethyl (MMHg) 127–130, 133–134, 136–140, 143, 144, 331
Metabolic
  activity 142
  rate 101
Metabolites (*see* Xenobiotic, metabolites)
Metals (*see also* individual metals) 26, 129, 219
  assimilation efficiency 162, 170–178, 181
  bacteria-associated 162, 165, 167
  bioaccumulation factors 169
  bioconcentration factors 169
  detoxification 179
  dissolved, conditional stability constant 157
  dissolved, free-ion activity 154–155, 169–170, 179
  dissolved, uptake rates 170–171, 173–178, 181
  distribution coefficient (K_D) 159–161, 170, 181
  efflux rates 170–171, 173–178, 181
  in sludges 60
  neutral complexes 130–131, 158
  oxide phases 144
  phytoplankton-associated 162, 166, 169, 171–173, 179
  prey-associated 168–169, 173, 179–180
  reproductive toxicity 179
  toxicity test exposure duration 165, 180
Metal ion concentration 129
Metalloids 132, 142
Metallothionein 179
Metapopulation 26, 332
Method transparency 74
Methylated species 127
Microbial bioluminescence 35
Microelectrodes (for pH) 163
Midge (*see* Chironomid, *Cryptochironomus* sp.)
Migration 261, 287, 290–291
Milk 220
Mineral oils 26
Ministerial declaration on the North Sea 21
Mink (see *Mustela vison*)
Minke whale (see *Balaenoptera acutorostrata*)
Mirex 112
Mite 176, 178
Mitochondria 114
Mixed function oxygenase system (*see* Cytochrome P–450 monooxygenase complex)
Mixtures (chemical) 330
Molecular diffusion 102
Mollusk (*see also* specific molluscan species) 98, 107, 111
Monobutyltin 117
Monte Carlo technique 275, 279

Morbillivirus 225–226
Morone saxatilis (striped bass) 173
Mud flats 302–303
Multiple stressors 327
Multiple xenobiotic resistance (MXR) 59
Mummichogs (see *Fundulus heteroclitus*)
Muscle 111, 115, 117
  cardiac 112
Musk compounds 58–59
Mussel
  feeding rates 35
Mussel (see *Mytilus edulis* and *M. galloprovincialis*)
Mustela erminea (stoat) 232
Mustela furo (domestic ferret) 232
Mustela nivalis (weasel) 232
Mustela putorius (polecat) 232
Mustela vison (mink) 227–232
Mustelids (family of ferrets) 227–228, 232
Mysid 173–174
Mystacides sp. (caddisfly) 176, 178
Mytilus edulis (mussel, blue or edible) 118, 164, 172, 176–177
Mytilus galloprovincialis (mussel, blue) 171

N

National Environmental Protection Act (NEPA) 326
National Marine Monitoring Plan (NMMP) 33
Natural and synthetic estrogens (*see also* Endocrine disrupting agents) 49–52
  effects of 50
  fate in wastewater treatment 50–51
  sources of 51
Natural killer (NK) cell 226, 230
Natural resource damage assessment (NRDA) 299–304, 307
Natural selection 261, 264
Neanthes arenaceodentata (polychaete) 167, 172
Neomysis mercedis (mysid) 173–175
Nereis succinea (polychaete) 171–172
Nereis virens (polychaete) 105
Nesting (of heron) 279
Netted dog whelk (see *Hinia reticulata*)
Neuse and Pamlico River systems 91
New Bedford Harbor (NBH) 253–255, 332
New York/New Jersey (NY/NJ) Harbor Estuary 302, 307, 333
New Zealand fur seal (see *Arctocephalus forsteri*)
Nickel (Ni) 35, 134, 139–140, 157, 160
Nitrites 26
Nitrocera spinipes (harpacticoid copepod) 201

No observed adverse effect level (NOAEL) 226–231
Non-ionic surfactants (*see also* Nonylphenol-based surfactants) 192
Non-metric multidimensional scaling (MDS) 35
Nonylphenol 191, 194–195, 198, 202
Nonylphenol-based surfactants 52–56, 330
   chemical properties of 55
   concentrations in
      biosolids 60–61
      effluents 52–54
      sediments 55–56
   degradation of 52, 55–56
   effects of 54–56
   uses and sources of 52–54
North Sea 30
Northeastern Atlantic Ocean 30
Nucella lapillus (dog whelk) 194–196
Nucleophiles (from phase-one reactions) 109

O

Objectives
   chemical standards 30
   ecological standards 30
Occultatio 76–77
Ocean Dumping Ban 60, 326
Ocenebra erinacea (sting winkle) 198
Octanol-Water Partition Coefficient (K_{OW}) 47, 55, 102–103, 130
Odds
   posterior 88–89
   prior 86–89, 91
Oil Pollution Act (OPA) 326
Oligochaete (*see also* Annelid and specific species) 104, 179
Oncorhynchus mykiss (rainbow trout) 19
Orchestia sp. (amphipod) 176
Organic
   carbon 136, 140
   carbon coatings 161–162
   ligands 131, 140–141
   phase 140
Organisation for Economic Cooperation and Development (OECD) 20
Organohalogen 26, 98
Organometallic compounds 98
Organophosphate insecticide 110
Organophosphorus 26
Organotin (*see also* Monobutyltin, Dibutyltin, and Tributyltin) 26
Organotin Antifouling Paint Control Act 326
Oslo and Paris Commission (OSPAR)
   DYNAMEC group 23

Otter (see *Lutra lutra*)
Ouabain 142
Overwinter survival 260
Oxyanions 132
Oyster (see also *Crassostrea virginica* and *C. gigas*) 102, 181, 195, 198, 202

P

PAH (*see* Polycyclic aromatic hydrocarbons)
Palaemonetes pugio (grass shrimp) 105, 180, 195, 200
Palaemonetes sp. (grass shrimp) 105, 163
Paracentrotus lividus (purple sea urchin) 195
Parasites 261, 264, 266
Parophrys vetula (English sole) 78–80, 93
Particulate organic content (POC) 128, 134–135, 137, 140, 144
   sediment 136
Passaic River 302–303, 317, 333
Passive
   accumulation 130
   uptake 130
PCB (*see* Polychlorinated biphenyls)
PCDD (*see* Polychlorinated dibenzo-*p*-dioxins)
PCDF (*see* Polychlorinated dibenzofurans and Furans)
Pathways, active 142
Pelagic food webs 129
Pentachlorophenol 30, 109, 115, 118
Perfusion 141–143
Persistent organochlorine (POC) compounds 218, 224–225, 229, 233, 237, 331–332
Personal care products 58–59, 330
Pesticide 26, 98, 110
   lipophilic 78
Petroleum hydrocarbons 26, 102, 118
Pfiesteria-like organism (PLO) 92
Pfiesteria piscicida (dinoflagellate) 90–93, 330
pH 163, 169–170
Phantom midge (see *Chaoborus punctipennis*)
Pharmaceuticals 56–58, 330
   fate in the environment 57–58
Pharmacokinetics
   half-life 87
Phenol 106
Phenolic group 109
Phenylethylenes 191
Phoca hispida (ringed seal) 225
Phoca sibirica (Baikal seal) 225
Phocoena phocoena (harbor porpoise) 234
Phocoenidae dalli (Dall's porpoise) 234–235
Phosphoadenosyl phosphosulfate (PAPS) 109
Phylogenetic diversity 24

Physiological shifts 332
Phytoplankton (*see also* Algae, Diatom) 129, 132–133, 144, 170–171, 173, 176–177, 223
Phytosterols 51
Pilings 304
Pilot whale (see *Globicephala melaena*)
Pimephales promelas (fathead minnow) 157
Pinniped (*see* Seal)
Platichthys flesus (flounder) 191–193, 232
Pleuronectes vetulus (English sole) 78
Pleuronectes yokohamae (Yokohama sole) 191
Polar bear (see *Ursus maritimus*)
Polecat (see *Mustela putorius*)
Polychaete (*see also* Annelid and specific species) 102, 104–105, 163, 167, 171–172
  cast formation 35
Polychlorinated biphenyls (PCBs) 48–49, 79, 102, 104–107, 219–220, 224–226, 232–236, 252–253, 219, 230, 330–332
  dioxin-like toxicity 49, 225
  individual congeners 48, 107
  role in endocrine disruption 49
Polychlorinated dibenzo-*p*-dioxins (PCDDs) 219, 319, 233
Polychlorinated dibenzofurans (PCDFs) 219, 319, 233
Polycyclic aromatic hydrocarbons (PAHs) 78–80, 93, 99, 100, 106–107, 109, 111, 115, 118, 318, 330
Polyurethane foam
  as a BDE source 62–63
Pontoporeia hoyi (amphipod) 99, 117
Popper (Sir Karl) 77, 80
Population
  carrying capacity 287–290, 294–295
  decline 227
  demographic projections 332
  density (human) 327
  dynamics 250, 266, 274
  ecology 77, 247–272
  genetics 264
  growth rate 3, 250, 257
  matrix models 248, 257, 265, 332
  size 250
Pore water 99, 104, 140, 167
Postexposure mortality 77
Potamocorbula amurensis (Asian clam) 162, 165, 167, 171, 173–178
Precautionary Principle 18, 21–23, 329, 333
Precipitate explanation 75
Predator, metal transfer to 168
Predatory fish 133
Predicted
  no effect concentration (PNEC) 20, 24, 202

environmental concentration (PEC) 20, 202
Primary consumers 134
Primary production
  of world 327
Principle
  Polluter Pays 18
  subsidiarity 18
Priority pollutants 42
Probabilistic risk assessment 290, 329
Probability
  blindness 76
  conditional 84–85, 90
  density function 275, 281, 283, 285, 290
  explanation of 82–84
  posterior 81, 86, 88, 90
  prior 81, 86, 88
  temporal 279
Problem formulation 1, 219–221
Product testing 19
Productive bottoms 307
Protein 107, 115, 133
Pygmy right whale (see *Caperea marginata*)
Pyrite 140

R

Racial profiling 85
Radioisotopes 161
Rainbow trout 168
Recruitment 291
Redistribution 144
Reference dose 230, 237
Reproductive failure 198, 225
Residency time (of herons) 279
Retinoids (hepatic) 227
Rhepoxynius abronius (amphipod) 166
Ringed seal (see *Phoca hispida*)
Riparian corridors 307
Risk assessment
  definition 1
  deterministic 19
  marine 23
  population-based 273–296, 332
  probabilistic 274, 290, 329
  prospective 19, 23
  relative 8
  retrospective 19
Risk 274
  characterization 20
  factor 79
  ranking 8
River Basin Management Plan 31
River Pollution Prevention Act (RPPA) 18
RNA 115

Ruditapes decussatus (Tunisian clam) 194

S

Safety factor 23
San Francisco Bay 162, 172–173
Saturation 143
Scientific ethics 91
Scrobicularia plana (peppery furrow shell) 198
Scombrus sp. (mackerel) 233
Sea birds 220
Seal (*see also* Baltic gray seal, Harbor seal, Ringed seal, New Zealand fur seal) 229–230
Sediment 104–105, 118, 160, 221–223, 162, 165, 167, 177
    chemical extractions 163
    chemistry 140
    iron oxyhydroxides 164, 169–170
    nutritional constraints 162
    organic matter 99, 176, 301
    quality guidelines 6, 329
    quotient value 6
    toxicity tests 166
        contaminated, uncontaminated food 165–166
        dietary metals exposure 159–161
    sulfides (*see also* Acid volatile sulfides) 164, 167
    triad 7
    water interface 138
Selenastrum capricornutum (green algae) 169
Selenium (Se) 127–129, 133–134, 142, 144, 158, 162, 171–173, 175–176
Sewage effluent 191
Sewage sludge 59–63, 330
    brominated diphenyl ethers in 61–63
    contaminants in 60–63
    disposal of 60
    land-application of 60
    nonylphenols in 60–61
    regulation of 60
Sex steroid 191
Sheepshead minnow (see *Cyprindon variegatus*)
Sherlock Holmes 74
Shoreline structure 301, 304
Sialis valeta (predatory insect) 168, 177
Silent Spring 18, 325
Silicon 26
Silver (Ag) 129–130, 133, 138–140, 157–158, 160, 167, 172–173, 176, 179
Simultaneously extractable metal (SEM) 138
    AVS model 140
    AVS paradigm 138–139

AVS ratio 139
Snow crab (see *Chionectes opolio*)
Social biases 77
Solubilization
    during digestion 134, 144
    from food or sediment 143
    studies 136–137
Solution to Pollution is Dilution Paradigm 326
Sorption properties 135
Speciation (chemical) 24, 330
Species sensitivity distribution (SSD) 23, 25, 202–203, 207
Spontaneous generalization 75
Stenella coeruleoalba (Mediterranean striped dolphin) 225
Stoat (see *Mustela erminea*)
Stomach 99, 111–112, 115, 117
Stressors
    multiple 6, 59
    response relationships 251, 266
Striped bass (see *Morone saxatilis*)
Structure-Activity Relationships (SARs) 199–200, 207
Sturgeon (see *Acipenser* sp.) 173
Substances
    List I (black list) 26
    List II (gray list) 26
Sulfotransferase 109
Superfund Amendments and Reauthorization Act 326
Superfund site 253–254
Superoxide anion (O_2^-) 107
Surface microfilm 224
Surfactants (*see also* Nonylphenol-based surfactants) 164–165
Suspended particulate matter, nutritional constraints 162

T

Technical guidance document (TGD) 20
Tees Estuary (U.K.) 34,35
Temora longicornus (Calanoid copepod) 171–172
Temperature 101
Tetrachlorobenzene 115–116
Tetrachlorodibenzo-*p*-dioxin (TCDD) 220, 228
Thalassiosira pseudonana (diatom) 181
Thallium (Tl) 179
Theory
    dependence 76
    immunization 75
    life history 250, 259
    tenacity 76
Thiessen polygons 305

Thiol groups 140
Threshold effect 78, 230–231, 275
Tides 301
Tin (Sn) 158
Tisbe battagliai (Harpacticoid copepod) 33, 195
Tissue residue guidelines (TRG) 224
Tobacco 80
Toxic equivalents (TEQs) 220, 225, 228, 230–231, 233–237
Toxic equivalency factor (TEF) 228
Toxicity
    experiments 134
    exposure ratio 19
    identification evaluation (TIE) 206–207
    tests 136
Toxicity reference values (TRVs) 218, 225, 228–230, 234, 236–238, 285–286, 290
Toxicity response function 285
Toxics characterization methods 329
Toxics reduction inventory (TRI) 45
Toxins (natural) 225
Transboundary contaminants (*see also* Atmospheric transportation) 43
Transitional waters (estuaries) 30
Treaty of Rome 16
Tributaries 307
Tributyltin 110, 113, 115, 117, 190, 194–195, 198–199, 202–203, 205
p-Trichlorobenzoic acid 115
Triclosan 58
Trophic
    level 144
    orders 173
    position 133
    structure 133
    transfer 132–134
Tursiops truncates (bottlenose dolphin) 225, 227, 234–235

U

Uncertainty 21, 224, 229, 231, 237
Uncertainty factor (UF) 229, 236
Unionid clam (see *Anadonta grandis*)
United Nations Conference on the Environment 18
Uptake 97–98, 100–106, 118, 144, 167, 174–175, 179
    rate or rate constant 100, 102, 170–172, 178
    clearance coefficient 104
Uridine diphospho-D-glucose (UDPG) 110
Ursus maritimus (polar bear) 220

V

Vanadium (V) 158
Vegetation 301
Vitamin A 226–227
Vital rates 77, 250
Vitellogenin 191

W

Water
    depth 304
    quality criteria 5, 329
Water Effect Ratio (WER) 155–156
Water Framework Directive (WFD) 16, 30
    parameters
        biological 31
        chemical and phyochemical 31
        hydromorphological 31
Weasel (see *Mustela nivalis*) 232
Weedlots 302
Weight-of-evidence 74–75, 80, 93, 229
Wetlands 300, 304, 307, 317
Whale (*see* specific whale species)
    baleen 234–237
    beaked 234–236
    mysticete plankton-feeding 233
    open ocean, toothed (odontocete) 233
White-sided dolphin (see *Lagenorhynchus* sp.)
Windermere Humic Acid Model (WHAM) 157

X

Xenobiotics
    general metabolism/biotransformation 103, 106–115, 118
    phase-one reactions 106–108, 118
    phase-two reactions 106, 108–111, 118
    metabolites 98, 110–115, 117, 330
    metabolizing enzyme 253

Z

Zinc (Zn) 26, 35, 129, 133–134, 139–140, 142, 154, 160, 164, 166–167, 171–173, 175, 177–178
Zooplankton 133, 173, 179, 232